Interest in groundwa... increased in recent years. Hydrogeologists and geologists are now actively exploring the role of subsurface fluids in such fundamental geologic processes as crustal heat transfer, hydrocarbon migration, earthquakes, diagenesis, and metamorphism.

Groundwater in Geologic Processes first develops the basic theory of groundwater motion, solute transport, and heat transport. The second section applies flow and transport theory in a generalized geologic context and focuses on particular geologic processes and environments.

The systematic presentation of theory and application makes this book ideal for graduate-level hydrogeologists and geologists with backgrounds in calculus and introductory chemistry. It will also be an invaluable reference for professiona

Two week

GROUNDWATER IN GEOLOGIC PROCESSES

One

GROUNDWATER IN GEOLOGIC PROCESSES

STEVEN E. INGEBRITSEN

United States Geological Survey

Menlo Park, California

WARD E. SANFORD

United States Geological Survey

Reston, Virginia

CAMBRIDGE
UNIVERSITY PRESS

PUBLISHED BY THE PRESS SYNDICATE OF THE UNIVERSITY OF CAMBRIDGE
The Pitt Building, Trumpington Street, Cambridge, United Kingdom

CAMBRIDGE UNIVERSITY PRESS
The Edinburgh Building, Cambridge CB2 2RU, United Kingdom
40 West 20th Street, New York, NY 10011-4211, USA
10 Stamford Road, Oakleigh, Melbourne 3166, Australia

First published 1998
First paperback edition 1999

Printed in the United States of America

Typeset in Times Roman

Library of Congress Cataloging-in-Publication Data
Ingebritsen, S. E.
Groundwater in geologic processes / Steven E. Ingebritsen, Ward E.
Sanford.
p.   cm.
Includes bibliographical references.
1. Hydrogeology.   I. Sanford, Ward E.   II. Title.
GB1003.2.I54   1998
551.49 – dc21   97-18014
                                                CIP

A catalog record for this book is available from the British Library

ISBN 0 521 49608 X hardback
ISBN 0 521 66400 4 paperback

Contents

Preface

Historically, interest in groundwater and other subsurface fluids has largely been confined to a few specific disciplines in the earth sciences, notably ground-water hydrology, soil physics, engineering geology, petroleum geology, and petroleum engineering. These disciplines have tended to be "applied" in nature, with practitioners concentrating on the immediate and practical problems of water supply, water quality, mine dewatering, deformation under structural loads, and the location and recovery of fluid hydrocarbons. This situation has begun to change in recent years, and both hydrogeologists and geologists are now actively exploring the role of subsurface fluids in such fundamental geologic processes as sedimentary basin evolution, diagenesis, ore deposition, and tectonism. The growing interest in the role of groundwater in geologic processes is evidenced by Penrose Conferences on "Basin-Scale Flow and Transport" (1991) and "Fluid-Volcano Interactions" (1992), by a number of symposia at recent annual meetings of the American Geophysical Union and the Geological Society of America, and by the birth of an annual, informal "Hubbert Quorum" on groundwater and geologic processes. The activity in this hybrid discipline is also highlighted in a collection of papers published by the U.S. National Research Council (*The Role of Fluids in Crustal Processes*, Bredehoeft & Norton (1990)).

We perceived a need for a more comprehensive treatment of this body of inquiry. The National Research Council volume noted above is a collection of seminal research papers by various workers in the field. In contrast, we have attempted an integrated and systematic accounting, suitable for use as an adjunct to more-or-less general curricula in hydrogeology and geology, or as the primary text for courses on groundwater and geologic processes. The books that come closest to addressing the intended audience (*Fluids in the Earth's Crust* by Fyfe, Price, & Thompson, 1978; *Geology and Water* by Chapman, 1981; *Geodynamics* by Turcotte & Schubert, 1982; and *Physical and Chemical*

Hydrogeology by Domenico & Schwartz, 1990) in fact touch on only a few of the couplings between geology and groundwater.

This book describes how the the occurrence and movement of groundwater affects a wide range of geologic processes, such as ore deposition, petroleum migration, upper-crustal heat transfer, earthquakes, diagenesis, and metamorphism. We have defined "groundwater" in a broad sense, as any subsurface, aqueous fluid, including those of meteoric, connate, and magmatic origin. We originally considered naming this book *Fluids and Geologic Processes*. The term "fluids" is more inclusive but also tends to imply full consideration of the most important nonaqueous crustal fluid, magma. Here, although we consider magma as a source of heat, fluids, and solutes to groundwater systems, we do not treat the dynamics of the magma itself.

The organization of this book is intended to facilitate use by both students and professionals. Chapters 1 to 3 develop the basic theory of groundwater motion, solute transport, and heat transport. These three chapters comprise the theoretical core of the book, and we expect that most course instructors would elect to assign them during the first few weeks of a quarter or semester. Chapter 4 applies flow and transport theory in a generalized geologic context, and Chapters 5 to 10 focus on particular geologic processes or environments. Practicing earth scientists who are interested in particular geologic problems might turn to the latter part of the book first and then refer back to Chapters 1 to 3 as necessary to review pertinent theory.

We expect that when this book is used as a textbook, it will most often be in the context of a graduate-level course, and that most of the students will already have taken an introductory groundwater course. However, we hope that it will also prove accessible to earth science students who have an understanding of calculus and introductory chemistry and physics but lack previous formal training in hydrogeology.

"Everything is coupled"

In order to treat the role of groundwater in geologic processes, we must first develop the pertinent theory, including equations of groundwater motion (Chapter 1) and descriptions of solute and heat transport. The coupling of groundwater flow with solute transport (Chapter 2) is usually included in standard developments of groundwater theory, because of its importance to water-quality and contamination issues. However, the coupling with heat transport (Chapter 3) is often neglected. In most hydrogeologic practice it is fairly standard (and reasonable) to assume that groundwater flow takes place in an isothermal, nondeforming medium. For the geologic applications considered here, these assumptions often are not tenable.

Over the large spatial scales that pertain in many geologic applications, the assumption of isothermal flow (and thus constant fluid properties) is often poor. Given a typical geothermal gradient of 25°C/km, temperatures in many large sedimentary basins may be expected to vary by 100°C or more. For pure water, a 100°C variation in temperature can cause a 4% variation in fluid density (greater than the density difference between freshwater and seawater) and a 600% variation in fluid viscosity (equivalent to a six-fold variation in hydraulic conductivity). In moderate- to high-temperature hydrothermal systems, much larger variations in fluid properties occur. For example, there is a three-fold variation in the density of pure liquid water between room temperature and its critical point (374°C, 22.06 MPa), and in multiphase (liquid–steam) systems, density variations range up to several orders of magnitude.

Futhermore, over the large time scales that pertain in many geologic problems, deformation often will be substantial. In this book our theoretical treatment of coupled groundwater flow and transport is fairly complete, but the description of the coupling between groundwater flow and deformation is rudimentary. We generally neglect deformation except in deriving *specific storage* (Chapter 1.5.2) and in discussing sediment dewatering (Chapter 4.1), the "stress-heat flow paradox" of the San Andreas fault (Chapter 4.3.5), and earthquakes (Chapter 8). This is in accordance with standard hydrogeologic practice, but for some geologic problems a more general and rigorous treatment of coupled flow and deformation based on poroelastic theory is needed (Biot, 1941; Rice and Cleary, 1976; Roeloffs, 1996). In some environments the deformation rate will actually exceed the groundwater flow rate. Thus, for instance, in the rapidly deforming *accretionary prisms* of sediment that overlie subducting oceanic plates, the net direction of fluid flow relative to fixed coordinates may be very different from the flow direction relative to the sedimentary matrix (Chapter 4.1.5). Deformation can also reduce porosity, thereby increasing fluid pressure, decreasing the effective stress, and reducing the amount of work required for tectonism. Deformation-related forcing of groundwater flow is particularly effective in low-permeability strata (Chapter 4.1.2).

From the theoretical point of view, then, a unique aspect of the study of groundwater and geologic processes is the importance of the couplings of groundwater flow with heat transport and deformation, in addition to solute transport. Another unusual consideration (relative to standard hydrogeologic practice) is that every one of these couplings may be important to a given problem. As an example, let us consider hydrothermal circulation near mid-ocean-ridge (MOR) hydrothermal vents (Chapter 7.9). We can assume that the fluid flow is governed by some form of Darcy's law, except perhaps very near some of the vents, where flow rates may be high enough that some energy is lost to turbulence. We would need to invoke a relatively complex, multiphase form of Darcy's law (Chapter 3.1.3), because large gradients in salinity between vents indicate that active *phase separation* (Chapter 7.4.1) occurs between a relatively dense, saline brine and a less saline vapor. Large gradients in salinity and temperature (approximately 2–400°C) dictate that any complete model of the system must include both solute and heat transport. Furthermore, we must expect that the flow systems are highly transient; the exceptionally high rates of heat discharge from individual vents can only be explained as the result of rapid crystallization and cooling of large volumes of magma (Chapter 7.2.1), so that the intensity and spatial distribution of heat sources must vary with time. We would also expect that precipitation and dissolution of minerals causes continuous variations in porosity and permeability, because the extreme variations in fluid composition and temperature make for a highly reactive chemical environment. As a result of these transient phenomena, deformation enters the picture: As permeability, flow rates, and temperatures wax and wane, near-vent rates of thermomechanical deformation are likely large enough to substantially affect permeability (Germanovich and Lowell, 1992). On a longer time scale, plate movement away from the MOR itself (another mode of deformation) will also influence the overall pattern of fluid circulation.

Although we can recognize the probable importance of each of these couplings, we must invariably neglect some of them in our analyses. There is as yet no quantitative model that can fully describe MOR hydrothermal circulation, or other complex, transient systems. We account, at most, for one or two of the couplings in each analysis, hoping to capture the essence of the system. For any but near-surface conditions, efforts to describe more than one coupling at a time (e.g., fluid flow–solute transport–heat transport) are still in their infancy.

Acknowledgments

The notion of a book on this topic was originally inspired by a course that we developed for colleagues at the U.S. Geological Survey, and several of the chapters were refined on the basis of short courses held at the Survey's National Training Center and elsewhere. We extend our gratitude to our co-instructors Chris Neuzil and Evelyn Roeloffs, to the many course participants, and to numerous colleague reviewers. The following people reviewed all or part of one or more chapters; unless otherwise indicated, they are members of the USGS: Barbara Bekins, Jim Bischoff, Hedeff Essaid, Lisl Gaal (University of

Minnesota), Dan Hayba, Paul Hsieh, Blair Jones, Lenny Konikow, Robert Lowell (Georgia Institute of Technology), Les Magoon, Craig Manning (UCLA), Brian McPherson (New Mexico Institute of Technology), Casey Moore (University of California–Santa Cruz), Chris Neuzil, Mark Person (University of Minnesota), Evelyn Roeloffs, Elizabeth Rowan, Martha Scholl, Claire Tiedeman, Fiona Whitaker (University of Bristol), Warren Wood, and Mary Lou Zoback. Classes taught by Paul Hsieh (at Stanford University) and Roy Haggerty (at Oregon State University) tested this book in manuscript form; they found and corrected a number of errors in the presentation. We also gratefully acknowledge John Bredehoeft, who first stimulated our interest in the geological aspects of hydrogeology. A special thanks is also owed to David R. Jones for drafting or redrafting nearly all of the figures.

Finally, we owe a debt of thanks to the USGS itself, and specifically to our immediate supervisors, for allowing us to invest time in this effort. In our view the USGS has a strong organizational interest in the topic of groundwater in geologic processes. This view is borne out to some extent by examination of the chapter titles, several of which (e.g., ore deposits, hydrocarbons, earthquakes) directly correspond to important USGS programs. Others that show no direct correspondance (e.g., geothermal processes) nevertheless represent topical areas that are significant to one or more USGS programs.

List of symbols

Listed are common symbols in alphabetical order, first in the Latin, then the Greek alphabets. Other, less frequently used symbols are defined where they appear in the text. Dimensions given in square brackets are M, mass, L, length, t, time, E, energy, T, temperature, and C, electrical charge.

A = area $[L^2]$

A_o = radiogenic heat production measured near the Earth's surface $[E/L^3\text{-}t]$

b = thickness $[L]$

c = isochoric heat capacity $[E/m\text{-}T]$

C = aqueous concentration $[M/L^3]$

C_d = dimensionless concentration [dimensionless]

C_{eq} = concentration at equilibrium with respect to a given mineral $[M/L^3]$

C_i = concentration in inflowing groundwater $[M/L^3]$

C_o = concentration in outflowing groundwater $[M/L^3]$

C_p = concentration in precipitation $[M/L^3]$

C_s = sorbed concentration (Chapter 2) $[M/M]$

 = solute concentration in inflowing surface water (Chapter 9) $[M/L^3]$

C_o = reference concentration or concentration at an initial time $[M/L^3]$

CEC = the exchange capacity of a porous medium $[M/M]$

d = depth $[L]$

d_m = mean or median grain diameter (particle size) $[L]$

D = hydrodynamic dispersion (Chapters 2, 4) $[L^2/t]$

 = hydraulic diffusivity (Chapter 7) $[L^2/t]$

D'_{ij} = component of diffusion coefficient matrix for a multicomponent system $[L^2/t]$

D_k = Damkohler number [dimensionless]

D_L = coefficient of longitudinal dispersion $[L^2/t]$

D_m = coefficient of molecular diffusion in a porous medium $[L^2/t]$

D'_m = coefficient of molecular diffusion per unit area of effective porosity $[L^2/t]$

D_T = coefficient of transverse dispersion $[L^2/t]$

D_w = coefficient of molecular diffusion in open water $[L^2/t]$

e = activation energy $[E/M]$

E = impelling force $[ML/t^2]$

E_j = gradient in the electrochemical potential of the jth ion $[E/M\text{-}L]$

F = Faraday's constant $[C/M]$

g = gravitational acceleration $[L/t^2]$

h = hydraulic head $[L]$

H = enthalpy $[E]$

k = intrinsic permeability $[L^2]$

$k_i^{\cdot}$ = kinetic rate constant for a given mineral i (i = A, B, C) [variable]

k_r = relative permeability [dimensionless]

K = hydraulic conductivity $[L/t]$

K_{ac} = equilibrium constant for an aqueous complex [variable]

K_d = linear sorption distribution coefficient $[L^3/M]$

K_{eq} = ion-exchange distribution coefficient [variable]

K_i = equilibrium constant for mineral i (i = A, B, min) [variable]

K_m = medium thermal conductivity $[E/L\text{-}T]$

K_r = a reaction-rate constant [variable]

L = characteristic length or distance $[L]$

m = empirical exponent in a kinetic rate equation [dimensionless]

m_s = the molality of a species $[M/M]$

M = a point source of mass $[M]$

n = porosity [dimensionless, L^3/L^3]

n_e = effective porosity [dimensionless]

Nu = Nusselt number [dimensionless]

P = pressure $[M/L\text{-}t^2]$

Pe = Peclet number [dimensionless]

q = volumetric flow rate per unit area (specific discharge or Darcy velocity) $[L/t]$

q_a = advective flux of a solute $[M/L^2\text{-}t]$

q_d = diffusive flux of a solute $[M/L^2\text{-}t]$

q_h = conductive heat flux per unit area $[E/L^2\text{-}t]$

q_h^* = reduced conductive heat flux per unit area $[E/L^2\text{-}t]$

q_i = diffusive flux of the ith ion $[M/L^2\text{-}t]$

Q = total volumetric flow rate $[L^3/t]$

Q_e = volumetric flow rate of evaporating basin water $[L^3/t]$

Q_i = volumetric flow rate of inflowing groundwater [L^3/t]
Q_o = volumetric flow rate of outflowing groundwater [L^3/t]
Q_p = volumetric flow rate of precipitation [L^3/t]
Q_s = source or sink of solute (Chapter 2) [M/L^3-t]
   = volumetric flow rate of inflowing surface water (Chapter 9) [L^3/t]
r = radial distance (Chapter 1) or a radius of curvature (Chapter 6) [L]
R = general source/sink term for mass, chemical reactions, or heat [variable]
   = gas constant (Chapter 6) [E/M-mol-T]
   = D/H or $^{18}O/^{16}O$ ratio (Chapter 10) [dimensionless]
R_c = rate of calcite dissolution [variable]
R_f = retardation factor [dimensionless]
R_p = rate of porosity development [L^3/L^3-t]
Ra = Rayleigh number [dimensionless]
s = storage coefficient (storativity) [dimensionless]
s_s = specific storage [1/L]
s_0 = specific surface [L^2/L^3]
S = volumetric saturation [L^3/L^3]
S_y = specific yield [dimensionless]
t = time [t]
t_d = dimensionless time [dimensionless]
T = temperature [T]
   = transmissivity (Chapters 1 and 8) [L^2/t]
v = average linear velocity (seepage velocity) [L/t]
V = volume [L^3]
V_e = evaporated basin volume (Chapter 9) [dimensionless]
x = mass fraction steam (Chapter 7) or H_2O or CO_2 in an H_2O–CO_2 mixture (Chapter 10) [dimensionless]
x_d = dimensionless distance in the x direction
X = Lagrangian x coordinate [L]
X' = translational transformation variable [L]
X_a = exchange site for ion a on a mineral surface [dimensionless]
Y = Lagrangian y coordinate [L]
z = elevation above a datum, elevation head, or vertical Cartesian coordinate [L]
z_i = ionic charge of the ith ion [dimensionless]
Z = Lagrangian z coordinate [L]

α = coefficient of thermal expansion (Chapters 3 and 10) [1/T, $\Delta L/LT$]
   = contact angle (Chapter 6)

α_L = longitudinal dispersivity [L]

α_T = transverse dispersivity [L]

β = fluid compressibility [Lt^2/M]

γ = change in fluid density per change in solute concentration [dimensionless]

γ_m = activity coefficient of aqueous species m (Chapter 2) [dimensionless]

Γ = a volumetric source or sink of fluid (Chapter 4) [L^3/L^3/t]

δ = dirac delta function (Chapter 2) [dimensionless]

    = [(R_{sample}/R_{SMOW}) − 1] × 1, 000 (Chapter 10)

ΔCa_s = mass of dissolved calcite required to reach equilibrium [M/L^3]

Δ_x = discretization length in the x direction for a computational grid [L]

ε = strain [dimensionless, $\Delta L^3/L^3$]

θ = angle between failure plane and greatest principal stress

λ = radioactive decay constant [1/t]

λ' = loss-by-diffusion constant (Chapter 4) [1/t]

μ = dynamic viscosity [M/L-t]

μ_i = chemical potential of the ith ion (Chapter 2) [E/M]

ν = stoichiometric coefficient in a chemical reaction [dimensionless]

ρ = density [M/L^3]

ρ_f = density of freshwater (Chapter 9) [M/L^3]

ρ_o = reference fluid density [M/L^3]

ρ_s = density of saltwater (Chapter 9) [M/L^3]

σ = one of three mutually orthogonal principal stresses [M/L-t^2]

    = surface tension (Chapter 6) [M/t^2]

σ_e = effective normal stress on an arbitrarily oriented plane ($\sigma_n - P_f$) [M/L-t^2]

σ_n = normal stress on an arbitrarily oriented plane [M/L-t^2]

σ_T = total overburden pressure or lithostatic load on a horizontal plane [M/L-t^2]

σ_1 = greatest principal stress [M/L-t^2]

σ_2 = intermediate principal stress [M/L-t^2]

σ_3 = least principal stress [M/L-t^2]

τ = tortuosity (Chapter 2) [L/L]

    = shear stress (Chapter 8) [M/L-t^2]

τ_0 = cohesion [M/L-t^2]

ϕ = electrostatic potential (Chapter 2) [E/C]

    = fluid potential per unit mass (Chapter 7) [E/M]

    = angle of internal friction (Chapter 8); tan ϕ is also known as the *coefficient of friction*

Subscripts

Unless otherwise locally redefined, these subscripts have the following meanings. They mainly apply to the description of multiphase systems. For example, ρ_g, ρ_o, and ρ_s are the densities of gas, oil, and steam, respectively.

atm	refers to atmospheric conditions
c	refers to capillary effects
f	refers to the fluid mixture in place (either a single phase or a two-phase mixture)
g	refers to gas
m	refers to the porous medium
o	refers to oil
r	refers to the rock
s	refers to steam
v	refers to void space (porosity)
w	refers to liquid water

1

Groundwater flow

The primary coupling between groundwater flow and deformation, solute transport, and heat transport is through *Darcy's law*. Groundwater flow rates calculated by Darcy's law are used to describe fluid flow through a deforming volume, solute transport by advection and mechanical dispersion, and heat transport by advection. In this book we will generally assume that some form of Darcy's law adequately describes rates of fluid flow. It thus seems appropriate to begin with a discussion of Darcy's law and associated parameters such as *fluid potential, hydraulic head, porosity, hydraulic conductivity*, and *permeability*. We will then introduce the concept of the *water table* and *confined* and *unconfined* groundwater systems, briefly discuss the *elemental volume* or macroscopic approach that is implicit in the definition of hydrogeologic parameters, and proceed to derive the groundwater flow equation by considering conservation of mass and the production of fluid from "storage" in a geologic medium (introducing *specific storage* and *specific yield*). Finally, we will consider the assumptions embodied in various forms of the groundwater flow equation. It will become clear that the commonly invoked forms of the equation, posed in terms of hydraulic head, are too simplified or incomplete to apply to many geologic problems. Nonetheless, they need to be examined and understood as a common point of reference and departure. In succeeding chapters we will develop the coupled theory of solute transport (Chapter 2) and heat transport (Chapter 3) and then describe applications of flow and transport theory to geologic processes (Chapters 4 to 10).

1.1 Darcy's law

In 1856 the French engineer Henry Darcy published the results of a set of experiments aimed at developing design parameters for sand water filters. The experimental apparatus, shown schematically as Figure 1.1, allowed him to vary the length (L) and cross-sectional area (A) of a sand-packed column and also

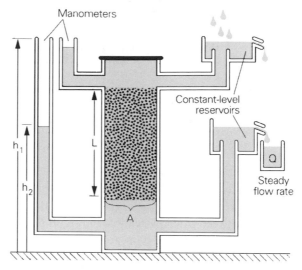

Figure 1.1 Schematic diagram of the apparatus used in Henry Darcy's sand-filter experiments.

the elevations of constant-level water reservoirs connected to the upper (h_1) and lower (h_2) boundaries of the column. Under steady flow conditions, the volumetric flow rate (V_w/t) through the column (Q) was positively correlated with A and $(h_1 - h_2)$ and inversely correlated with L. By introducing a constant of proportionality K, Darcy's experimental results can be summarized as

$$Q = \frac{KA(h_1 - h_2)}{L}. \tag{1.1}$$

We arrive at what has become known as *Darcy's law* by dividing both sides by A, writing $(h_1 - h_2)/L$ in differential form by defining $(h_1 - h_2) \rightarrow dh$ and $L = (L_1 - L_2) \rightarrow dL$, and introducing a minus sign to indicate that flow is in the direction of decreasing h, so that

$$q = -K\left(\frac{dh}{dL}\right), \tag{1.2}$$

where q denotes the volumetric flow rate per unit area and has units of L/t $(V_w/t/L^2 = L^3/t/L^2 = L/t)$. The constant of proportionality K, named *hydraulic conductivity*, must also have dimensions of L/t, because dh/dL is dimensionless. Subsequent laboratory-column experiments conducted using a variety of fluids revealed that K expresses a combination of fluid and solid properties. The flow rate is actually proportional to the specific weight of the fluid, ρg, inversely proportional to the dynamic viscosity of the fluid, μ, and proportional to a property of the solid medium, k, which is called *intrinsic permeability*. Thus

$$K = \frac{k\rho g}{\mu}, \tag{1.3}$$

where k has units of L^2. Theoretical considerations (Hubbert, 1940) and experiments with sand or glass beads of uniform diameter d_m (Hubbert, 1956) further revealed that, for granular porous media, q, K, and k are proportional to d_m^2.

The volumetric flow rate per unit area, q, is variously called the *volumetric flow rate, specific discharge*, or *Darcy velocity*. Despite the velocity-like units, q is not an actual fluid velocity, because the fluid is flowing only through void (pore) spaces, not through the total cross-sectional area A. An estimate of the actual fluid velocity is obtained by dividing q by that fraction of the total area or volume that is composed of connected void space, that is,

$$v = \frac{q}{\left(\frac{V_v}{V_t}\right)} = \frac{q}{n_e}, \tag{1.4}$$

where v is called the *average linear velocity* or *seepage velocity*, V_v is the volume of connected void space, V_t is the total volume, and the ratio V_v/V_t defines the *effective porosity*, n_e. The average linear velocity is an estimate of the mean of the pore-scale fluid velocity. It represents, for example, the mean velocity at which a conservative (nonreacting) solute would move through Darcy's experimental column (Figure 1.1). In Darcy's experiments, $n_e = 0.38$ (Darcy, 1856), so $v = 2.6q$.

Darcy's law in the form of Eq. (1.1) or (1.2) is a simple *linear transport law*, completely analogous to the expressions that describe conduction of heat and electricity or the diffusion of a solute. For example, an analogous form of Fourier's law for conductive heat flow is

$$Q_h = -K_m A\left(\frac{dT}{dz}\right), \tag{1.5}$$

where Q_h is heat flow, K_m is thermal conductivity, and dT/dz denotes the vertical temperature gradient. Similarly, Ohm's law for conduction of electrical current can be written

$$i = -K_e A\left(\frac{dV}{dL}\right), \tag{1.6}$$

where i is electrical current, K_e is electrical conductivity (the reciprocal of resistivity), and V is (temporarily) redefined here as voltage. Fick's first law of diffusion (Eq. 2.1) has the same form, relating the diffusive flux of a solute to its concentration gradient through a constant of proportionality.

Combining these linear transport laws with *continuity equations* that express mass or energy conservation results in *diffusion equations* that describe the transient flow of groundwater, heat via conduction, electricity, and solute transport via diffusion. In Chapter 1.5 we use this approach to derive the diffusion equation for groundwater flow. The full-fledged diffusion equations for electricity and conductive heat flow are not particularly relevant to the purposes of this book. However, those readers with some background in electrical or heat-flow theory should recognize and benefit from the analogy. Furthermore, one should be aware that many analytical solutions to diffusion-type equations are available in the heat-flow literature, as are many solutions to *advection–diffusion equations* similar to those that describe the coupled processes of groundwater flow and solute or heat transport (e.g., Carslaw and Jaeger, 1959; see Chapters 2 and 3 for solute and heat transport, respectively). With an appropriate substitution of parameters, the solutions in the heat-flow literature can be directly applied to groundwater flow and transport problems.

We regard Darcy's law as fundamentally empirical, although it may be derived by averaging more fundamental equations of fluid motion known as the *Navier–Stokes equations* over a representative volume of porous medium and assuming laminar, steady flow or negligible inertial terms (e.g., Hubbert, 1956).

1.1.1 The limits of Darcy's law

Darcy's law has been tested empirically over a fairly wide range of conditions. Experimental results indicate that it fails at sufficiently high volumetric flow rates. Above a certain threshold flow rate, significant amounts of energy begin to be lost to turbulence and, as a result, Darcy's law overpredicts the flow rate associated with a given value of dh/dL. Such flow rates are exceedingly rare in the subsurface but can occur very near to a wellbore or in areas of cavernous porosity, such as occur in carbonate rocks and lava flows.

The upper limit for application of Darcy's law is usually estimated on the basis of the dimensionless *Reynold's number*, $Re = \rho q L/\mu$, where ρ is fluid density, q is the volumetric flow rate per unit area, μ is the dynamic viscosity of the fluid, and L is some characteristic length. In granular porous media, L is commonly related to the grain-size distribution (although it is also sometimes taken as $k^{1/2}$ (Ward, 1964), where k is the intrinsic permeability), and the transition to non-Darcian (nonlinear) flow appears to take place at $Re \sim 5$ (Bear, 1979; Freeze and Cherry, 1979). In fractured media the transition to non-Darcian flow can be estimated in the context of a parallel-plate model of fracture porosity ($k \sim Nb^3/12$, where N is the fracture spacing; Snow, 1968). In this approach, b would be taken as the fracture aperture, q is replaced by v, the average linear velocity through the fractures, and the critical Re value is about 1,000 (Vennard and Street, 1975). Applying the porous-medium approach, evaluating ρ (1,000 kg/m^3) and μ (1.1 × 10^{-3} Pa-sec) at 15°C, and taking d_m as the median grain size (d_{50}) of a coarse sand (0.001 m), we find that Darcy's law would begin to overpredict flow rates for $q \geq 5.5 \times 10^{-3}$ m/s or 1.7×10^5 m/yr.

Although Darcy's law fails at high flow rates, a low flow rate or low permeability limit has not been demonstrated. Apparent departures from a linear relation between a hydraulic gradient dh/dL and the volumetric flow rate q (Eq. 1.2) were reported in the literature as early as 1899, mainly from experiments under low-gradient (e.g., King, 1899) and/or low-permeability conditions. Many of these experiments have since been shown to have been flawed or incorrectly interpreted (e.g., Olsen, 1965; Neuzil, 1986). However, all experiments with intrinsic permeabilities less than about 10^{-15} m^2 have imposed very large values of dh/dL (Neuzil, 1986), rather than values representative of in situ conditions (dh/dL generally ≤ 1), because experiments with in situ gradients are impractically slow. Thus it is still reasonable to speculate that Darcy's law might not strictly apply in extremely low permeability environments. In most geologic media, low permeability is associated with extremely small pore and fracture openings, such that a large fraction of the pore water occurs in close proximity to hydrophilic solid surfaces. There are significant variations in some water properties with distance from solid surfaces (e.g., Mitchell, 1993); for example, there is a more rigid short-range order or "structure" among the water molecules nearer to a solid. If the narrowest connected openings are dominated by such "structured" water, there might be a finite threshold value of dh/dL below which Darcy's law does not apply (that is, $q = 0$ for some finite range of $dh/dL > 0$).

1.1.2 Driving forces for groundwater flow

In hydrogeologic practice the driving force for groundwater flow is generally expressed in terms of a parameter called *hydraulic head* or simply *head*. This is the same quantity indicated by h_1 and h_2 in Henry Darcy's laboratory manometers (Figure 1.1). The nature of hydraulic head was clarified by Hubbert (1940), who derived it from basic physical principles after stating complacently that " ... to adopt [head] empirically without further investigation would be like reading the length of the mercury column of a thermometer without knowing that temperature was the physical quantity being indicated."

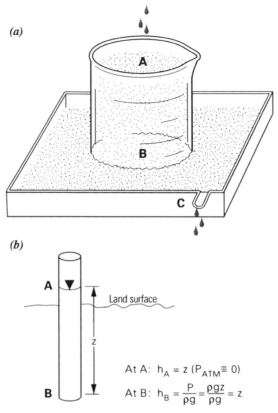

Figure 1.2 (a) A simple experiment demonstrating that groundwater does not necessarily flow from areas of high pressure to areas of low pressure. Water is added at a steady rate to an open-ended sand-filled column partly embedded in a sand-filled tray. Flow is from point A (P = atmospheric $\equiv 0$) to point B ($P > 0$) to point C ($P = 0$). (b) Components of hydraulic head ($h = P/\rho_f g + z$) illustrated with reference to a *piezometer*, a tube that is open to the atmosphere at the top and to groundwater flow at the bottom. The pressure datum ($P = 0$) is taken as atmospheric pressure, and the bottom of the tube is the reference elevation datum where $z = 0$. At point A, then, $h_A = 0 + z = z$. The pressure at point B is determined by the weight of the overlying column of water, $\rho_w g z$, so that $h_B = \rho_w g z / \rho_w g + 0 = z$. The relation $h_A = h_B = z$ defines *hydrostatic* (nonflowing) conditions within the piezometer tube.

Intuitively, one might tend to think of groundwater as flowing from areas of high pressure to areas of low pressure. This idea is readily debunked by consideration of the pressure distribution in a static column of water ($P = \rho g d$, where d is the depth below the water surface) or, alternatively, by simple experiments such as the one diagramed in Figure 1.2a, where flow is from $P = 0$ at point

A to $P > 0$ at point B to $P = 0$ at point C. As Hubbert (1940) demonstrated, groundwater flow is actually from areas of high energy to areas of low energy.

Two fundamental forms of energy are of interest in this context: *kinetic energy* (KE) and *potential energy* (PE). Kinetic energy is associated with motion, and the kinetic energy of a unit volume of liquid water is given by $\rho_w v^2/2g$. Most readers will recognize that this quantity is very similar to the *velocity head* term of *Bernoulli's equation*. In a typical groundwater environment the kinetic energy is negligibly small relative to the total potential energy (KE is typically at least 10^{10} less than PE) and can be neglected.

Potential energy is associated with the work required to move from one place to another in a conservative force field, that is, a field in which the work done in moving from point A to point B does not depend on the path taken. In a groundwater context the most important conservative force fields are gravity and pressure. The *gravitational potential energy* of a unit volume of liquid water is $\rho_w g z$, where z is its height above an arbitrary datum and $\rho_w g$ is its *specific weight*. The *pressure potential energy* per unit volume is simply the pressure P, a force per unit area. For a fluid with variable density (ρ not a constant), flow is proportional to the gradient in the quantity $(P + \rho_f g z)$ (see Chapters 2.3 and 3.1.3). For an *incompressible fluid* (ρ constant), one can divide through by the specific weight to obtain the quantity $(P/\rho_f g + z)$, which defines the *hydraulic head h*. Both the *pressure head* ($P/\rho_f g$) and the *elevation head* (z) have units of length, and the total hydraulic head h can be equated to the water level observed in a manometer (Figure 1.1) or well (Figure 1.2b). (See also Problem 1.1.) In this case the driving force for groundwater flow is the head gradient.

Linear flow laws such as Darcy's law describe what are termed *direct flow phenomena* (Mitchell, 1993), in which the driving force and flow are of like types (head gradients and Darcian fluid flow, concentration gradients and solute flux, temperature gradients and heat flux). Often, simultaneous flows of different types occur when only one driving force is acting, for example, fluxes of solutes and heat via fluid flow driven by gradients in hydraulic head. There are also *coupled flow phenomena* in which gradients of one type (e.g., concentration gradients) cause flows of another type (e.g., fluid flow). Some direct and coupled flow phenomena are summarized in Table 1.1, where direct flow phenomena appear on the main diagonal and coupled flow phenomena appear in the off-diagonal positions. This relatively complete listing of flows and forcings suggests a more general Darcy-type flow law that could be written (still in one dimension) as

$$q = -K \left(\frac{dh}{dL} \right) - \frac{K_C}{C} \left(\frac{dC}{dL} \right) - K_T \left(\frac{dT}{dL} \right) - K_E \left(\frac{dV}{dL} \right), \qquad (1.7)$$

Table 1.1 *Direct and coupled flow phenomena. Direct flow phenomena appear on the main diagonal and coupled flow phenomena appear in the off-diagonal positions[a]*

| | Gradient | | | |
Flow	Hydraulic head	Chemical concentration	Temperature	Electrical
Fluid	Darcian flow (*Darcy's law*)	chemical osmosis	thermoosmosis	electroosmosis
Solute	streaming current	diffusion (*Fick's first law*)	thermal diffusion (*Soret effect*)	electrophoresis
Heat	isothermal heat transfer	Dufour effect	heat conduction (*Fourier's law*)	Peltier effect
Current	streaming current	diffusion and membrane potential	thermoelectricity (*Seeback effect*)	electric conduction (*Ohm's law*)

[a] After Mitchell (1993).

where K_C is osmotic conductivity, C is solute concentration, K_T is thermo-osmotic conductivity, dT/dL is the temperature gradient, K_E is electroosmotic conductivity, and V is voltage.

In this book we will generally neglect the coupled or indirect driving forces and assume that groundwater flow is driven solely by gradients in hydraulic head or, more generally, by gradients in the quantity $(P + \rho_f gz)$. However, *chemical osmosis*, in particular, appears to be important in some regional-scale groundwater systems. Certain argillaceous media seem to act as semipermeable membranes that retard movement of ions relative to liquid water (e.g., Neuzil, 1995). Water flows across these layers toward regions of high concentrations to lessen the concentration gradient, so that high hydraulic heads coincide with high solute concentrations. In a comprehensive study of the Dunbarton basin of South Carolina and Georgia, hydraulic heads approximately 130 m above hydrostatic were attributed to chemical osmosis in Triassic sediments with pore-fluid salinities of 12,000 to 19,000 mg/liter (Marine and Fritz, 1981).

1.2 Crustal permeability

Permeability is unquestionably the crucial hydrologic parameter. Unfortunately, it is often a very difficult parameter to evaluate and apply in a meaningful fashion, especially over the enormous space and time scales that apply in many geologic problems.

The measured permeability of common geologic media varies by an almost inconceivable 16 orders of magnitude, from values as low as 10^{-23} m^2 in intact crystalline rock, intact shales, fault gouges, and halite to values as high as 10^{-7} m^2 in well-sorted gravels. Furthermore, permeability appears to vary with the scale or method of measurement, particularly in media where fracture permeability is significant. For example, in fractured rocks the relatively large-scale permeabilities measured by in situ wellbore testing are commonly 10^3 times higher than the smaller-scale permeabilities measured on the associated drill core. Finally, in many geologic problems permeability must be regarded as a time-dependent parameter, being created or destroyed over time by mineral dissolution and precipitation, by changes in *effective stress* (Chapter 8.1) that result in consolidation or *hydraulic fracturing* (Chapter 4.1.3), and by thermo-elastic effects.

1.2.1 Permeability versus porosity

The terms porosity (n) and permeability (k) are often used interchangeably by nonhydrogeologists. This confusion is understandable, because there is a strong positive correlation between the two quantities in many porous and fractured geologic media. For well-sorted, unconsolidated porous media, this correlation is often expressed by the *Kozeny–Carman equation* (Carman, 1956),

$$k = \frac{n^3}{\left[5s_0^2(1-n)^2\right]},$$ (1.8)

where s_0 is the solid surface exposed to the fluid per unit volume of solid material. The Kozeny–Carman relation was developed by solving the *Navier–Stokes equations* for a system of parallel capillary tubes; in this model, s_0 can readily be related to the hydraulic radii of the capillaries. Alternatively, s_0 can be related to some mean particle size d_m. For spheres of constant diameter d, $s_0 = 6/d$.

However, the positive correlation between n and k does not hold for some important classes of geologic media, namely clays, clay-rich materials, and volcanic tuffs. Clays and unwelded ash-flow or air-fall tuffs tend to have very large porosities ($\sim$0.50) but relatively low permeabilities. There is a weak positive correlation between n and k for clays themselves, but clays as a group tend to be about 10^6 times less permeable than sands despite having higher porosities (n is only $\sim$0.35 for a fairly well-sorted sand). Within ash-flow tuff units there is often a negative correlation between n and k (Figure 1.3). The top and bottom of an ash flow cool relatively rapidly, retaining their original high porosities ($\sim$0.50). However, the permeability of the unwelded material is

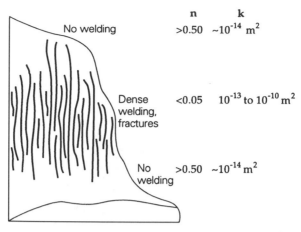

Figure 1.3 Cross section through a hypothetical ash-flow tuff unit showing typical values of porosity (n) and permeability (k). The thickness of individual ash-flow tuff sheets ranges from a few meters to over 300 m. Tertiary ash-flow tuffs are widespread in the western United States, particularly in the Basin and Range province. After Winograd (1971).

relatively low, because the pores are fairly small and not well-connected. The interior of the flow cools relatively slowly and, in the process, loses much of its original porosity, so that the final value of porosity is quite low (<0.05). However, the interior also fractures during cooling, and the interconnected fractures transmit water very effectively despite the low overall porosity. The net result of the cooling history is that flow interiors typically have up to 10^4 times higher permeability than flow tops and bottoms (Figure 1.3; Winograd, 1971) despite their much lower porosities (0.05 versus 0.50).

1.2.2 Heterogeneity and anisotropy

In general, permeability exhibits extreme spatial variability or *heterogeneity*, both among geologic units and within particular units. We have already mentioned a typical 10^3-fold variation between core-scale and in situ–scale values in fractured crystalline rocks (Brace, 1980, 1984) and up to 10^4-fold variation in in situ–scale values within ash- flow tuff units (Winograd, 1971). Similarly large variations have been measured within particular soil units (Mitchell, 1993), and even larger variations in in situ permeability have been inferred between basalts near the surface of the east rift zone of Kilauea volcano ($k \sim 10^{-10}$ to 10^{-9} m^2) and compositionally identical rocks at 1- to 2-km depth ($k \sim 10^{-16}$ to 10^{-15} m^2) (Ingebritsen and Scholl, 1993).

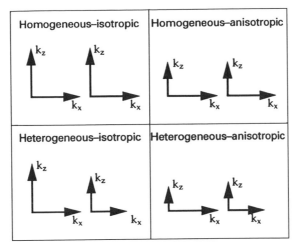

Figure 1.4 The four possible combinations of heterogeneity and isotropy. In a homogeneous, isotropic medium, permeability does not vary with position or with direction at a single point. In a homogeneous, anisotropic medium, permeability does not vary with position but does vary with direction at a point ($k_x > k_z$ in the case shown here). In a heterogeneous, isotropic medium, permeability varies with position but not with direction. Finally, in a heterogeneous, anisotropic medium, permeability varies with both position and direction. After Freeze and Cherry (1979).

Permeability is also generally an *anisotropic* or direction-dependent property. Thus in hydrogeologic practice, two pairs of adjectives are used to characterize the permeability of geologic media. A unit is either *homogeneous* or *heterogeneous* and either *isotropic* or *anisotropic* (Figure 1.4). The homogeneous, isotropic medium is convenient fiction; in reality, no geologic material is homogeneous with respect to permeability. Usually what is meant by "homogeneous" is that permeability is assumed to be randomly distributed about some mean value, or that permeability at the *representative elementary volume* (REV) level (Chapter 1.4) does not change from one REV to the next. The heterogeneous, anisotropic medium is the general case. Analytical solutions can often be obtained for groundwater flow in homogeneous, isotropic media, whereas equations describing flow in heterogeneous, anisotropic media must generally be solved numerically.

The most important cause of anisotropic permeability is sedimentary and volcanic layering. Such layers exhibit anisotropy at multiple scales, ranging from small-scale anisotropy caused by the horizontal orientation of platey minerals to larger-scale anisotropy caused by (for example) alternating clay and sand layers. The general effect of sedimentary layering is that permeabilities parallel to bedding tend to be much higher than the permeabilities orthogonal to bedding.

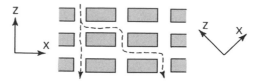

Figure 1.5 Schematic diagram of a two-dimensional medium illustrating how off-diagonal terms in the hydraulic-conductivity tensor (Eqs. 1.9 to 1.11) come into play when the Cartesian coordinate axes are not aligned with the principal directions of permeability. If we adopt the coordinate system on the right-hand side, then a hydraulic gradient in the z direction will cause components of flow in both x and z directions.

By solving Darcy's law for a series of stacked layers (e.g., Maasland, 1957) it can be shown that the equivalent vertical permeability k_z of n layers is given by the harmonic mean of the layer permeabilities, $k_z = b_t / (\sum_{i=1}^{n} b_i / k_i)$, where b_i and k_i are the thickness and permeability of the ith layer and b_t is the total thickness of the stacked layers. The equivalent horizontal permeability k_x of i layers is given by the arithmetic mean, $k_x = \sum_{i=1}^{n} k_i (b_i / b_t)$. This implies that the equivalent vertical permeability of stacked layers will tend to be controlled by the lower-permeability layers, whereas the equivalent horizontal permeability will tend to be controlled by the higher-permeability layers. For example, consider a sedimentary sequence composed of equal thicknesses of isotropic, intact shale ($k_z \sim k_x \sim 10^{-20}$ m^2) and isotropic sandstone ($k_z \sim k_x \sim 10^{-14}$ m^2). The equivalent vertical permeability of the entire sequence will be 2×10^{-20} m^2, whereas the equivalent horizontal permeability will be 0.5×10^{-14} m^2.

The anisotropy of permeability motivates introduction of a three-dimensional form of Darcy's law. In order to simplify the notation we will write this three-dimensional form in terms of hydraulic conductivity (K) rather than permeability (k); recall that $K = k \rho_f g / \mu$. Consider the schematic two-dimensional medium shown in Figure 1.5 and suppose that we adopt the coordinate system shown on the left-hand side. If a unit head gradient $\partial h / \partial x = 1$ is imposed on this system, with $\partial h / \partial z = 0$, then

$$q_x = -K_x \quad \text{and} \quad q_z = 0.$$

Similarly, if a unit head gradient $\partial h / \partial z = 1$ is imposed on this system, with $\partial h / \partial x = 0$, then

$$q_x = 0 \quad \text{and} \quad q_z = -K_z.$$

However, if we adopt the coordinate system shown on the right-hand side, a unit gradient in the z direction will cause flow in both the x and z directions.

Thus a more general, three-dimensional form of Darcy's law is

$$q_x = -K_{xx}\frac{\partial h}{\partial x} - K_{xy}\frac{\partial h}{\partial y} - K_{xz}\frac{\partial h}{\partial z}$$

$$q_y = -K_{yx}\frac{\partial h}{\partial x} - K_{yy}\frac{\partial h}{\partial y} - K_{yz}\frac{\partial h}{\partial z} \qquad (1.9)$$

$$q_z = -K_{zx}\frac{\partial h}{\partial x} - K_{zy}\frac{\partial h}{\partial y} - K_{zz}\frac{\partial h}{\partial z}$$

or, in vector notation,

$$\begin{pmatrix} q_x \\ q_y \\ q_z \end{pmatrix} = - \begin{pmatrix} K_{xx} & K_{xy} & K_{xz} \\ K_{yx} & K_{yy} & K_{yz} \\ K_{zx} & K_{zy} & K_{zz} \end{pmatrix} \begin{pmatrix} \partial h/\partial x \\ \partial h/\partial y \\ \partial h/\partial z \end{pmatrix}. \qquad (1.10)$$

We can further simplify the notation by introducing the vector operator ∇ to indicate $\mathbf{i}\partial/\partial x + \mathbf{j}\partial/\partial y + \mathbf{k}\partial/\partial z$ and writing

$$q = -\overline{K}\,\nabla h, \qquad (1.11)$$

where boldface type indicates a vector quantity and the overline indicates that K is now a second-order tensor. The data implications of the nine spatially variable coefficients $\overline{K}(x, y, z)$ are rather overwhelming, particularly when one considers the difficulty of determining a single value of "homogeneous, isotropic" permeability. Fortunately, because $\overline{K}(x, y, z)$ is symmetric and has principal directions, the coordinate axes can frequently be aligned with the principal directions of the $\overline{K}$ tensor, so that the off-diagonal terms of Eqs. (1.9) to (1.11) become zero. For example, in layered media the coordinate axes can often be aligned with bedding.

1.2.3 Scale dependence

Values of permeability are often determined by three methods, each of which measures or infers permeability at a different volume-averaged scale. Laboratory tests measure permeability at the drill-core scale, sometimes using methods as simple as that shown in Figure 1.1 but often using more sophisticated apparatus to make transient (non-steady-state) measurements, impose large head gradients, and/or replicate in situ pressure and temperature conditions. Regardless of the exact experimental design, the volume of material sampled in laboratory tests is generally very small, almost always $\ll 1$ m^3.

In situ or wellbore tests are done by pumping from a wellbore at a steady rate while monitoring hydraulic head, by monitoring the loss of fluid from a wellbore at a fixed head value, or by changing the hydraulic head in a wellbore and monitoring its recovery. The head response may be monitored in the perturbed well only (single-well tests) or in the perturbed well and one or more nearby *observation wells* (multiwell tests). The response of a well or wells is matched with solutions to the *diffusion equation* (Chapter 1.5.3) under various assumptions about boundary conditions and the nature of the well–aquifer system (for a description of wellbore test methods, see, e.g., Earlougher, 1977; Driscoll, 1986; Fetter, 1994). The volume of material sampled by such in situ tests varies with the size and duration of the perturbation and with the hydraulic properties of the medium, but it generally ranges from <10 m^3 (for a single-well test in a low-permeability medium) to perhaps $>10^5$ m^3 (in a high-permeability medium).

There is no direct way to measure larger-scale permeabilities, but larger-scale "regional" values are often inferred from the results of numerical-modeling experiments. In such experiments the unknown values of regional-scale permeability are varied so that the numerical simulation results match known values of hydraulic head, rates of groundwater flow, solute concentrations, or temperatures. Regional permeability values inferred on this basis are applied to volumes ranging from perhaps 10^2 m^3 to $>10^3$ km^3.

On the basis of measurements and inferences at these different and often nonoverlapping scales, permeability often appears to be a scale-dependent property. The apparent scale dependence is most pronounced in crystalline rocks, where permeability is largely provided by discrete fractures. For such rocks there is typically a 10^4- to 10^6-fold variation among laboratory-scale measurements at a particular site (Brace, 1980), and the mean in situ value is typically about 10^3 higher than the mean value determined by laboratory tests (Brace, 1980, 1984; Clauser, 1992). This is because the cores for laboratory analysis (typically $\sim$0.1-m long) tend to come from mechanically sound (unfractured) sections of the wellbore. The in situ tests sample a longer section of wellbore (usually >10 meters) and encounter more of the natural heterogeneity of the medium.

Laboratory and in situ permeability measurements on argillaceous rocks often agree to within an order of magnitude (Brace, 1980; Neuzil, 1994), but there are possible exceptions. For example, numerical-modeling experiments suggest that the Cretaceous Pierre Shale of the North American midcontinent, which has a laboratory and in situ–scale permeability of $\sim$$10^{-20}$ m^2 (e.g., Neuzil, 1994), may have a regional-scale permeability of $\sim$$10^{-16}$ m^2 in South Dakota (Bredehoeft and others, 1983). However, the same Pierre Shale appears to have a regional-scale permeability of $\leq$$10^{-19}$ m^2 in the Denver basin of Colorado, Nebraska, and Wyoming (Belitz and Bredehoeft, 1988), consistent with laboratory-scale measurements. Perhaps the higher regional-scale permeability in South Dakota is due to large-scale fractures, that is, fractures on too large a scale to be encountered by in situ tests that sample relatively small volumes of this low-permeability material. If this is the case, we must speculate that such fractures do not contribute significantly to regional-scale permeability in the Denver basin area to the southwest.

In studies of basin-scale flow and heat transport in the Uinta basin, Utah, and in the North Slope basin, Alaska, calibration with thermal data was used to estimate regional-scale permeabilities (see Chapter 4.3.3). The Uinta basin workers found that temperature gradients are depressed by groundwater recharge near the high-elevation margin of the basin and enhanced by upflow of groundwater in the central part of the basin (Chapman and others, 1984). By simulating groundwater flow and heat transport, they determined that the thermal observations can be reproduced if the two uppermost sedimentary units (a total thickness of 1–3 km) are assigned a permeability of about 5×10^{-15} m^2 (Willet and Chapman, 1987). This value is significantly less than the mean values of permeability determined by in situ hydraulic testing of the same rocks (about 10^{-12} m^2). In the case of the North Slope basin, heat flow is depressed in the foothills of the Brooks Range and enhanced near the Arctic Ocean, in the vicinity of the Prudhoe Bay oil fields (Deming and others, 1992). Deming (1993) found that the thermal observations were well matched by assigning horizontal permeabilities (k_x) that are equivalent to the arithmetic mean of the in situ (drillstem-test) values for a particular unit and vertical permeabilities (k_z) equivalent to the harmonic mean of the in situ values (see Chapter 1.2.2).

The preponderance of the evidence thus suggests that any trend towards higher permeability at larger scales does not continue between in situ and "regional" scales. In two of the most thoroughly studied cases (Denver and North Slope basins) there is fairly good agreement between in situ permeability measurements and the inferred regional-scale permeabilities. In one case (South Dakota), regional permeability values greatly exceed in situ values; in another (Uinta basin) the regional values are significantly less than the in situ values. Furthermore, Clauser's (1992) compilation of crystalline-rock data shows reasonably good agreement between the overall in situ and regional-scale permeability ranges. We have already suggested that regional-scale permeabilities higher than in situ values (e.g., the South Dakota example) might be due to undersampling of large-scale, regionally significant fractures. Regional-scale permeabilities lower than in situ values (e.g., the Uinta basin) might also be caused by sample bias; in situ hydraulic data are perhaps more likely to be obtained from high-permeability layers, whereas the lower-permeability layers would tend to control regional-scale flow (Willett and Chapman, 1987). An alternative explanation for some cases in which in situ permeabilities greatly exceed the "thermally reasonable" regional values is that the in situ tests have sampled permeable fracture networks that are discontinuous on a regional scale.

1.2.4 Depth dependence

At any particular site, and for a uniform lithology, permeability is likely to decrease more-or-less systematically with depth. Depth itself is not the activating factor; the depth dependence is due mainly to loss of porosity through increasing confining pressure and *effective stress* (Chapter 8.1) and to temperature- and pressure-dependent diagenetic and metamorphic processes. In practice it may be hard to distinguish among these various effects. Furthermore, the general decrease in permeability with depth is not necessarily monotonic. It may be temporarily reversed by the presence of permeable geologic structures or strata at depth, by *anomalous fluid pressures* (Chapter 4.1.2) that decrease the effective stress, or by *hydraulic fracturing* (Chapter 4.1.3).

There is an important conceptual difference between the apparent scale dependence of permeability discussed in the previous section and the depth dependence considered here. We have suggested that much of the scale dependence might in fact be due to inadequate sampling of the medium. Insofar as this is the case, the scale dependence is not a real phenomenon, but rather an artifact of our inadequate sampling capabilities. In contrast, the general depth dependence of permeability is due to physical and chemical processes common to most geologic settings.

At depths such that temperatures are significantly elevated above near-surface ambient conditions, permeability tends to be reduced by a suite of metamorphic reactions that are collectively termed *diagenesis* or *hydrothermal alteration*. The term *diagenesis* refers to postdepositional physical and chemical changes in sedimentary strata and is generally applied in a low-temperature context (see Chapter 10 for greater detail). The moderate- to high-temperature reactions termed *hydrothermal alteration* often decrease permeability by decreasing porosity, as relatively low-density minerals (e.g., clays, zeolites) replace higher-density ones (e.g., feldspars). In the Cascade Range of the northwestern United States, substantial reductions in the permeability of volcanic rocks seem to be associated with temperatures $>50°C$ (Blackwell and Baker, 1988). We have already mentioned (Chapter 1.2.2) the large difference in in situ permeability between basalts near the surface of the east rift zone of Kilauea volcano ($\sim25°C$; 10^{-10} to 10^{-9} m^2) and compositionally identical rocks at 1- to 2-km depth (50 to 200°C; 10^{-16} to 10^{-15} m^2). At Kilauea it is not clear whether pervasive intrusion by less-permeable basaltic dikes or hydrothermal alteration is the primary cause of the lower permeability at depth. Both are probably important and are related by the role of recently emplaced dikes as heat sources that help to drive alteration.

Permeability–depth curves are often invoked in models of regional-scale groundwater flow to represent the effects of *effective stress*–dependent consolidation. To represent the permeability of the Dakota Sandstone in a model of the North American midcontinent, Belitz and Bredehoeft (1988) chose $\log(\log k / k_o) \propto$ depth, where k_o is the near-surface permeability. Their choice was motivated by observed correlations between $\log k$ and n and between $\log n$ and depth, and it leads to sandstone permeability variations of about 3 orders of magnitude over a depth range of about 4 km. To represent the permeability of cross sections across the San Andreas fault in California (Chapter 4.3.5), Williams and Narasimhan (1989) chose $\log k = -0.2 \times$ depth(km) $- 15$, so that permeability varied from 10^{-15} m^2 at the surface to 10^{-18} m^2 at 15-km depth, roughly matching the range of in situ permeabilities measured in fractured (10^{-15} m^2) and intact (10^{-18} m^2) crystalline rock in nearby drillholes.

Both theoretical models (e.g., Gangi, 1978) and laboratory results (e.g., Morrow and others, 1994) suggest that the permeability of fractured crystalline rocks will generally decrease with increasing confining pressure or *effective stress* (Chapter 8.1). Although early data compilations by Brace (1980, 1984) showed little evidence of depth-dependent permeability in crystalline rocks, most of Brace's in situ data were from <500 m depth. More recent data indicate a dependence on confining pressure that varies with rock type. The in situ permeability of highly fractured granitic rocks from the Black Forest of Germany does not appear to vary with depth, whereas less-fractured gneissic rocks from the same region do show depth-dependent in situ permeability (Stober, 1996). Similarly, the laboratory-scale permeabilities of relatively fracture-free basalts from the deep Kola (Russia) and KTB (Germany) drillholes are sensitive to confining pressure, whereas highly fractured granodiorite gneisses from the same holes are less pressure sensitive (Morrow and others, 1994). In the KTB hole, significant in situ permeability (10^{-17} m^2) is found at >9 km depth (Huenges and others, 1997).

The laboratory-scale permeability of deep drillhole core samples is generally both lower and more pressure sensitive than the permeability of similar rocks sampled at surface outcrops (Morrow and Lockner, 1994). The lack of pressure sensitivity in the outcrop samples may result from near-surface unloading and weathering, which causes near-surface fracture openings to be less uniform than those at depth and therefore less easily closed.

1.2.5 Time dependence

Because of ongoing deformation, dissolution and precipitation of minerals, and other metamorphic processes, permeability is also a time-dependent property. The transient nature of permeability has long been recognized by economic geologists, who see evidence of episodic fracture creation and healing in fossil hydrothermal systems (e.g., Titley, 1990). However, hydrogeologists have rarely incorporated time-dependent permeabilities in quantitative analyses of groundwater flow and transport.

As is the case with depth, time itself is not the activating factor. Nonetheless, it is useful to develop some appreciation of the time scales over which various geologic processes are likely to affect permeability. Some geologic processes (e.g., compaction of sediments) cause a gradual evolution of permeability, whereas others (e.g., hydrofracturing, earthquakes) act very rapidly. The examples that we cite in this section include compaction (really a form of effective stress dependence; Chapter 8.1), carbonate dissolution (*diagenesis*; Chapter 10), hydraulic fracturing (again effective stress dependence) and tectonic strain (Chapter 8.6).

The reduction of pore volume during sediment burial within evolving sedimentary basins causes compaction-driven groundwater flow and also modifies the permeability structure. Porosity reduction due to compaction is frequently described by Athy's (1930) law

$$n = n_0 e^{-bd}, \tag{1.12}$$

where n_0 is the porosity at the land surface, b is an empirical constant, and d is depth. The depth dependence of porosity and permeability in subsiding basins can be complicated by metamorphic reactions and by the development of *overpressured* zones in which fluid pressures significantly exceed hydrostatic (e.g., Bethke, 1986), thereby inhibiting compaction. Laboratory-scale shale permeabilities from the U.S. Gulf Coast vary from about 10^{-18} m^2 near the surface to about 10^{-20} m^2 at 5-km depth (Neglia, 1979), and the natural subsidence rate for Gulf Coast–type basins is 0.1 to 10 mm/yr (Sharp and Domenico, 1976). Thus we can infer that in this environment it takes perhaps 10^5 to 10^7 years for the permeability of a subsiding packet of shale to decrease by a factor of ten.

Water–rock reactions can cause permeability to evolve much more rapidly. Simulations of groundwater flow and calcite dissolution in coastal carbonate aquifers (Chapter 10.3.2) suggest significant changes in porosity and permeability over time scales of 10^4 to 10^5 years (Sanford and Konikow, 1989a, 1989b). Analyses of near-surface silica precipitation in hydrothermal upflow zones indicate that at high temperatures ($\sim$300°C) even large (1 mm) fractures can be effectively sealed in 10 years (Lowell and others, 1993). Although such chemical deposition is a major cause of permeability reduction in hydrothermal systems, the opening and closing of crack networks due to thermoelastic response to heating and cooling can cause permeability to evolve even more rapidly (Germanovich and Lowell, 1992).

Hydraulic fracturing and earthquakes cause essentially instantaneous changes in permeability. Hydraulic fracturing occurs when pore-fluid pressure exceeds the sum of the least principal stress and the tensile strength of the surrounding rock. In general, this is likely to occur only in low-permeability geologic media. The lithostatic load ($P_f = \rho_r g d$) sets an upper bound for failure, and common causes of overpressuring ($P_f \gg \rho_f g d$) include rapid burial of low-permeability sediments (as along the Gulf Coast), thermal pressurization, or some source of fluids at depth. The effects of hydraulic fracturing on the permeability structure can be simulated by arbitrarily increasing permeability when pore pressures are sufficient to induce failure (e.g., Bredehoeft and Ingebritsen, 1990) or through more rigorous simulations of the physics of fracture growth (e.g., Renshaw, 1996). The fact that earthquakes can instantaneously change permeability on a regional scale was vividly demonstrated by the **M** 7.1 Loma Prieta (California) earthquake of 1989 (Chapter 8.6.1; Rojstaczer and Wolf, 1992). Immediately after the earthquake, many streams within about 50 km of the epicenter experienced a 10-fold increase in streamflow. Simultaneously, groundwater levels dropped 20 m or more in some nearby highland areas. Both effects can be attributed to an abrupt 10-fold increase in near-surface (0–300 m depth) permeability, likely caused by coseismic movement along favorably oriented and once-cemented fractures.

1.2.6 Some limiting values

The permeability of geologic media largely determines the feasibility of some important geologic processes (e.g., advective solute and heat transport) as well as their economic potential in terms of ore genesis, hydrocarbon extraction, and water supply. We will conclude this section by introducing some useful limiting values of permeability (Figure 1.6).

At sufficiently low permeabilities, solute transfer occurs mainly by diffusion (Chapter 2.1.1) and heat transport occurs mainly by the analogous process of heat conduction (Chapters 3.1.4 and 7.1). These processes do not depend directly on the rate of fluid flow. For high enough flow rates, *advective transport* by the flowing fluid becomes significant. The effectiveness of advective solute transport is proportional to the average linear velocity $v(q/n)$, and an analytical solution for one-dimensional flow and transport (Chapter 3.4.1) shows that advection will substantially affect the concentration distribution for $q \geq 10^{-13}$ m/s, assuming a length scale of a few hundred meters and that $n \sim 0.1$. We can relate this flow rate to a hydraulic conductivity of $> 10^{-13}$ m/s by further assuming $dh/dL \ll 1$. Under standard temperature and pressure conditions this value of hydraulic conductivity translates to an intrinsic permeability

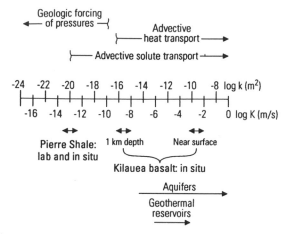

Figure 1.6 Range of permeabilities observed in geologic media showing certain process-limiting values and other selected values discussed in the text. Permeability (k, m^2) and hydraulic conductivity ($K = k\rho_f g/\mu$, m/s) are related for water density ρ_w and viscosity μ_w at 15°C.

of $>10^{-20}$ m^2 (Figure 1.6). The effectiveness of advective heat transport is proportional to the quantity $c_f \rho_f q_f$, where c_f is the heat capacity of the fluid. Analytical solutions show that, over a similar length scale, advection of heat by liquid water can begin to substantially affect the temperature distribution at flow rates on the order of 10^{-9} m/s. For $dh/dL \ll 1$, this translates to hydraulic conductivities $\geq 10^{-9}$ m/s or permeabilities $\geq 10^{-16}$. Numerical simulations of hydrothermal systems confirm that advective heat transport generally becomes more effective than heat conduction for permeabilities $\geq 10^{-16}$ m^2 (Norton and Knight, 1977; Manning and others, 1993; Ingebritsen and Hayba, 1994). In certain ore-forming environments, such as epithermal systems, both solute and heat transport is dominated by advection, at least episodically, so that we can infer permeabilities $\geq 10^{-16}$ m^2.

Thus in most media we can assume that advective solute transport is important, whereas some media (e.g., shales) have low enough permeabilities that we can neglect advective heat transport. The roughly 10^4 disparity between the permeability thresholds for significant advection-dominated solute transport ($\sim 10^{-20}$ m^2) and significant advection-dominated heat transport ($\sim 10^{-16}$ m^2) is due to the relative efficiency of conduction as a heat-transport mechanism versus diffusion as a solute-transport mechanism (see Chapter 3.4.1).

When permeabilities are sufficiently low, what has been termed *geologic forcing* can strongly influence hydraulic heads (Chapter 4.1; Neuzil, 1995). That is, the fluids produced by such processes as subsidence and sedimentation,

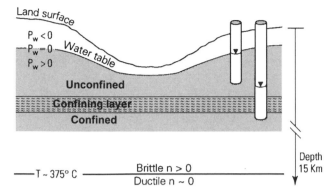

Figure 1.7 Schematic diagram of a hydrogeologic system showing the domain of interest between the water table and the brittle–ductile transition and defining the *water table* and *confined* and *unconfined* groundwater systems. Below the brittle–ductile transition, ductile deformation presumably collapses porosity, so that $n \sim 0.0$.

petroleum generation, pressure solution, crustal deformation, heating, and mineral dehydration can significantly influence the head distribution. The rate of fluid production by various geologic processes is such that they can be important when moderately large regions of a flow domain ($L > 100$ m) are composed of or bounded by material with $k \le 10^{-17}$ m^2.

Comparatively high permeabilities are needed for economic production of fluid from wells. Even a poor aquifer – for example, one that could support a single-household domestic well – requires a permeability of at least 10^{-14} m^2. Commercially exploited geothermal reservoirs (Chapter 7.7) generally consist of zones with $k \ge 10^{-13}$ m^2. Exceptionally permeable aquifer materials, such as well-sorted gravels, have permeabilities of up to 10^{-7} m^2.

1.3 Conceptualizing groundwater systems

Having thoroughly explored Darcy's law and the associated hydrogeologic parameters in Chapters 1.1 and 1.2, we will now introduce some less-quantitative but equally important hydrogeologic concepts, including the *water table* and *confined* and *unconfined* systems.

In this book we are generally concerned with the depth range between the *water table* and the brittle–ductile transition in the midcrust (which might occur at $\sim$15-km depth in a region with a normal geothermal gradient). The water table is defined as the surface where pressure in the liquid phase, $P_{\rm w}$, is equivalent to atmospheric pressure, $P_{\rm atm}$ (Figure 1.7). Below the water table $P_{\rm w} > P_{\rm atm}$ and, less obviously, above the water table $P_{\rm w} < P_{\rm atm}$, because air and water coexist in the pore space. The air pressure above the water table $P_{\rm air} \sim P_{\rm atm}$, requiring

$P_w < P_{atm}$ in order to balance the surface tension acting along the curved air–water interfaces (see Chapter 6.2.1 for a discussion of surface tension effects).

Under most hydrogeologic conditions the water table is a subdued replica of the topography, as is shown schematically in Figure 1.7. Even under extreme conditions, depths to the water table from the land surface generally amount to < 500 m. Large excursions from the land surface may occur in regions of moderate to great topographic relief, high permeability, and/or in arid climates (e.g., Kilauea volcano, Hawaii, the High Lava Plains of Oregon, and the Basin and Range province of the western United States). Most regional-scale groundwater flow is driven by variations in hydraulic head caused by topographic relief.

The groundwater flow system immediately below the water table is an *unconfined* system, within which the total hydraulic head (Chapter 1.1.2) roughly coincides with the elevation of the top of the water-bearing unit (i.e., with the water table; see Figure 1.7). Less-permeable horizons below the water table act as *confining layers* (Figure 1.7), and the total hydraulic head h in the underlying *confined* systems is greater than the elevation at the base of the overlying confining layers ($h = P_w/\rho_w g + z > z$). The mathematical description of unconfined groundwater systems is complicated by the fact that the water table acts as a moving upper boundary and by the fact that water flows across this moving boundary in response to variable head gradients. The particular complication caused by water-table conditions is one that we will generally be able to ignore in this book, because most of the processes that we are concerned with occur at depth ranges such that it is reasonable to assume some degree of confinement.

Another basic hydrogeologic concept is that the pore or void space is saturated with (dominantly) aqueous fluids between the water table and midcrustal depths. Hydrogeologists tend to call these aqueous fluids "groundwater," whereas geologists and geophysicists tend to call them simply "fluids." With increasing depth there is in fact an increasing likelihood that the pore fluid will be dominated by something other than groundwater of recent meteoric origin, for example, by connate waters; metamorphic, magmatic, or mantle-derived fluids; or by hydrocarbons. However, regardless of the origin of the fluids, the "groundwater" theory described in Chapters 1 to 3 is generally applicable so long as some form of Darcy's law describes the flow.

A final basic assumption is that all nonductile geologic media have some finite ability to transmit water via Darcian flow. Since modern equipment has decreased the feasible measurement limit to perhaps 10^{-23} m^2, nearly all geologic media have been shown to have finite permeability. Whether that permeability is significant with respect to a particular problem will depend on the process(es) under consideration (Chapter 1.2.6).

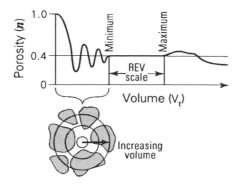

Figure 1.8 Porosity (*n*) as a function of averaging volume. At a particular point ($V_t \sim 0$), the value of *n* is either 0 or 1. The computed value of *n* stabilizes as it is averaged over progressively larger volumes. Over a representative elementary volume, or REV (Bear, 1972), the value is essentially constant. After Hubbert (1956).

1.4 The continuum approach

We do not describe Darcian groundwater flow and transport at a microscopic level, that is, at the level at which individual molecules or even the details of pore-fracture geometries are important. Instead, we describe flow and transport phenomena at a macroscopic level, using averaged properties. The domain of interest consists of both solids and void space filled with one or more fluids, and in nearly all cases we do not know the distribution of solid–fluid boundaries well enough to use classical fluid-mechanics approaches. Key parameters such as permeability (Chapter 1.2) and dispersivity (Chapter 2.1.3) can be regarded as effects of small-scale solid-fluid geometries that we cannot map and thus wish to ignore.

To demonstrate that some key medium properties are only definable on a macroscopic scale, let us consider porosity ($V_v / V_t = n$). At any microscopic point in a domain, porosity will be either 0 in the solid material or 1 in a pore space (Figure 1.8). As one averages over progressively larger volumes, the computed value of *n* will fluctuate over a progressively smaller range. If the medium is sufficiently homogeneous, the volume-averaged value of *n* will eventually become nearly constant (e.g., at a value of 0.4 in Figure 1.8). The volume range over which the average *n* value remains constant has been termed the *representative elementary volume*, or REV range (Bear, 1972, 1979). In a homogeneous material, the REV range could be arbitrarily large. However, all geologic media have some larger-scale heterogeneity, and as one continues to average over larger volumes, *n* will eventually depart from its REV-scale average. The variation at $>$REV scale will be relatively smooth, because the averaging volume has become large. The diagram shown as Figure 1.8 would apply equally as well to volumetric saturation ($V_f/V_v = S_f$), a key parameter in descriptions of multiphase flow (see, e.g., Chapter 3.1). Like porosity, volumetric saturation would be 0 or 1 at any microscopic point, nearly constant in the REV range, and smoothly varying at $>$REV range.

It is important to consider when an REV-based or *continuum* approach is justified. The REVs must be large relative to the scale of microscopic heterogeneity (e.g., grain size in a granular porous medium) but small relative to the entire domain of interest. The appropriate REV size will vary dramatically depending on the problem under consideration and the nature of the geologic medium. For example, the minimum REV size needed to represent permeability in a well-sorted coarse sand ($d_{50} \sim 0.001$ m) would be about 10^{12} times smaller than the minimum REV size needed to represent permeability in a granite where flow is dominated by fractures spaced 10 m apart.

The REV or continuum approach is implicit in the derivation of the groundwater flow equation in the next section and in the derivation of heat- and solute-transport equations in Chapters 2 and 3. Furthermore, the REV concept is relevant to numerical solution of these equations. Both the flow and transport equations are usually solved numerically over spatially discretized problem domains. In this context the sizes of the discrete volume elements implicitly define the REVs for purposes of solution. One must therefore attempt to select volume-element sizes that are within the likely REV ranges for the natural system under consideration, so that their size does not influence the solution.

1.5 The groundwater flow equation

It has been said that

The physics of flow through porous media is embodied in a flow equation that distills into one short mathematical notation all the essential truths of groundwater flow.

(Freeze and Back, 1983)

We are now in a position to derive this groundwater flow equation, having discussed most of the essential parameters (*fluid potential, hydraulic head, hydraulic conductivity, permeability,* and *porosity*) in Chapters 1.1 and 1.2 and the continuum or REV approach in Chapter 1.4. We will proceed through the derivation by considering conservation of mass for an REV and the production of fluid from "storage" in a geologic medium (introducing *specific storage* and *specific yield*), critically evaluating the assumptions required at various steps.

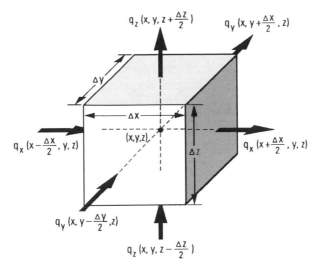

Figure 1.9 The control volume used to derive the continuity equation (Eq. 1.15). The vertical flow vectors (bold) denote volumetric flow rates per unit area through the $\Delta x \Delta y$ faces at distances of $\pm \Delta z/2$ from the central point x, y, z.

1.5.1 Conservation of mass

The groundwater flow equation is derived on the basis of conservation of mass. Consider a finite control volume fixed in space and centered on a point with spatial coordinates given by (x, y, z) (Figure 1.9). The change in mass stored within the volume over time increment Δt must equate to the difference between the mass flowing into the control volume and the mass flowing out of the control volume ($\Delta \text{mass} = \text{mass}_{\text{in}} - \text{mass}_{\text{out}}$). That is,

$$
\Delta[n\rho_f S_f \Delta x \Delta y \Delta z] = \left[\rho_f q_x \left(x - \frac{\Delta x}{2}, y, z \right) \Delta y \Delta z \Delta t \right. \tag{1.13}
$$

$$
+ \rho_f q_y \left(x, y - \frac{\Delta y}{2}, z \right) \Delta x \Delta z \Delta t
$$

$$
+ \left. \rho_f q_z \left(x, y, z - \frac{\Delta z}{2} \right) \Delta x \Delta y \Delta t \right]
$$

$$
- \left[\rho_f q_x \left(x + \frac{\Delta x}{2}, y, z \right) \Delta y \Delta z \Delta t \right.
$$

$$
+ \rho_f q_y \left(x, y + \frac{\Delta y}{2}, z \right) \Delta x \Delta z \Delta t
$$

$$
+ \left. \rho_f q_z \left(x, y, z + \frac{\Delta z}{2} \right) \Delta x \Delta y \Delta t \right],
$$

where S_f is the volume fraction of the void space that is occupied by fluid (V_f/V_v), $\Delta[n\rho_f S_f \Delta x \Delta y \Delta z]$ is the change in fluid mass stored in the volume $\Delta x \Delta y \Delta z$ over time increment Δt, and (for example) $q_x(x - \Delta x/2, y, z)$ is the volumetric inflow rate in the x direction across the $\Delta y \Delta z$ face of the control volume located at $x - \Delta x/2$ distance from point (x, y, z). Recall that q is the volumetric flow rate per unit area $(L^3/t/L^2)$ given by Darcy's law (Chapter 1.1, Eq. 1.2). This quantity must be multiplied by the fluid density ρ_f to arrive at a mass flow rate per unit area $(M/t/L^2)$, by the area of the face $\Delta y \Delta z$ to arrive at a total mass flow rate (M/t), and finally by the time increment Δt in order to arrive at units of mass (M). (The alert reader will notice a troubling inconsistency in Eq. (1.13). On the left-hand side we have introduced a saturation variable, S_f, to allow for the possibility of multiple fluid phases. However, we have not yet introduced a form of Darcy's law suitable for multiphase flow, so that the q terms on the right-hand side must describe a single fluid phase. In fact, multiphase forms of Darcy's law are not introduced until Chapters 3.1.3 and 6.4, where we describe steam–water and water–oil–gas flow, respectively. In this chapter we introduce the saturation variable for heuristic purposes, but we will generally assume that $S_f = 1$, so that the inconsistency between the two sides of the equation will eventually disappear.)

Rearranging Eq. (1.13) by grouping the flux through opposite faces and taking the limit as $\Delta t \to 0$ gives

$$\frac{\partial(n\rho_f S_f \Delta x \Delta y \Delta z)}{\partial t} = -\left[\rho_f q_x\left(x + \frac{\Delta x}{2}, y, z\right)\Delta y \Delta z \right. \tag{1.14}$$
$$\left. - \rho_f q_x\left(x - \frac{\Delta x}{2}, y, z\right)\Delta y \Delta z\right]$$
$$- \left[\rho_f q_y\left(x, y + \frac{\Delta y}{2}, z\right)\Delta x \Delta z \right.$$
$$\left. - \rho_f q_y\left(x, y - \frac{\Delta y}{2}, z\right)\Delta x \Delta z\right]$$
$$- \left[\rho_f q_z\left(x, y, z + \frac{\Delta z}{2}\right)\Delta x \Delta y \right.$$
$$\left. - \rho_f q_z\left(x, y, z - \frac{\Delta z}{2}\right)\Delta x \Delta y\right].$$

Then dividing both sides by $\Delta x \Delta y \Delta z$ and taking the limit as Δx, Δy, and $\Delta z \to 0$ gives

$$\frac{\partial(n\rho_f S_f)}{\partial t} + \frac{\partial(\rho_f q_x)}{\partial x} + \frac{\partial(\rho_f q_y)}{\partial y} + \frac{\partial(\rho_f q_z)}{\partial z} = 0, \tag{1.15}$$

which is known as the *continuity equation* for flow though a porous medium.

The continuity equation is a rather general statement of conservation of mass that involves few assumptions about the nature of the fluid or the geologic medium. From this point onward, though, the derivation of commonly invoked forms of the groundwater flow equation proceeds through a series of steps that require increasingly restrictive assumptions about the fluid, the medium, and/or the nature of the flow system. Several of these assumptions are aimed at allowing the groundwater flow equation to be posed in terms of a single dependent variable: *hydraulic head* (Chapter 1.1.2).

We will proceed by invoking the chain rule to expand the flow rate terms of the continuity Eq. (1.15), for example,

$$\frac{\partial(\rho_f q_x)}{\partial x} = q_x \frac{\partial \rho_f}{\partial x} + \rho_f \frac{\partial q_x}{\partial x}. \tag{1.16}$$

By assuming that the fluid density ρ_f does not significantly vary in space, we can then substitute

$$\frac{\partial(\rho_f q_x)}{\partial x} = \rho_f \frac{\partial q_x}{\partial x}, \quad \frac{\partial(\rho_f q_y)}{\partial y} = \rho_f \frac{\partial q_y}{\partial y}, \quad \text{and} \quad \frac{\partial(\rho_f q_z)}{\partial z} = \rho_f \frac{\partial q_z}{\partial z} \tag{1.17}$$

into Eq. (1.15), arriving at

$$\frac{\partial(n\rho_f S_f)}{\partial t} + \rho_f \left[\frac{\partial q_x}{\partial x} + \frac{\partial q_y}{\partial y} + \frac{\partial q_z}{\partial z} \right] = 0. \tag{1.18}$$

By invoking a form of Darcy's law appropriate for a single *incompressible fluid* and assuming that the Cartesian coordinates x, y, and z are aligned with the principal directions of the hydraulic conductivity tensor $\overline{K}$ (Chapter 1.2.2), we can make another significant set of substitutions:

$$q_x = -K_x \frac{\partial h}{\partial x}, \quad q_y = -K_y \frac{\partial h}{\partial y}, \quad \text{and} \quad q_z = -K_z \frac{\partial h}{\partial z}, \tag{1.19}$$

where h is hydraulic head. Substituting (1.19) into (1.18) and dividing through by ρ_f results in an equation

$$\frac{1}{\rho_f} \frac{\partial(n\rho_f S_f)}{\partial t} = \left[\frac{\partial \left(K_x \frac{\partial h}{\partial x} \right)}{\partial x} + \frac{\partial \left(K_y \frac{\partial h}{\partial y} \right)}{\partial y} + \frac{\partial \left(K_z \frac{\partial h}{\partial z} \right)}{\partial z} \right], \tag{1.20}$$

in which the right-hand ("flux") side is posed in terms of the single dependent variable h. Posing the left-hand ("storage") side of the equation in terms of h requires some further manipulation and definition of terms.

1.5.2 The storage term

The time-derivative or mass-storage term of the continuity equation, $\partial(n\rho_f S_f)/\partial t$, indicates that the mass of fluid stored at any point in the system is affected by temporal changes in porosity n, fluid density ρ_f, and volumetric fluid saturation S_f. The mass of fluid in storage increases with each of these quantities.

For a *confined* (Chapter 1.3) case in which the pore space is fully saturated by a single fluid phase, $S_f = 1$, and we can use the chain rule to expand the left-hand side of Eq. (1.20):

$$\frac{1}{\rho_f}\frac{\partial(n\rho_f)}{\partial t} = \frac{1}{\rho_f}\frac{\partial(n\rho_f)}{\partial h}\frac{\partial h}{\partial t}. \tag{1.21}$$

Then the entire groundwater flow equation can be written in terms of hydraulic head by simply defining a new parameter, *specific storage*,

$$S_s = \frac{1}{\rho_f}\frac{\partial(n\rho_f)}{\partial h}, \tag{1.22}$$

to represent the change in the volume of fluid stored in a unit volume of porous medium per unit change in hydraulic head. The specific storage has units of 1/L.

This approach to deriving specific storage (Eqs. 1.21 and 1.22) does not provide much insight into its physical nature, but the specific storage can in fact be related to two physically recognizable quantities, the uniaxial compressibility of the porous medium α and the compressibility of the fluid β. Differentiating the hydraulic head ($h = P/\rho_f g + z$) for a control volume fixed in space ($\partial z = 0$) gives

$$\partial h = \frac{1}{\rho_f g}\partial P, \tag{1.23}$$

and substituting the right-hand side of Eq. (1.23) into (1.22) gives

$$S_s = g\frac{\partial(n\rho_f)}{\partial P}. \tag{1.24}$$

Using the chain rule to expand the derivative in Eq. (1.24) then results in

$$S_s = \rho_f g\frac{\partial n}{\partial P} + ng\frac{\partial \rho_f}{\partial P} = \rho_f g\left[\frac{\partial n}{\partial P} + \frac{n}{\rho_f}\frac{\partial \rho_f}{\partial P}\right], \tag{1.25}$$

where we recognize

$$\frac{\partial n}{\partial P} \equiv \alpha \tag{1.26}$$

and

$$\frac{1}{\rho_f}\frac{\partial \rho_f}{\partial P} \equiv \beta, \tag{1.27}$$

so that

$$s_s = \rho_f g(\alpha + n\beta). \tag{1.28}$$

By assuming one stress (usually vertical) held constant and no deformation in the two perpendicular directions, s_s can also be expressed rigorously in terms of certain bulk moduli that Biot (1941) used to describe the elastic behavior of a fluid-saturated porous medium (see Green and Wang, 1990).

Numerical values of s_s are generally $\ll 10^{-3}$/m. In hydrogeologic practice the storage properties of a confined unit (Chapter 1.3) are often described in terms of the *storage coefficient* or *storativity*

$$s = bs_s, \tag{1.29}$$

where b is the thickness of a unit or layer. Because s_s has dimensions of 1/L, the storage coefficient is dimensionless. For confined units the storage coefficient is generally <0.005.

In unconfined or water-table single-phase systems (Chapter 1.3) the amount of fluid present in the system will vary with water-table elevation as well as with the porosity and fluid density. Under unconfined conditions the amount of fluid added or removed to the system per unit change in hydraulic head will be relatively large, amounting to a substantial fraction of the porosity n, and the storativity is defined by

$$s = bs_s + S_y, \tag{1.30}$$

where S_y, the *specific yield*, accounts for the changes in saturation due to movement of the water table. As the water table moves down (for example), the release of water due to pressure relief (defined by bs_s) is instantaneous, but the release of water from desaturating material (expressed by S_y) is gradual and depends upon the rate of drainage. This *delayed yield* phenomenon makes the storage behavior of unconfined systems relatively complex (cf. Neuman, 1979), but, as noted previously (Chapter 1.3), in this book we will generally be concerned with confined systems. The storativity of unconfined systems generally ranges from 0.05 to 0.25.

1.5.3 Various forms of the groundwater flow equation

We can now substitute Eqs. (1.21) and (1.22) into Eq. (1.20) to write, for a confined system

$$s_s \frac{\partial h}{\partial t} = \left[\frac{\partial \left(K_x \frac{\partial h}{\partial x} \right)}{\partial x} + \frac{\partial \left(K_y \frac{\partial h}{\partial y} \right)}{\partial y} + \frac{\partial \left(K_z \frac{\partial h}{\partial z} \right)}{\partial z} \right], \tag{1.31}$$

a form of the groundwater flow equation that is posed in terms of the single dependent variable hydraulic head. Let us review the assumptions that we made to arrive at this point.

We have assumed that the fluid density ρ_f does not vary significantly in space, both in simplifying the expanded flow rate terms of the continuity equation (Eqs. 1.15–1.18) and by invoking a form of Darcy's law appropriate for the single-phase flow of an incompressible fluid (Eqs. 1.19–1.20). We have further assumed that variations in fluid properties have a negligible effect on the hydraulic conductivity tensor $\overline{K}$, which is a function of both medium and fluid properties (e.g., $K_x = k_x \rho_f g / \mu$), and that its principal directions are aligned with the spatial coordinates x, y, and z.

There are also basic inconsistencies in the way that fluid-medium interactions were defined in developing the several essential components of Eq. (1.31): Darcy's law (Chapter 1.1) appears to describe groundwater flow rates with respect to a (possibly deforming) solid matrix rather than fixed spatial coordinates; the continuity equation (1.15) was developed for a nondeforming control volume with fixed spatial coordinates; and the derivations of specific storage (Eqs. 1.21–1.28) implied one-dimensional elastic deformation of the control volume itself.

Equations of groundwater motion similar to Eq. (1.31) were first derived by Jacob (1940), and the issue of the appropriate reference frame (fixed in space versus moving with the fluid) was widely discussed in the embryonic hydrogeologic community between 1940 and circa 1966. Questions about the validity of Jacob's formulation were in fact raised on the basis of the conflicting assumptions about deformation of the control volume (e.g., De Wiest, 1966). Cooper (1966) addressed many of these concerns by deriving equations of groundwater flow in both fixed and deforming coordinates, showing that Jacob's (1940) equation was generally appropriate so long as the relative velocity of the fluids was much larger than the velocity of the solids in a fixed reference frame.

Although equations such as (1.31) are widely applicable in hydrogeologic practice, there are both engineering (e.g., foundation settling) and geologic (e.g., accretionary prisms) applications where the velocity of the solid phase is significant and must be explicitly included in equations of groundwater motion. A more fundamental problem with respect to many geologic applications is that the fluid density (in particular) and viscosity cannot realistically be treated as constants, but in fact must be allowed to vary in space and time as a function of temperature, pressure, and sometimes solute concentration (e.g., Chapters 2.3 and 3.1).

Returning to the basic continuity equation (1.15), assuming single-phase fully saturated conditions, inserting a more general form of Darcy's law, and using the vector operator ∇ to indicate $\mathbf{i}\partial/\partial x + \mathbf{j}\partial/\partial y + \mathbf{k}\partial/\partial z$ gives rise to

$$\frac{\partial(n\rho_f)}{\partial t} = \nabla \cdot \left[\frac{\bar{k}\rho_f}{\mu_f}(\nabla P + \rho_f g \nabla z) \right], \tag{1.32}$$

a more general form of the groundwater flow equation that accounts for the effects of variable fluid properties by calculating fluxes in terms of the forces acting on the fluid $(\nabla P + \rho_f g \nabla z)$ rather than hydraulic head, by allowing hydraulic conductivity to vary with fluid density and viscosity, and by posing changes in mass storage in terms of changes in the fundamental controlling parameters, porosity and fluid density, rather than specific storage and hydraulic head. For most purposes of this book it is useful to consider Eq. (1.32), rather than more evolved variants such as Eq. (1.31), to be the fundamental groundwater flow equation. In Chapter 3.1 we will illustrate the extension of (1.32) to multiphase systems.

In hydrogeologic practice, however, it is common (and generally reasonable) to invoke even less general versions of Eq. (1.31). We will conclude this chapter by reviewing some common simplifications and the assumptions that they involve.

For a system with homogeneous but still anisotropic permeabilities (Chapter 1.2.2), K_x, K_y, and K_z no longer vary in space and can be moved outside of the spatial derivatives, giving

$$s_s \frac{\partial h}{\partial t} = K_x \frac{\partial^2 h}{\partial x^2} + K_y \frac{\partial^2 h}{\partial y^2} + K_z \frac{\partial^2 h}{\partial z^2}. \tag{1.33}$$

For a homogeneous and isotropic system, $K_x = K_y = K_z = K$, so that

$$\frac{\partial h}{\partial t} = \frac{K}{s_s} \left[\frac{\partial^2 h}{\partial x^2} + \frac{\partial^2 h}{\partial y^2} + \frac{\partial^2 h}{\partial z^2} \right], \tag{1.34}$$

a form that is well known to mathematicians as the *diffusion equation*. In Eq. (1.34) the term K/s_s is referred to as the *hydraulic diffusivity*. Under steady-state conditions, $\partial h/\partial t = 0$, and (1.34) reduces to

$$\frac{\partial^2 h}{\partial x^2} + \frac{\partial^2 h}{\partial y^2} + \frac{\partial^2 h}{\partial z^2} = 0, \tag{1.35}$$

another well-known equation in mathematics called *Laplace's equation*.

To treat media consisting of horizontal or subhorizontal layers, the form shown as Eq. (1.34) is often reduced to two spatial dimensions by invoking the storativity (Eq. 1.29) rather than the specific storage, defining an analogous depth-averaged property called *transmissivity* ($T = Kb$) to describe flow resistance, and writing

$$\frac{\partial h}{\partial t} = \frac{T}{s}\left[\frac{\partial^2 h}{\partial x^2} + \frac{\partial^2 h}{\partial y^2}\right], \tag{1.36}$$

which is sometimes called the *aquifer equation*. The *leakance* of fluid in the z direction between subhorizontal aquifers can be described with (1.36) by adding another term using a simple version of Darcy's law, and this *quasi-three-dimensional* approach to layered systems is employed in such widely used computer codes as the U.S. Geological Survey's MODFLOW (McDonald and Harbaugh, 1988). Although the hydraulic conductivity K will vary over the thickness of an aquifer, Eq. (1.36) is also often convenient to use in aquifer analysis, because analyses of in situ hydraulic tests (Chapter 1.2.3) usually lump variability in K into a single value of transmissivity T.

Equations 1.34 through 1.36 can often be solved analytically, especially for problems that can be reduced to one or two spatial dimensions. Many analytical solutions to the diffusion equation and Laplace's equation under various boundary conditions can be found, for example, in Carslaw and Jaeger (1959). (Chapters 2.1.6 and 3.2 discuss boundary conditions for hydrogeologic problems.) More general forms of the groundwater flow equation involving heterogeneous permeabilities and variable fluid properties require numerical solution.

Problems

1.1 The following field notes were taken at a nest of piezometers installed side by side at a single site:

	Piezometer		
	a	b	c
Ground-surface elevation (masl)	450	450	450
Depth of piezometer tube (m)	150	100	50
Depth to water in piezometer (m)	27	47	36

Let A, B, and C refer to the points of measurement of piezometers a, b, and c, and calculate **(a)** fluid pressure at A, B, C (N/m^2); **(b)** pressure head at

A, B, C (m); (c) elevation head at A, B, C (m); (d) hydraulic head at A, B, C (m); and (e) the hydraulic gradients from A to B and from B to C, assuming that the horizontal distances between A, B, and C are negligible. Finally, (f) propose a scenario that could explain the vertical flow directions indicated by the data. See Figure 1.2b for components of hydraulic head illustrated with reference to a single piezometer. (Freeze and Cherry, 1979.)

1.2 Show that the equivalent vertical permeability k_z of n layers is given by the harmonic mean, $k_z = b_t/(\sum_{i=1}^{n} b_i/k_i)$, where b_i and k_i are the thickness and permeability of the ith layer and b_t is the total thickness of the stacked layers. Use the fact that, from Darcy's law, the steady flow rate $q_z = k_1 \rho_w g \Delta h_1/b_1 = k_2 \rho_w g \Delta h_2/b_2 = \cdots = k_n \rho_w g \Delta h_n/b_n$, where Δh_i is the hydraulic head drop across the ith layer. (See Chapter 1.2.2.)

1.3 Under what conditions (if any) is Belitz and Bredehoeft's (1988) description of the permeability of the Dakota Sandstone [$\log(\log k/k_o) \propto$ depth] (Chapter 1.2.4) consistent with the Kozeny–Carman equation (Chapter 1.2.1)?

1.4 The following assumptions are implicit in one or more of the forms of the groundwater flow equation listed below. Next to each assumption, list the forms to which it applies, using the index numbers to the left of the equations. (See Chapter 1.5.)

Assumptions:
homogeneous medium_____
isotropic medium_____
two-dimensional flow_____
one-dimensional flow_____
steady-state conditions_____
constant fluid density_____

(1) $\quad S_s \dfrac{\partial h}{\partial t} = \left[\dfrac{\partial \left(K_x \frac{\partial h}{\partial x} \right)}{\partial x} + \dfrac{\partial \left(K_y \frac{\partial h}{\partial y} \right)}{\partial y} + \dfrac{\partial \left(K_z \frac{\partial h}{\partial z} \right)}{\partial z} \right]$

(2) $\quad S_s \dfrac{\partial h}{\partial t} = K_x \dfrac{\partial^2 h}{\partial x^2} + K_y \dfrac{\partial^2 h}{\partial y^2} + K_z \dfrac{\partial^2 h}{\partial z^2}$

(3) $\quad S_s \dfrac{\partial h}{\partial t} = K \left[\dfrac{\partial^2 h}{\partial x^2} + \dfrac{\partial^2 h}{\partial y^2} + \dfrac{\partial^2 h}{\partial z^2} \right]$

(4) $\dfrac{\partial^2 h}{\partial x^2} + \dfrac{\partial^2 h}{\partial y^2} + \dfrac{\partial^2 h}{\partial z^2} = 0$

(5) $\dfrac{\partial h}{\partial t} = \dfrac{T}{s} \left[\dfrac{\partial^2 h}{\partial x^2} + \dfrac{\partial^2 h}{\partial y^2} \right]$

(6) $\dfrac{\partial^2 h}{\partial r^2} + \dfrac{1}{r} \dfrac{\partial h}{\partial r} = \dfrac{s}{T} \dfrac{\partial h}{\partial t}$

(7) $\dfrac{\partial (n \rho_{\mathrm{f}})}{\partial t} = \nabla \cdot \left[\dfrac{\bar{k} \rho_{\mathrm{f}}}{\mu_{\mathrm{f}}} \left(\nabla P + \rho_{\mathrm{f}} g \nabla z \right) \right]$

2

Solute transport

The transport of dissolved chemical mass, or solutes, in groundwater comprises a significant component of many geologic processes. Many forms of metasomatism and diagenesis, for example, are driven by chemical reactions whose components are supplied by flowing groundwater. In addition, accumulations of hydrocarbons and metal-rich ore deposits are greatly controlled by the transport of chemicals by flowing groundwater. Groundwater also plays a role in the formation and dissolution of evaporite deposits. Detailed examples of how groundwater affects metasomatism, diagenesis, evaporites, and the formation of ore and hydrocarbon deposits will be discussed in later chapters. In this chapter we introduce the basic principles of solute transport in groundwater. As the chapter progresses we discuss the governing equations, techniques for solving these equations, variable-density flow and transport, and finally, multi-constituent reactive transport.

In this chapter, all solute components will be assumed to be miscible with the aqueous phase. Although miscible transport is adequate for describing many geologic processes, many others, such as hydrocarbon migration and steam transport in hydrothermal systems, involve the movement of two or more fluid phases. Multiphase systems require additional considerations that will be left for discussion in Chapter 6 for hydrocarbon phases and Chapters 3 and 7 for the steam phase.

2.1 Governing equations

As is the case for groundwater flow, solute transport can be described mathematically by combining principles of mass balance with expressions that relate the fluxes of solute to fundamental driving forces. Solutes move by molecular diffusion, advection, and hydrodynamic dispersion. Sources and sinks of solutes

35

(e.g., chemical reactions) can also be incorporated into solute-transport equations. If multiple solute-transport equations are written for different solutes, and these equations are coupled with the groundwater flow equation (Chapter 1.5), they can be used to quantify many of the reactive-transport processes that occur in the subsurface.

2.1.1 Molecular diffusion

In Chapter 1 we introduced Darcy's law as the flux of fluid resulting from a gradient in fluid potential or hydraulic head (Eq. 1.2). An analogous situation exists for solute transport – diffusive flux of chemical mass results from a gradient in the chemical potential, or *concentration*. This flux of chemical mass is directly proportional to both the concentration gradient and a coefficient of molecular diffusion. This is known as *Fick's first law* and can be expressed for open, single-phase water as

$$q_d = -D_w \frac{dC}{dx}, \tag{2.1}$$

where q_d is the diffusive flux, D_w is the *coefficient of molecular diffusion* in free or open water, C is the concentration of the molecule or ion, and x is the direction along the concentration gradient. Note the negative sign; the coefficient is always positive, so the flux of chemical mass is down the concentration gradient. The physical process driving molecular diffusion is simply the random motion of ions in solution. Ions in a region of higher concentration will eventually mix with ions in a region of lower concentration to create an equal distribution in space.

Different ions have different coefficients of molecular diffusion, as illustrated in Table 2.1. The coefficients in Table 2.1 are self-diffusion coefficients measured using isotopes in single-salt solutions. These tracer-diffusion coefficients show a general trend of decreasing value with both increasing charge and decreasing ionic radius. They do not take into account any interactions between the different types of ions present in solution. Such *binary interactions* and higher-order interactions will be discussed in Chapter 2.4.

Equation 2.1 expresses Fick's law in terms of D_w for open water. In the subsurface, diffusion occurs within a porous medium, and the coefficient of diffusion must be expressed in somewhat different terms. The presence of a solid phase restricts the area through which a solute can diffuse, and the tortuosity of the flow path increases the distance over which the solute must diffuse. Greenkorn and Kesslar (1972) express the coefficient of diffusion in a porous

Table 2.1 *Diffusion coefficients in water for some ions at 25°C[a]*

Cation	D_w $(10^{-10}\,m^2/s)$	Anion	D_w $(10^{-10}\,m^2/s)$
H^+	93.1	OH^-	52.7
Na^+	13.3	F^-	14.6
K^+	19.6	Cl^-	20.3
Rb^+	20.6	Br^-	20.1
Cs^+	20.7	I^-	20.0
Li^+	10.3	HS^-	17.3
		HCO_3^-	11.8
Mg^{2+}	7.05	HSO_4^-	13.3
Ca^{2+}	7.93	NO_2^-	19.1
Sr^{2+}	7.94	NO_3^-	19.0
Ba^{2+}	8.48	$H_2PO_4^-$	8.46
Ra^{2+}	8.89		
Mn^{2+}	6.88	CO_3^{2-}	9.55
Fe^{2+}	7.17	SO_4^{2-}	10.7
Cr^{3+}	5.94	HPO_4^{2-}	7.34
Fe^{3+}	6.07	CrO_4^{2-}	11.2

[a] After Li and Gregory (1974).

medium as

$$D_m = \frac{n_e}{\tau} D_w, \tag{2.2}$$

where n_e is the *effective porosity* and τ is the *tortuosity*. The effective porosity is that pore space through which molecules or fluid can move continuously, and it excludes any dead-end pore space. The tortuosity is defined here as the distance along a travel path that a molecule must follow between two points, divided by the shortest distance between those two points. In a porous medium the tortuosity is always greater than 1. The effect of porosity and tortuosity is that D_m is always less than D_w. Typical diffusion coefficients for geologic media range from about 10^{-11} to 10^{-10} m²/s.

Fick's first law (Eq. 2.1) combines with a mass-balance equation to produce *Fick's second law*. The mass-balance equation can be derived from principles of mass conservation in a *representative elementary volume* (Chapter 1.4), where [mass flux into the volume] − [mass flux out of the volume] = [rate of change of mass inside the volume] (for example Chapter 1.5.1). When one substitutes

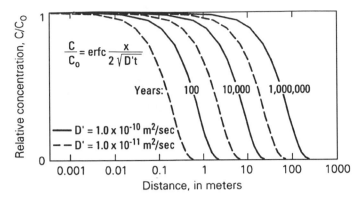

Figure 2.1 Rates of *molecular diffusion* illustrated by solutions of the one-dimensional diffusion equation with diffusion coefficients typical of geologic media.

Fick's first law in for the mass flux, one obtains *Fick's second law*:

$$D_m \nabla^2 C = n_e \frac{\partial C}{\partial t},\tag{2.3}$$

where the ∇^2 operator is $\partial^2/\partial x^2 + \partial^2/\partial y^2 + \partial^2/\partial z^2$. Equation 2.3, often called the *diffusion equation* in mathematics, is directly analogous to the groundwater flow equation of Chapter 1 and the heat conduction equation of Chapter 3. The particular form of the diffusion equation in (2.3) assumes that porosity does not vary with respect to space or time and that the coefficient of molecular diffusion does not vary with respect to space or concentration. One can solve Eq. (2.3) for a one-dimensional system with typical geological distances and coefficients (Figure 2.1). The results show that because the typical magnitudes of D_m in geologic media are relatively small, it can take from thousands to millions of years for chemicals to migrate significant distances. Thus diffusion alone cannot account for the transport of chemical mass over the long distances or relatively short time frames required by many geologic processes.

2.1.2 Advection

Another important transport mechanism for solutes is *advection*. Advection is simply the movement of chemical mass with flowing groundwater, whereby the solutes move at the same mean velocity as the groundwater and no concentration gradient is required for transport to occur. The one-dimensional advective transport equation can be written as

$$\frac{\partial C}{\partial t} = -v_x \frac{\partial C}{\partial x},\tag{2.4}$$

where v_x is the *average linear velocity*, or *seepage velocity*, in the x direction. Equation (2.4) can be understood intuitively as describing the translation of a concentration distribution in the x direction at a velocity v_x. The negative sign indicates that forward advection of a positive concentration gradient (concentration increases in the x direction) will lead to a decrease in concentration at that point. Although a concentration gradient is not required for advective transport to occur, in the absence of a gradient ($\partial C / \partial x = 0$) advection will not change the concentration at a point in space ($\partial C / \partial t = 0$). Another way to describe this relation is by applying a transformation variable, X', defined as a translation along the x direction at velocity v_x:

$$X' = x + v_x t. \tag{2.5}$$

Then, using the chain rule, Eqs. (2.4) and (2.5) can be combined to show that

$$\frac{\partial C}{\partial X'} = 0. \tag{2.6}$$

Equation (2.6) simply states that the concentration does not change at a point that is moving with the groundwater at seepage velocity v_x. Thus transport of mass in Eqs. (2.4) and (2.6) is by advection only.

In order to solve the advection equation (2.4), fluid fluxes and velocities must first be calculated using Darcy's law. Recall from Chapter 1 that the *average linear* or *seepage velocity*, v (Eq. 1.4), is different from the *Darcy velocity* or *specific discharge*, q (Eq. 1.2). The seepage velocity is equal to the specific discharge divided by the effective porosity and can be written as

$$v = \frac{q}{n_e} = -\frac{\overline{K}}{n_e} \nabla h, \tag{2.7}$$

where K is the hydraulic conductivity, h is the hydraulic head, boldface indicates a vector quantity, and ∇ represents the vector operator $\mathbf{i}\frac{\partial}{\partial x} + \mathbf{j}\frac{\partial}{\partial y} + \mathbf{k}\frac{\partial}{\partial z}$. To obtain a mass flux of solute (mass per unit area per time) by advection, q_a, one must multiply the seepage velocity by the concentration:

$$q_a = vC. \tag{2.8}$$

2.1.3 Mechanical dispersion

The heterogeneity of porous media creates groundwater velocity fields that are highly complex at the pore, outcrop, and regional scale (Figure 2.2). These heterogeneities create a variance in the groundwater velocity around the average linear velocity or seepage velocity, v, that we defined in Eq. (2.7).

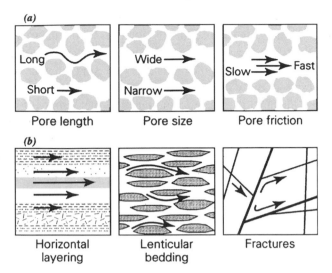

Figure 2.2 Physical mechanisms that cause *mechanical dispersion* on the (a) pore and (b) field scale.

The variations in the velocity field create an indirect transport process called *mechanical dispersion*. Because of mechanical dispersion, a concentration front that is originally sharp will spread out or *disperse* as it is advected with the groundwater.

Mechanical dispersion, like molecular diffusion, operates to disperse chemical mass in groundwater. Thus, mathematically, both processes can be invoked by a similar coefficient in a transport equation. These processes differ in that molecular diffusion depends solely on a concentration gradient to operate, is dominant at lower velocities, and can be treated as a scalar constant; mechanical dispersion, in contrast, depends on advection to operate and dominates at higher velocities; its rate coefficient is generally treated mathematically as a second-order tensor. At the pore scale, diffusion and mechanical dispersion are interconnected and can only be artificially separated (Bear, 1972). The combined effects of mechanical dispersion and molecular diffusion are called *hydrodynamic dispersion*. The coefficient of hydrodynamic dispersion can be defined in Cartesian tensor notation as

$$\overline{D}_{ij} = \alpha_{ijkl}\frac{v_k v_l}{|v|} + D'_{\mathrm{m}}, \quad i, j, k, l = 1, \ldots n_{\mathrm{d}} \tag{2.9}$$

(Scheidegger, 1961), where $\overline{D}_{ij}$ is the coefficient of hydrodynamic dispersion, α_{ijkl} is the *dispersivity* of the porous medium (a fourth-order tensor), v_k and v_l are the spatial components of the flow velocity, n_{d} is the number of spatial

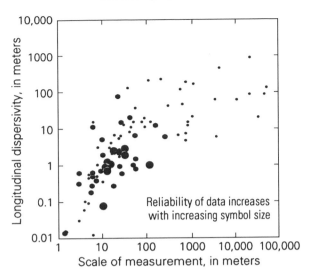

Figure 2.3 Values of *longitudinal dispersivity* as a function of the scale at which they were measured. After Gelhar (1986).

dimensions, D'_m is D_m/n_e, and $|v|$ is the magnitude of the velocity vector. The dispersion coefficient of (2.9) has 81 components in three-dimensional space (Scheidegger, 1961), but in an isotropic medium the dispersivity reduces to just two components: α_L, the dispersivity of the medium parallel to groundwater flow, and α_T, the dispersivity of the medium transverse to flow. For example, under two-dimensional transport conditions in a one-dimensional flow field, Eq. (2.9) reduces to

$$D_{xx} = D_L = \alpha_L v_x + D'_m \tag{2.10a}$$

and

$$D_{yy} = D_T = \alpha_T v_x + D'_m, \tag{2.10b}$$

where D_L and D_T are the longitudinal and transverse dispersion coefficients. For a more thorough description of the mathematics of dispersion, see Bear (1961, 1972) and Konikow and Grove (1977).

In applying the dispersion equations to real problems, it is necessary to determine appropriate values for dispersivity. Laboratory column experiments usually yield values on the order of centimeters. However, field-scale calibrations have yielded values on the order of meters to kilometers. These discrepancies can be attributed in part to the scale at which the heterogeneities occur and the distance of transport (Figure 2.3). A trend clearly exists for larger dispersivity values at larger scales of measurement. This scale dependence is not present in the dispersion coefficients described by Eqs. (2.9) and (2.10). The large dispersivity values seen at larger scales ($\gg 100$ m) are in part artifacts of incomplete or biased field sampling and numerical-model calibrations. Sampling through wells always results in mixing of waters from within the formation, which tends to average concentrations and hence exaggerate dispersivity estimates. The use of larger dispersivities can create smoothly varying concentrations over relatively large distances. Such concentration distributions are oversimplified and reflect our ignorance of the detailed velocity field. Because calculations of the extent of chemical reactions are sensitive to values of concentration, using overly large dispersivities in simulations of reactive transport can lead to erroneous results.

As more measurements are made in larger-scale ($\gg 100$ m) systems, estimates of dispersivity tend to decrease. At the Otis Air Force Base at Cape Cod, for example, the U.S. Geological Survey (USGS) ran a large tracer experiment in glacial outwash sands and gravels, collecting tens of thousands of samples down-gradient of the release point (LeBlanc and others, 1991). Spatial moments analysis of the data gave a value for longitudinal dispersivity, α_L, of about 1.0 meter, and a value for transverse dispersivity, α_T, of about 0.02 meters (Garabedian and others, 1991). These values were for a relatively homogeneous sand aquifer; estimates from more heterogeneous aquifers, which are much more difficult to characterize, are usually greater. One notable result from the USGS tracer test was the large ratio of longitudinal to transverse dispersivity. Such large ratios were suggested much earlier by laboratory results, which, however, also suggested that molecular diffusion is a significant component of the transverse dispersion over a relatively large velocity range (Perkins and Johnston, 1963). Both at the Cape Cod site (Garabedian and others, 1991) and at a sand aquifer at Borden, Canada (Sudicky, 1986), the transverse dispersion appears to be small, on the order of the bulk diffusion coefficient. Other field studies have suggested much higher values of transverse dispersivity, with ratios of longitudinal to transverse dispersivity near unity (Robertson, 1974). Many of these latter studies were, however, model calibrations using sparse field data, and thus have much higher levels of uncertainty.

2.1.4 Mass-balance equation

The total solute flux at any given point will be a sum of the fluxes of advection, molecular diffusion, and mechanical dispersion. This total flux can be substituted into a mass-balance equation to obtain the solute transport equation, just as we substituted Darcy's law for groundwater flux into a mass-balance equation to obtain the groundwater flow equation in Chapter 1.5. Upon doing this substitution one obtains for a general case

$$\nabla \cdot (n_e \rho \overline{D} \nabla C) - \nabla \cdot (n_e \rho v C) + Q_s = \frac{\partial(n\rho C)}{\partial t}, \qquad (2.11)$$

where $\overline{D}$ is the hydrodynamic dispersion tensor (second-order, discussed below), C is the concentration in terms of mass fraction, and Q_s is an outside source (+) or sink (−) of solute. The solute source or sink could be through a net loss or gain of a fluid with a finite concentration or through any number of chemical reactions. For systems with constant density and a constant effective porosity that is equal to the total porosity, Eq. (2.11) reduces to

$$\nabla \cdot (\overline{D} \nabla C) - \nabla \cdot (vC) + Q_s = \frac{\partial C}{\partial t}, \qquad (2.12)$$

where C can be expressed either as a mass fraction or mass per unit volume.

In multidimensional problems the dispersion tensor contains several components. For example, in a two-dimensional transport problem under isotropic conditions the tensor will have four components: two diagonal terms and two cross-dispersion terms. In two Cartesian coordinates, Eq. (2.12) would be expanded as

$$\frac{\partial}{\partial x}\left(D_{xx}\frac{\partial C}{\partial x}\right) + \frac{\partial}{\partial x}\left(D_{xy}\frac{\partial C}{\partial y}\right) + \frac{\partial}{\partial y}\left(D_{yx}\frac{\partial C}{\partial x}\right) + \frac{\partial}{\partial y}\left(D_{yy}\frac{\partial C}{\partial y}\right) \qquad (2.13)$$

$$- \frac{\partial}{\partial x}(v_x C) - \frac{\partial}{\partial y}(v_y C) = \frac{\partial C}{\partial t},$$

where the dispersion coefficients are defined as

$$D_{xx} = \alpha_L \frac{v_x^2}{|v|} + \alpha_T \frac{v_y^2}{|v|} + D'_m, \qquad (2.14a)$$

$$D_{yy} = \alpha_L \frac{v_y^2}{|v|} + \alpha_T \frac{v_x^2}{|v|} + D'_m, \qquad (2.14b)$$

and

$$D_{xy} = D_{yx} = (\alpha_L - \alpha_T)\frac{v_x v_y}{|v|}. \qquad (2.14c)$$

These expressions simply describe one way that hydrodynamic dispersion has been modeled. As discussed in Chapter 2.1.3, field measurements do not always produce constant dispersivity values that fit the context of these equations, but to date no other expression for dispersion has been adopted as a useful general alternative.

2.1.5 Chemical reactions

A general equation for solute transport (Eq. 2.12) includes a term for the external source or sink of a solute. This term, Q_s, can be the gain or loss of a solute due to a chemical reaction, R, and thus a reactive-transport equation can be written as

$$\nabla \cdot (\overline{D} \nabla C) - \nabla \cdot (vC) + R = \frac{\partial C}{\partial t}. \tag{2.15}$$

In this section we will consider R to be only a single-species reaction; multicomponent reaction-transport systems will be treated in Chapter 2.5. The R term is a rate of solute production or loss; thus, for each individual equation, R must take the form of a rate equation. Examples include radioactive decay, a kinetic rate equation, equilibrium-controlled sorption, or kinetically controlled sorption.

To illustrate this mathematical construction, we will give two examples: radioactive decay and equilibrium-controlled linear sorption. *Radioactive decay* can be expressed in a rate equation as

$$\frac{dC}{dt} = -\lambda C, \tag{2.16}$$

where the rate of decay (dC/dt) is a simple linear function of concentration scaled by the decay constant λ. The *half-life* of the species is related to the decay constant by $t_{1/2} = (\ln(2))/\lambda$.

In the second example, rather than undergoing decay, dissolved ions become attached to or released from mineral surfaces. This process is referred to as *adsorption* or simply *sorption*. Linear sorption can be expressed as the equilibrium reaction

$$C_s = K_d C \tag{2.17}$$

when we assume that the sorbed concentration, C_s, is directly proportional to the aqueous concentration, C, and K_d is the *distribution coefficient*. More realistic sorption or ion exchange requires a more complex treatment, but from this simple case we can develop a rate equation

$$\frac{dC_s}{dt} = K_d \frac{dC}{dt}. \tag{2.18}$$

We can then describe the transport of a decaying, sorbing solute by first writing

$$\nabla \cdot (\overline{D} \nabla C) - \nabla \cdot (vC) - \frac{\rho_b}{n} \frac{\partial C_s}{\partial t} - \lambda C - \frac{\lambda \rho_b}{n} C_s = \frac{\partial C}{\partial t}, \tag{2.19}$$

where ρ_b is the bulk density of the porous medium and C_s is the sorbed concentration in mass of sorbed solute per unit mass of porous medium. If we then define a *retardation factor*

$$R_f = 1 + \frac{\rho_b K_d}{n},\qquad(2.20)$$

substitute Eqs. (2.17) and (2.18) into Eq. (2.19), and then rearrange Eq. (2.20) in terms of K_d and substitute it into the new Eq. (2.19), we obtain

$$\nabla \cdot (\overline{D}\nabla C) - \nabla \cdot (vC) - R_f \lambda C = R_f \frac{\partial C}{\partial t}.\qquad(2.21)$$

The velocity of a concentration front is slowed or retarded by a multiple equal to the retardation factor. Solutions to the solute transport equation (2.21) will be illustrated in the next section, along with the effects of these simple single-species reactions on solute concentrations during transport.

2.1.6 Initial and boundary conditions

The solution of the solute transport equation requires a known velocity field and additional site-specific information in the form of *initial conditions* (for transient problems) and *boundary conditions* describing the solute conditions at the boundaries of the problem domain. Many problems invoke an initial condition in which a constant concentration is specified everywhere in space.

Boundary conditions for any second-order partial differential equation can be expressed mathematically in three different ways. For the solute transport equation, the type I, or *Dirichlet*, boundary condition specifies that the concentration is constant at that boundary. An example would be a simple diffusion problem where the concentration at one boundary will not change (for example, where one boundary represents the ocean floor). The type II, or *Neumann*, boundary condition specifies that the gradient in concentration is constant at that boundary. From Fick's first law (Eq. 2.1), we see that this also can be interpreted as a constant diffusive flux through the boundary. One common example would be a zero flux (zero gradient) boundary, for example, at a solid boundary where no solute can come in or out. The type III, or *mixed*, boundary, is a combination of the first two and specifies that a weighted sum of the concentration and flux remains constant at the boundary. The practical meaning is that the total advective and dispersive fluxes for the boundary are known. This type of condition is useful for specifying a flux of a fluid with an associated concentration. Under such a condition, dispersion is allowed to occur at the boundary and the concentration there is not rigorously fixed.

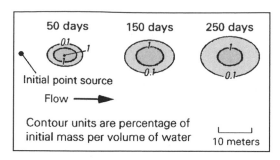

Figure 2.4 Spreading of an instantaneously injected tracer pulse due to hydrodynamic dispersion. In this example, longitudinal dispersivity is 0.5 meters and transverse dispersivity is 0.25 meters.

One solution to the solute transport equation is the problem where a point source of mass M is injected into an infinite porous medium at initial time zero in a linear velocity field. The mass is then simultaneously advected by the velocity field and dispersed through the porous medium. The initial condition for this problem is stated mathematically as

$$C(x, y, z, t = 0) = M \cdot \delta(x, y, z), \tag{2.22}$$

where $\delta(x, y, z)$ is the *Dirac delta function*, signifying a value of unity at coordinates $(0, 0, 0)$ and zero everywhere else. The boundary conditions dictate a zero solute concentration at infinite distance in all directions. The solution to the transport equation (2.12), assuming these conditions and no other sources or sinks, is (Baetsle, 1969)

$$C(x, y, z, t) = \frac{M}{8(\pi t)^{3/2}\sqrt{D_{xx}D_{yy}D_{zz}}} \tag{2.23}$$

$$\times \exp\left(-\frac{X^2}{4D_{xx}t} - \frac{Y^2}{4D_{yy}t} - \frac{Z^2}{4D_{zz}t}\right),$$

where X, Y, and Z are the moving coordinates following the center of mass of solute and are defined as: $X = x - v_x t$, $Y = y - v_y t$, and $Z = x - v_z t$. One solution to this problem is illustrated in Figure 2.4.

Another useful solution to the solute transport equation includes reactions (Eq. 2.21) and applies to a system with an initial concentration of $C = 0$ throughout, a unidirectional seepage velocity, v_x, and continuous source of solute at concentration C_0 at an inlet boundary. The downstream boundary at infinite distance from the inlet is kept at zero concentration. The analytical

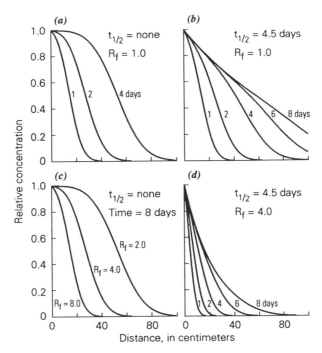

Figure 2.5 Solutions of the advection–dispersion equation for single-species transport with reactions in a one-dimensional, steady-state flow field: (a) no decay or sorption, (b) decay and no sorption, (c) sorption and no decay, and (d) decay and sorption. For all solutions, $v_x = 12.5$ cm per day and $\alpha_L = 1.5$ cm. Modified from Goode and Konikow (1989).

solution to this problem (Bear, 1972) is

$$C = \frac{1}{2}C_0 \exp\left(\frac{v_r x}{2D_r}\right)\left[\exp(-x\beta)\cdot\mathrm{erfc}\left(\frac{x - t\sqrt{v_r^2 + 4\lambda D_r}}{2\sqrt{t D_r}}\right)\right. \tag{2.24}$$

$$\left. + \exp(x\beta)\cdot\mathrm{erfc}\left(\frac{x + t\sqrt{v_r^2 + 4\lambda D_r}}{2\sqrt{t D_r}}\right)\right],$$

where D_r is equal to D_{xx}/R_f, v_r is equal to v_x/R_f, and $\beta^2 = (v_r^2/4D_r^2) + \lambda/D_r$. If $\lambda = 0$ and $R_f = 1$, Eq. (2.24) reduces to the analytical solution for a conservative (nonreactive) solute presented by Ogata and Banks (1961). In Figure 2.5a the solution for a conservative solute distribution is shown. Figure 2.5b shows that if radioactive decay is occurring, a steady-state concentration distribution will develop at relatively long times. A comparison of Figures 2.5a and 2.5c shows that the retardation factor is simply a scale factor for the seepage velocity. Figure 2.5d illustrates the combined effects of decay and sorption.

2.2 Numerical solution techniques

The complexity of the analytical solution to a simply defined transport problem
(Eq. 2.24) illustrates the limited applicability of analytical solutions to real-
world problems where heterogeneity and multidimensionality are the norm.
With the advent of high-speed computing, numerical techniques have proven to
be very efficient for solving partial differential equations, including the equations used to described groundwater flow and transport. To date, the most
popular numerical solution techniques have been finite differences and finite
elements, each of which have particular advantages and disadvantages when
applied to solute transport in groundwater. A description of these methods is
outside the scope of this book, but such descriptions have been developed in
detail elsewhere (e.g., Wang and Anderson, 1982; Huyakorn and Pinder, 1983).

Partial differential equations can be divided into classes based on their form
(for example, Smith, 1978). These classes include elliptic, parabolic, hyperbolic, and hybrid or mixed equations. Whereas the transient diffusion equation
is of the *parabolic* type and the wave equation is of the *hyperbolic* type, groundwater transport equations have both a diffusive and an advective component and
are thus of a mixed type that ranges from parabolic to hyperbolic depending
upon the values of the parameters. A parabolic-type equation (e.g., the groundwater flow equation) is relatively easy to solve with standard methods, whereas a
hyperbolic equation is more difficult to solve. An equation describing advection-
dominated groundwater transport will be closer to a hyperbolic equation. Using
realistic transport parameters usually results in solute transport equations being
more hyperbolic (often with sharp concentration gradients) and heat transport
equations being more parabolic (often with more diffuse temperature gradients). Partly for this reason our discussion of numerical methods and stability
is confined mainly to this chapter.

The tendency of standard numerical methods to produce oscillations around
a sharp front caused by advection of solutes can be predicted by examination
of the *Peclet number*. A dimensionless form of the one-dimensional solute
transport equation can be written as

$$\frac{1}{Pe}\frac{\partial^2 C_d}{\partial x_d^2} - \frac{\partial C_d}{\partial x_d} = \frac{\partial C_d}{\partial t_d}, \tag{2.25}$$

where C_d is dimensionless concentration, x_d is dimensionless distance, t_d is
dimensionless time, and Pe is the Peclet number (dimensionless) defined as
$v_x L/D_{xx}$. In the context of one-dimensional flow with negligible molecular
diffusion, D_{xx} is equal to $\alpha_L v_x$, and in the context of finite numerical methods
the characteristic length, L, can be defined as the grid spacing, Δx. The Peclet

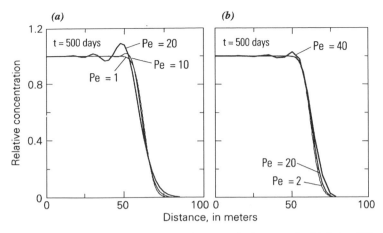

Figure 2.6 Effect of the Peclet number on the numerical error in the (a) finite-difference and (b) finite-element methods applied to the one-dimensional solute transport equation with a relatively sharp concentration front. All calculations were made using values of $v_x = 0.1$ m/day, $\alpha_L = 0.1$ m, $\Delta t = 1$ day, and $\Delta x = \alpha_L Pe$.

number then becomes $\Delta x / \alpha_L$, representing the ratio of the grid spacing to the dispersivity. When the Peclet number exceeds a value of 2, the solution for Eq. (2.25) becomes relatively unstable using either finite differences or finite elements, whereas when the Peclet number is lower than a value of 0.5, the solution becomes stable for both finite differences and finite elements. In the range $0.5 < Pe < 2$, finite-element techniques are somewhat more stable for the same Peclet number (Gray and Pinder, 1976) (Figure 2.6). However, the finite-element method requires more computations per node, and thus its relative efficiency at handling sharp fronts is not always the determining factor in deciding which method to use.

The Peclet number criterion for a stable solution can be too severe for many solute transport simulations regardless of the solution technique. The relatively small dispersivities found in many natural systems lead to relatively steep concentration gradients, which in turn require fine grid spacing for a stable solution. In geologic process applications, length and time scales are often forbiddingly large compared to the required discretization. One proposed solution to this problem has been to use deforming grids that adjust to the concentration fronts, but these are difficult to adapt to two- and three-dimensional problems.

Particle-tracking algorithms have also been developed to solve the solute transport equation. Among these are the *random-walk method* and the *method of characteristics* (MOC). Both of these methods ease the Peclet number requirement, though both also have their own particular constraints. The random-walk method adds a random adjustment to the particle position that is proportional to the dispersivity, but in so doing adds a nonuniqueness to each solution. The method of characteristics solves the advection part of the equation by tracking particles along flow lines and handles the dispersion part of the equation with finite differences. One MOC code that has seen wide use for environmental problems was developed by the U. S. Geological Survey (Konikow and Bredehoeft, 1978).

Other solute transport codes that have been made available to the public by the USGS include SUTRA (Voss, 1984), a two-dimensional finite-element code that will simulate heat or solute transport in saturated or unsaturated conditions, and HST3D (Kipp, 1987), a three-dimensional finite-difference code that will simulate both heat and solute transport. The USGS codes MODFLOW (McDonald and Harbaugh, 1988) and MODPATH (Pollock, 1989) have been used extensively to describe groundwater flow and advective transport in the shallow subsurface.

2.3 Density-driven flow

The solute concentration influences the density of groundwater. This influence becomes significant when the resulting buoyancy forces are large enough to create fluid velocities of the same order of magnitude as those created by the boundary-condition-induced hydraulic forces. For a variable-density fluid, Darcy's law can be written as

$$q = -\frac{\overline{k}}{\mu}(\nabla P + \rho g \nabla z), \tag{2.26}$$

where P is the fluid pressure, $\overline{k}$ is the intrinsic permeability, μ is the dynamic viscosity, and z is the elevation above a standard datum. In most cases we can realistically represent the density of an isothermal fluid as a linear function of one of the conservative major dissolved constituents (e.g., chloride). Thus, the density can be defined as

$$\rho = \rho_0 + \gamma(C - C_0), \tag{2.27}$$

where γ is the ratio of the change in fluid density to change in concentration, C_0 is the reference concentration, and ρ_0 is the fluid density at the reference concentration. To fully describe solute transport in groundwater having variable density requires an equation of state, such as Eq. (2.27), and a groundwater flow equation that incorporates the variable density form of Darcy's law (Eq. 2.26). Such a flow equation was given in Chapter 1.5 (Eq. 1.32), and by putting Eq. (2.27) into (1.32) one can write a flow equation applicable specifically to variable-density solute transport:

$$\nabla \cdot \left(\frac{\rho \overline{k}}{\mu} \left(\nabla P + \rho g \nabla z \right) \right) = \rho \frac{\partial n}{\partial P} \frac{\partial P}{\partial t} + n \frac{\partial \rho}{\partial P} \frac{\partial P}{\partial t} + \gamma n \frac{\partial C}{\partial t}, \qquad (2.28)$$

where the terms on the right-hand side represent changes in mass over time through changes in porosity with respect to pressure, changes in fluid density with respect to pressure, and changes in fluid density with respect to concentration.

The governing system of equations for nonreactive variable-density solute transport thus includes equations for both solute transport (e.g., Eq. 2.12) and groundwater flow (e.g., Eq. 2.28). These equations are posed in terms of two dependent variables: solute concentration and fluid pressure. Fluid flow is a function of fluid density, which is related to solute concentration by an equation of state (e.g., Eq. 2.27). In turn, solute concentration is a function of transport, which is a function of fluid flow. Thus, the equations are coupled and must be solved simultaneously. This can be done numerically for a wide range of cases.

Numerical solutions to the variable-density solute transport equations generally encounter stability problems in cases with a high *Rayleigh number*. The Rayleigh number has generally been defined in terms of heat transport for buoyancy-driven convection (see Chapter 3.5.3). In the context of variable-density solute transport, the Rayleigh number indicates the relative strength of the buoyancy drive caused by solute-related density differences. In such cases the Rayleigh number can in general be defined as

$$Ra = \frac{K \gamma L}{D}, \qquad (2.29)$$

where K is the hydraulic conductivity, L is a characteristic length of the system, and D is the coefficient of dispersion or molecular diffusion. In the context of a numerical stability analysis, L represents the grid spacing. In numerical solutions to the variable-density equations, a high Rayleigh number will cause very small numerical errors to be propagated into macroscopic convection cells that are purely numerical artifacts. This problem can only be controlled by reducing the Rayleigh number, which can be done in one of two ways: by reducing the grid spacing to a point where the solution is often intractable or by changing the physical parameters to a point where the solution is often unrealistic. There is often no feasible way to accurately solve the equations for certain realistic physical conditions. In some cases, however, a compromise between grid spacing and parameter values will produce solutions that, although not completely realistic, will have certain meaningful attributes.

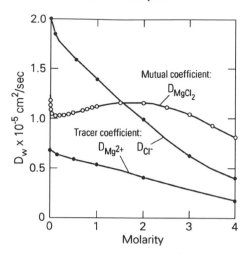

Figure 2.7 Experimental diffusion coefficients for Mg^{2+}, Cl^-, and $MgCl_2$ from Harris and others (1978) and Miller and others (1984). From Felmy and Weare (1991), reprinted with kind permission from Elsevier Science Ltd., Kidlington.

2.4 Multicomponent diffusion

Fick's first law (Eq. 2.1) states that the diffusive flux of an ion is the product of the concentration gradient and a constant parameter known as the diffusion coefficient. In reality this coefficient is not constant but is a function of both the concentration of the diffusing ion and the concentrations of all other ions in solution. The process of multicomponent diffusion can be derived from the theory of irreversible thermodynamics (DeGroot and Mazur, 1962; Haase, 1969). A dilemma arises when trying to apply tracer diffusion coefficients, such as those in Table 2.1, to real aqueous systems (Figure 2.7). For example, not only do Cl^- and Mg^{2+} diffusion coefficients vary with concentration, their ratio differs from 2, the value that would be required to maintain electrical neutrality. Moreover, it can be seen that the mutual $MgCl_2$ coefficient is not a linear combination of the individual coefficients. It is apparent that to accurately describe the process of multicomponent diffusion in water one must fully consider the interaction of other ions in solution. Here we present an introduction to multicomponent diffusion calculations with some examples of applications.

To understand the theory behind multicomponent diffusion, one must return to irreversible thermodynamics and the Onsager reciprocal relations (Miller, 1960). The *Onsager relations* come from statistical thermodynamics and state that each thermodynamic force within a system causes not only a finite flux of its own entity, but also of all other entities such that, for example, the force of A on ion B is equal and opposite to the force of B on ion A. In the context of diffusion the force of interest is the gradient in the electrochemical potential, which can be expressed as

$$E_j = -\left[\frac{\partial \mu_j}{\partial x} + z_j F \frac{\partial \phi}{\partial x}\right], \tag{2.30}$$

where E_j is the electrochemical-potential gradient of the jth ion, μ_j is the chemical potential of the jth ion, z is the ionic charge, F is Faraday's constant, and ϕ is the electrostatic potential. The system also requires electroneutrality, which can be specified as

$$\sum_{i=1}^{N} z_i J_i = 0, \tag{2.31}$$

where N is the total number of ions and J_i is the flux of the ith ion. By applying the electroneutrality constraint (Eq. 2.31) to Eq. (2.30), and by converting the chemical potential to concentrations, one can obtain a set of equations of the form

$$J_i = -\sum_{j=1}^{N} D'_{ij} \frac{\partial C_j}{\partial x}. \tag{2.32}$$

Equation (2.32) resembles Fick's first law except that there are now multiple diffusion coefficients for every ion, including off-diagonal coefficients relating the flux of each ion to the gradient of every other ion in solution. Models have been developed for multicomponent diffusion using an ion-pairing aqueous model for relatively dilute solutions (Lasaga, 1979) and using the virial-coefficient equation aqueous model of Pitzer (1973, 1975) for high concentrations (Felmy and Weare, 1991). Both models require incorporating laboratory measurements of the Onsager coefficients into off-diagonal diffusion coefficients for all of the pairs of ions under consideration.

One important calculation to come from such a model of multicomponent diffusion is a set of coefficients for seawater (Table 2.2). Felmy and Weare (1991) considered the six most concentrated ions in seawater, omitting chloride, because its flux can be calculated from the electroneutrality constraint once the other fluxes are known. Note that some of the off-diagonal terms are relatively large and many have negative signs (caused by the electroneutrality constraint). Also, on the main diagonal, the monovalent coefficients have values about twice the divalent ones (also reflecting electroneutrality). These values of coefficients are valid only for the exact composition of seawater, whereas, in general, each coefficient is a function of all the solute concentrations. In order to calculate diffusion in a spatial continuum where, for example, seawater is mixing with

Table 2.2 *Calculated matrix of diffusion coefficients, D'_{ij}, for seawater (all coefficients $\times 10^{-10} \, m^2/s)^a$*

Flux	$\frac{\partial C_{Na}}{\partial x}$	$\frac{\partial C_K}{\partial x}$	$\frac{\partial C_{Ca}}{\partial x}$	$\frac{\partial C_{Mg}}{\partial x}$	$\frac{\partial C_{SO_4}}{\partial x}$
J_{Na}	13.98	0.15	7.29	8.05	−7.54
J_K	0.07	17.93	0.23	0.26	−0.37
J_{Ca}	0.07	0.02	6.04	0.23	−0.46
J_{Mg}	0.23	0.08	0.76	5.75	−1.08
J_{SO_4}	−0.15	−0.26	−1.60	−1.13	8.33

^{a} After Felmy and Weare (1991).

a more dilute solution, the coefficients must be recalculated continually as concentrations change.

Such a continuum application was demonstrated by Felmy and Weare (1991), who examined what would happen if seawater were present in a nonreactive porous medium adjacent to a salt dome. To solve this problem in one dimension, they assumed negligibly low values of groundwater flow and substituted each diffusion equation into the mass balance equation to obtain a set of diffusion equations that must be solved simultaneously. Their equations were of the form

$$\frac{\partial (nC_i)}{\partial t} = \frac{\partial}{\partial x} \left[\frac{n}{\tau} \sum_{j=1}^{N} D'_{ij} \frac{\partial C_j}{\partial x} \right]. \tag{2.33}$$

The boundary condition at the edge of the salt dome was specified to be constant concentration with all species specified at seawater concentrations except Na and Cl, which were set close to halite saturation values. Porosity and tortuosity were both arbitrarily set to 1.0. Although other, more realistic, values could have been used for the above conditions, these were chosen to best illustrate the cross-coupling effects. Results of diffusion of sodium away from the salt dome after 1 and 3 million years are shown in Figure 2.8. The sodium profile indicates that the solute has been able to diffuse about 1 km in 3 million years. Chloride curves are very similar in shape to the sodium curves. For the other solutes (Figure 2.9), concentrations are normalized to seawater values in order to illustrate the presence of cross-coupling effects. Without the coupling effects the other relative solute concentrations would remain constant in space, but the off-diagonal coefficients cause additional interactions to occur among all of the ions.

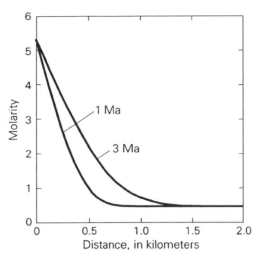

Figure 2.8 Calculated distribution of Na resulting from diffusion in a brine away from a hypothetical salt dome. Calculations were made by Felmy and Weare (1991) using a multicomponent diffusion model.

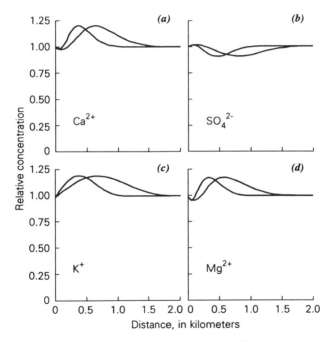

Figure 2.9 Calculated distributions of (a) calcium, (b) sulfate, (c) potassium, and (d) magnesium resulting from diffusion in a brine away from a hypothetical salt dome. Calculations were made by Felmy and Weare (1991) using a model of multicomponent diffusion.

2.5 Multicomponent reactive transport

In many geologic processes, chemical reactions are driven by a continual supply of reactive chemicals that are supplied in dissolved form by flowing groundwater. A complete mathematical description of such a process usually requires a transport equation for each of the solutes involved, as well as equations describing the pertinent chemical reactions. Solving such a system of equations requires a level of complexity that we have not yet addressed but will introduce here. Reactive-transport processes are important where there is a potential for fluid flow coexisting with spatial variations in the thermodynamic states of the system. Thermodynamic state variations can be classified as either variations in the compositions of the solid phase, variations in temperature or pressure, or variations in the composition of the aqueous phase. In reality, all three of these variations may exist simultaneously at one location in the subsurface, but often one variation is significantly more important than the others.

The three fundamental end-member types of reactive-transport environments are depicted in Figure 2.10. The first type of environment, where variations in mineralogy exist, leads to reaction zones or fronts that migrate through space and time. This is caused when water moves from one area, where it is in equilibrium with respect to one suite of minerals, to another area, where it comes into contact with a second mineral suite. The second type of environment is caused by a variation in temperature or pressure across a region where mineralogy and aqueous composition might otherwise be constant. The changes in temperature or pressure cause a concomitant change in chemical equilibrium as the water carrying the solutes moves through the temperature or pressure gradients. The result is dissolution or precipitation of mineral phases toward a new state of equilibrium. The third type of environment is caused by a variation in aqueous composition where mineralogy, temperature, and pressure might otherwise be constant. Two different waters could both be in equilibrium with respect to the same mineralogy, but upon their mixing the new mixed water will be out of equilibrium. This type of mixing disequilibrium is caused by the nonlinear nature of the *law of mass action* governing chemical reactions.

Another way to categorize reactive-transport systems is based on the mathematical nature of the chemical reaction equations and their coupling to transport equations (Rubin, 1983). In this approach (Figure 2.11), chemical reactions can first be classified by whether they are "sufficiently fast," to be described by an assumption of local equilibrium, or "insufficiently fast," occurring at rates similar to the fluid flow, such that they must be described by reaction rate equations, rather than mass-action equations. A second level of classification then depends on whether the reactions are *homogeneous* or *heterogeneous* – the former

(*a*) Isothermal reaction fronts

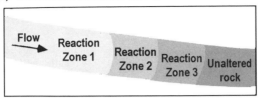

(*b*) Gradient reactions

Maximum rate of fluid
cooling and precipitation

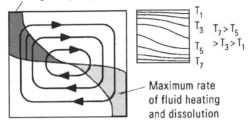

$T_7 > T_5$
$> T_3 > T_1$

Maximum rate
of fluid heating
and dissolution

(*c*) Mixing zone reactions

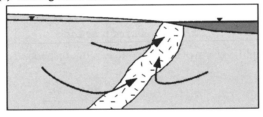

Figure 2.10 Three major types of reactive-transport environments (defined by Phillips, 1991) characterized by (a) introduction of a fluid that is out of equilibrium with existing mineral phases, (b) spatial gradients in temperature (T) or pressure, and (c) mixing of two different fluids, each of which is in equilibrium with the existing mineral phases before mixing. After Person and others (1996).

being, for our purposes, reactions between two different dissolved species, and the latter being reactions between dissolved species and solid phases. A third classification separates the heterogeneous reactions into "surface" and "classical" types. Surface reactions occur between aqueous species and ions sorbed or exchanged onto the surfaces of minerals, where the concentration of the surface species can vary and control the extent or rate of the reaction. The classical reactions are dissolution and precipitation, where the concentrations of the solid phases do not vary (they are assumed to equal a value of one by definition). In the remainder of this chapter we will briefly address the nature and implications of these different types of reaction-transport systems.

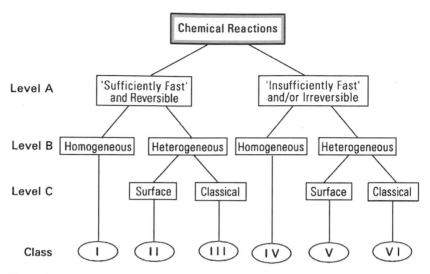

Figure 2.11 Classification of chemical reactions based on the mathematics of their coupling with solute transport equations. After Rubin (1983).

2.5.1 Rate-based reactions

Recall that a simplified form of a reactive advection–dispersion equation (Eq. 2.15) can be written

$$\nabla \cdot (\overline{D}\nabla C) - \nabla \cdot (vC) + R = \frac{\partial C}{\partial t}, \tag{2.34}$$

where R is the rate of production (or removal) by chemical reaction of the solute whose concentration is described by Eq. (2.34). The R term must be written for each reaction in which the given solute participates. For solute production by such rate-based reactions, one simply can substitute a rate-based equation for the R term. Such rate equations are usually descriptions of the kinetics of dissolution or precipitation of a mineral phase present in the system; homogeneous reactions are usually sufficiently fast so as not to warrant treatment with rate equations. Rate equations for dissolution or precipitation are often written in the form

$$R = \frac{\partial C}{\partial t} = -K_r \frac{s_0}{n}(C - C_{eq})^m, \tag{2.35}$$

where K_r is an empirical rate constant, s_0 is the surface area of mineral per volume of porous media, C_{eq} is the concentration of the solute at equilibrium conditions with respect to the mineral of interest, and m is an empirical exponent. Note that the reaction rate decreases as the concentration approaches equilibrium and that the rate is positive (a source of additional solute) when the aqueous concentration is less than the equilibrium level (resulting in dissolution). Data for rate constants are typically available for only the most common minerals. For those cases where rate data exist, the surface area term is often the parameter most difficult to quantify and apply to field-scale problems. Rate constants are often measured in the laboratory on fine-grained mineral samples that allow experimental results to be achieved over reasonable time frames. Scaling these high-surface-area systems to more realistic field conditions then becomes problematic.

In real systems a rate equation may be more complicated, such that there are multiple terms in parentheses, each representing a different solute. Also, a chemical element may be present in more than one reactive mineral, such that more that one R term must be written for each reaction. In such multicomponent systems, transport equations must also be written for each solute of interest. If the rate terms contain concentrations for solutes other than those for which that particular transport equation was written, then there will be transport equations with multiple terms that must be solved simultaneously. Inherent in the solution of this type of equation is a time-step restriction that is a direct function of the rate constant. This restriction requires that the time step be small relative to the change in the rate of the reaction. Otherwise the numerical solution may approach the chemical equilibrium too quickly for accurate reaction-rate calculations, or it may even overshoot equilibrium in severe cases.

Just as the advection–dispersion equation can be written in dimensionless form (Eq. 2.25), the reactive-transport equation can also be written in dimensionless form when a simple reaction-rate equation is appropriate. The dimensionless form becomes

$$\frac{1}{Pe} \frac{\partial^2 C_d}{\partial x_d^2} - \frac{\partial C_d}{\partial x_d} + D_k = \frac{\partial C_d}{\partial t_d}, \tag{2.36}$$

where D_k is the *Damkohler number* (developed in the chemical engineering field). A high Damkohler number indicates that the reaction rate of the solute at the mineral surface is fast relative to the advective transport of the solute, and a low Damkohler number indicates that the reaction rate is slow relative to advective transport. In the former case the rate-limiting process is delivery of the solute to the reaction site, whereas in the latter case the rate-limiting process is the reaction at the site. A similar Damkohler number exists for diffusion-only transport indicating relative rates of reaction to rates of diffusion.

2.5.2 Surface reactions

In Chapter 2.1.5 we discussed sorption as a single-species reaction coupled with solute transport. Here we will focus on multicomponent surface reactions – specifically, *competitive ion exchange*. Clay minerals have electrically nonneutral surfaces that attract charged ions (usually cations) in solution to occupy the available sites. When these sites become full, different ions will compete for the sites based on the characteristics of the ions and the clay. This process can have a large effect on aqueous concentrations, especially as groundwater that is out of equilibrium with the exchange sites moves through the formation. It is usually most important in relatively shallow, low-temperature, clay-rich environments.

To incorporate cation exchange into solute transport equations, one must also solve equations representing the equilibrium between the aqueous concentrations and surface concentrations. One can write a single exchange reaction as

$$\frac{1}{a}A^a + \frac{1}{b}BX_b \leftrightarrow \frac{1}{a}AX_a + \frac{1}{b}B^b, \tag{2.37}$$

where X is the exchange site (one site per ion) and a and b are the ionic charges of solutes A and B, respectively. Exchange reactions written in this way use the Gaines–Thomas (1953) convention. When A and B have the same charge, the exchange is said to be *homovalent*, and the equilibrium equation can be written as

$$K_{eq} = \frac{[AX_a][B^b]}{[BX_b][A^a]}, \tag{2.38}$$

where the terms in brackets represent concentrations for aqueous species and molar fractions for surface species. As an example, a Na–K exchange would be written as

$$K_{NaK} = \frac{[NaX][K]}{[KX][Na]}. \tag{2.39}$$

For *heterovalent* exchanges such as Na–Ca, the equilibrium equation would have the form

$$K_{NaCa} = \frac{[NaX][Ca]^{1/2}}{[CaX]^{1/2}[Na]}, \tag{2.40}$$

or, by the Gapon (1933) convention,

$$K_{NaCa} = \frac{[NaX][Ca]^{1/2}}{[CaX][Na]}, \tag{2.41}$$

where the [CaX] term now has an exponent of one. A third convention, the Vanselow (1932) convention, uses mole fractions in Eq. (2.40). All of these conventions are equivalent in homovalent exchange. In heterovalent exchange there are some differences, but they are often second-order. The equilibrium constant is usually assumed to be constant, but under most real conditions it is a function of the ionic strength of the solution.

For any given set of cations undergoing transport and exchange, the complete system can be described by $2N$ equations, where N is the number of cations and each cation exists in both the aqueous and surface phase. This equation set is composed of N transport equations, $N-1$ equilibrium exchange equations, and one exchange capacity equation. Although an additional exchange equation could be written, it would be redundant. The equation set can be written as follows:

N transport eqs.:

$$\nabla \cdot (\overline{D} \nabla C_i) - \nabla \cdot (v C_i) = \frac{\partial C_i}{\partial t} + \frac{\rho_b}{n} \frac{\partial C_{si}}{\partial t}, \quad i = A, B, \ldots N, \quad (2.42)$$

$N - 1$ equilibrium eqs.:

$$K_{eq} = \frac{[AX_a]^{1/a}[B^b]^{1/b}}{[BX_b]^{1/b}[A^a]^{1/a}}, \quad (2.43)$$

and one exchange capacity eq.:

$$CEC = [AX] + [BX] + \cdots + [NX], \quad (2.44)$$

where C_{si} is the equivalent fraction of solute on the exchange site, and where CEC is the *cation exchange capacity* (total number of exchange sites) of the porous medium, assumed to be a constant. For further details of modeling exchange reactions, see Appelo and Postma (1993).

2.5.3 Homogeneous reactions

In aqueous solutions a significant fraction of the dissolved species often occur as aqueous complexes, and an adequate thermodynamic description of the solution cannot be made without taking into account the *homogeneous reactions* among these complexes. Many complexes exist in solution as a combination of one cation and one anion, and they are thus often referred to as ion pairs. They usually react at a rate that is fast relative to fluid flow and heterogeneous reactions and are therefore nearly always represented by equilibrium equations following the *law of mass action*. In solving for the distribution of species in solution (known as speciation), one equation must be written for each aqueous complex that is included in the description of the system. According to the law of mass action, we can write the equation as

$$K_{ac} = \frac{\prod_{s=1}^{Ns} \gamma_s^{\nu s} m_s^{\nu s}}{\gamma_{ac} m_{ac}},\tag{2.45}$$

where K_{ac} is the equilibrium constant of dissociation of the complex, Ns is the number of reactant species present in the complex, γ is the activity coefficient of the species or complex, m is the molality of the species, s refers to the reactant species, ac refers to the aqueous complex, ν refers to the stoichiometric coefficient in the reaction, and the $\prod$ operator denotes continuous multiplication.

The composition of an aqueous solution is usually measured and described in terms of the concentrations of the dissolved elements present. These concentrations are measured and reported in terms of the total dissolved component and do not indicate how the element is speciated. A model for the aqueous system therefore must solve for a set of unknowns that includes both component species (defined here as the uncomplexed ion chosen to represent a given element, e.g., Ca^{2+}, Na^+, SO_4^{2-}) and aqueous-complex concentrations. Such a system will have $Nc + Nac$ unknowns, that is, the number of different component species, Nc, plus the number of aqueous complexes, Nac. Nac equations are given by Eq. (2.45). Nc equations are provided by writing one mass balance equation for each dissolved element in the form

$$M_T = M_{cs} + \sum_{m=1}^{Nac} \nu_c M_m,\tag{2.46}$$

where M_T is the total dissolved elemental concentration (in moles), M_{cs} is the molar concentration of the component species, M_m is the molar concentration of the aqueous complex, Nac is the number of aqueous complexes containing the element of interest, and ν_c is a stoichiometric coefficient indicating the number of moles of that element present per mole of complex. For natural waters with any significant acid–base reactions, one equation is also written for the dissociation of water; this accounts for the unknown value of the hydrogen ion, or pH. In systems with redox reactions, an equation is also required to balance either electrons or the valence state; this accounts for the extra unknown that is present in the specified redox pair. More detailed descriptions of writing and solving equations describing aqueous chemistry can be found in Stumm and Morgan (1981), Bethke (1996), and Langmuir (1997).

When homogeneous reactions are coupled with solute transport, Eqs. (2.45) and (2.46) must be solved together with the transport equation (Eq. 2.34). The nature of the equations are such that one transport equation (with no R term) can be written for each component. Once the component concentrations are known from the solutions to the transport equations, the values of M_T can be used in (2.46) so that Eqs. (2.45) and (2.46) can be solved independently of the transport. This is possible because the total-component concentrations are independent of the extent of the homogeneous reactions. Total-component concentrations have taken on other names in the literature such as reaction-invariants and tenads (Rubin, 1983). In systems where different total-component concentrations are linearly dependent during transport, only one transport equation is necessary for each independent set of linearly dependent components. This approach can significantly reduce computational time and storage (Sanford and Konikow, 1989a).

2.5.4 Heterogeneous reactions

Heterogeneous reactions can be of the classical or surface type and of the sufficiently fast or insufficiently fast type. Surface and insufficiently fast types have been considered in previous sections, leaving the "classical-fast" type to consider here. Rubin (1983) used this term to refer to dissolution or precipitation of minerals under the assumption that reaction rates were fast enough to satisfy local conditions of thermodynamic equilibrium. To include such reactions into a reactive-transport analysis that may already consider homogeneous reactions, one must simply add the mass-action equations for the minerals of interest. These equations are written in the form

$$K_{min} = \prod_{s=1}^{Ns} \gamma_s^{vs} m_s^{vs},\qquad(2.47)$$

where K_{min} is the equilibrium constant for the mineral of interest. An additional equation now exists for each mineral present in the porous medium. The additional corresponding unknowns are the R terms in Eq. (2.34); one R term is added representing the rate of reaction for each particular mineral. Each transport equation incorporates R terms only for minerals that are composed of the element being transported.

When heterogeneous reactions are present in a system with a mineralogy that varies in space, reaction fronts can develop (Figure 2.10a). These fronts become sharp when local equilibrium assumptions are made, leading to a moving boundary problem. In these cases the reaction front itself becomes a boundary that divides the system into two zones whose solutions require different sets of reaction equations. The upstream zone has water in equilibrium with its particular mineral assemblage, but as that water moves across the reaction front, it becomes out of equilibrium as soon as it enters the downstream zone. Under the local equilibrium assumption, dissolution or precipitation is instantaneous, so all the mineral transfer occurs at the front, causing the boundary between the two mineralogical zones to move as mass is etched from or accreted to the boundary. Standard discrete numerical solution techniques have difficulties representing such sharp fronts. Other algorithms such as continual grid refinement can be used to improve results, but these are difficult to implement in more than one spatial dimension. The presence or absence of the reactive mineral phases across the fronts must be explicitly accounted for.

2.5.5 Solution algorithms

Various numerical algorithms exist for coupling partial-differential transport equations to algebraic geochemical reaction equations. Figure 2.12 organizes these different algorithms based on the level of coupling between the algebraic and partial differential equations. In general, a more tightly coupled set of equations will be more stable and robust, whereas a more loosely coupled set of equations often requires less computer time and storage. Thus, highly nonlinear problems are better suited for tightly coupled algorithms and computationally intensive problems are better suited for loosely coupled algorithms. The technique of choice can be highly problem-dependent. Many problems can be formulated that are too large or too nonlinear to be solved with existing computer technology. Assumptions usually need to be made to scale the problem down to a tractable level.

The innermost region in Figure 2.12 represents the tightest coupling, where the reaction equations are substituted into the transport equations. Because there is a close dependence between the solutes through the reactions, all of the transport equations can then be solved simultaneously. One example of this is the treatment of *competitive ion exchange* by Sanford and others (1992). When total exchange capacities are high relative to solute concentrations, iterations between transport and exchange equations can easily lapse into nonconvergence. By using the chain rule to expand the terms representing the time rate of change of the sorbed concentrations, the equations could be solved simultaneously much more quickly and efficiently. Another example where tight coupling is required is the problem of multicomponent diffusion. The off-diagonal diffusion coefficients and the dependence of these coefficients on all solute concentrations virtually requires simultaneous solution.

The next two levels in Figure 2.12 represent degrees of separation of the reaction and transport equations. Here the reaction equations are solved as an isolated set, using the total aqueous concentrations from the transport equations from the previous iteration or time step as known values to constrain

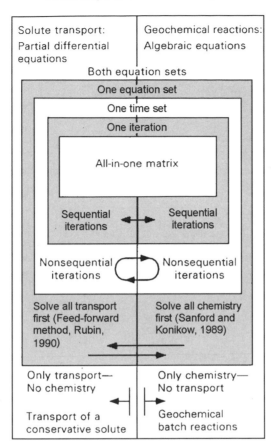

Figure 2.12 Levels of coupling in different algorithms for solving multiconstituent reactive solute transport. The innermost domain represents the tightest coupling of equations and the outermost domain the loosest coupling. Arrows indicate general direction of movement in the algorithm between solving the two sets of equations.

the geochemical solution. Mineral mass transfer is calculated based on equilibrium or kinetic conditions, and the reacted amount is treated as a source of solute in the transport equation via the R term. This is sometimes called the two-step or operator-splitting technique. Iterations can be performed between the solute concentrations and the mass transfer amounts in order to ensure a stable and accurate solution. Other models that are noniterative rely on an adequately small time step to ensure accuracy. In either case the set of geochemical reaction equations are the same ones that are solved by various geochemical codes such as PHREEQC (Parkhurst, 1995) or EQ3/6 (Wolery, 1992). This approach often allows one to link existing geochemical and transport codes to solve the system of equations being considered.

A further separation of the sets of equations during computations is possible under certain conditions. If the heterogeneous reactions are relatively few, the geochemical reactions can be solved first for the range of solute concentrations expected to be encountered during transport. The dissolution or precipitation rates can then be taken from either a look-up table or an interpolation scheme during the solution of the transport equations. This approach can be very computationally efficient (Sanford and Konikow, 1989a). An opposite approach is to solve all the transport equations first and then solve the geochemical reaction equations on the basis of those results. This can be done in some systems through the feed-forward method of Rubin (1990). Certain linear combinations can be made of component species such that all the R terms will fall out of the transport equation when written in terms of these new variables. The outermost region of Figure 2.12 represents total decoupling of the equations, where transport or geochemical reactions are assumed to be completely independent.

Fully coupled sets of reactive-transport equations, as described above, can be written to describe many of the geochemical transport processes occurring in the Earth's crust. Heat transport (discussed in Chapter 3) must also be considered for nonisothermal problems. Examples of fully coupled reactive transport for ore deposits and diagenesis will be given in Chapters 5 and 10.

Problems

2.1 A shale unit one kilometer thick ($L = 1$ km) is deposited in a marine environment such that the initial pore-fluid composition is equivalent to seawater. The shale is bounded above and below by more permeable sands. Relatively rapidly (in a geologic sense) the entire sequence is lifted above sea level, and the sands become freshwater aquifers. Assuming that the dissolved salts can transport out by simple (noncoupled) diffusion only, about how long would it take for the salinity concentrations in the shale formation to reach one tenth of the original seawater concentration? Assume a reasonable diffusion coefficient. Figure 2.13 provides appropriate solution curves (for half the shale unit).

2.2 Matrix diffusion can be a significant sink for solutes moving through fractured rock. This can pose a problem when radioactive isotopes are used as tracers for age dating. The equation describing steady-state diffusion and decay of a solute in a matrix, and its solution, can be written as

$$D_{\mathrm{m}} \frac{\partial^2 C}{\partial x^2} - \lambda n_{\mathrm{e}} C = 0 \quad \text{and} \quad C = C_{\mathrm{f}} \exp\left[-\sqrt{\frac{\lambda n_{\mathrm{e}}}{D_{\mathrm{m}}}} x\right],$$

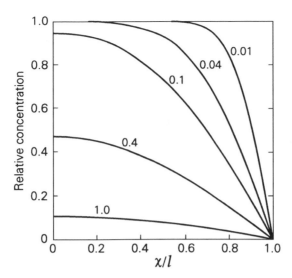

Figure 2.13 Concentration values in a porous medium $0 < X < L$ with no flow at $X = 0$, zero concentration at $X = L$, and a constant concentration initial condition. Curves represent values of $D_m t / L^2$. Modified from Carslaw and Jaeger (1959, p. 98).

where C_f is the concentration in the fracture, λ is the decay constant, and x is the distance from the fracture. Assuming a coefficient of diffusion of 10^{-11} m²/sec and a half-life of 5,700 years (carbon-14), how far into the matrix would one have to go in order to sample concentration levels 1/100th of that in the fracture?

2.3 Assume a unidirectional groundwater velocity field and that the longitudinal dispersivity of an unconsolidated sand unit is one meter and the transverse dispersivity is one centimeter. How low would the velocity have to be before molecular diffusion (10^{-10} m²/sec) would be greater than transverse dispersion? Than longitudinal dispersion?

2.4 A sediment core is taken from the shallow sea floor and concentration gradients are measured for the major ions (Ca, Mg, Na, K, Cl, and SO₄). All gradients appear to be zero accept for Ca and Mg. The dissolved calcium concentration at the seawater/sediment interface increases downward at 1 mmol/(liter-m) whereas the dissolved magnesium concentration decreases downward at 1 mmol/(liter-m). Use Fick's law (in the form of Eq. 2.32) and the data from Table 2.1 to calculate fluxes of all of the ions, assuming a porosity of 40% and a tortuosity of 2.0. Then use the data from Table 2.2 to calculate fluxes of all of the ions (the chloride flux can be obtained by an electrical balance). Compare the two results. Is there a significant difference?

2.5 Assume a unidirectional groundwater flow field and that local equilibrium conditions are valid. Groundwater that is undersaturated with respect to calcite is moving across a mineralogical boundary where it suddenly comes into contact with calcite (see Chapter 2.5.4 and Figure 2.10a). The inflowing seepage velocity of the groundwater is 10 m/day, the porosity changes from 20 to 30% as the calcite dissolves, and there is a 10 mg/liter increase in the concentration of total dissolved calcium across the boundary. At what speed is the boundary moving? *Hint:* Use the density of calcite and a simple mass-balance calculation.

3

Heat transport

In this chapter we derive general equations describing groundwater flow and heat transport, introduce several useful dimensionless parameters, and describe simple quantitative approaches to such specific processes as one-dimensional flow, buoyancy-driven flow, and heat pipes. In later chapters we will describe applications of the theory outlined here to regional-scale heat transfer (Chapter 4.3), ore genesis (Chapter 5), hydrocarbon maturation (Chapter 6), and geothermal systems (Chapter 7).

General models of hydrothermal circulation in the Earth's crust must accommodate a temperature range of 0°C (near the surface) to 1,200°C (the approximate basalt solidus). Corresponding fluid pressures might range from 0.1 MPa (atmospheric pressure) to 1,000 MPa (the approximate lithostatic pressure at 40-km depth). Over this pressure–temperature range, water can exist as liquid, vapor (steam), or supercritical fluid. On the macroscopic scale at which we model hydrologic systems, water can also be regarded as comprising a two-phase mixture of liquid and steam. We will consider heat as being *advected* by the moving fluid and *conducted* through the rock–fluid medium. "Conduction" refers to the transfer of kinetic, rotational, and vibrational energy according to a temperature gradient. Advection is generally more effective than conduction where volumetric rates of fluid flow are greater than 1 cm per year.

In pressure–temperature coordinates, the *vaporization* or *saturation curve* separates the liquid and steam regions (Figure 3.1a). Saturated liquid water and steam coexist along this curve. The energy content of the fluid differs dramatically across the vaporization curve due to the *latent heat of vaporization*. When the energy of the fluid is described in terms of energy per unit mass (internal energy, enthalpy, or entropy), rather than temperature, two-phase conditions define an area, rather than a single curve. For example, in the pressure–*enthalpy* diagrams of Figures 3.1b and 3.1c the breadth of the two-phase region corresponds to the latent heat of vaporization.

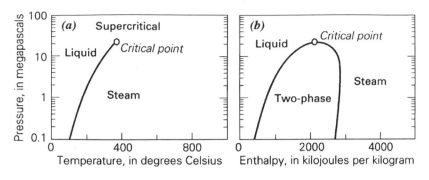

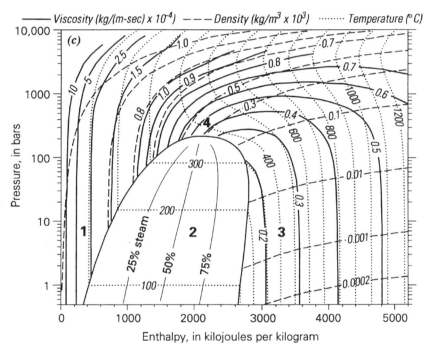

Figure 3.1 (a) *Pressure–temperature* and (b) *pressure–enthalpy* diagrams, showing thermodynamic regions; (c) pressure–enthalpy diagram showing contours of equal temperature, density, viscosity, and mass fraction steam. In (b) and (c) the curves bounding the two-phase region define the enthalpies of saturated steam and liquid water; they intersect at the *critical point* (220.55 bars and 2086. kJ/kg). In (c), the thermodynamic regions are denoted by (1) "compressed" liquid water; (2) two-phase; (3) superheated steam; and (4) supercritical fluid. On a microscopic scale, any particular water molecule will lie either on the saturation curves or fully outside the two-phase region. However, on the macroscopic scale at which we model hydrogeologic systems (Chapter 1.4), some *representative elementary volumes* may contain a mixture of phases, and these will lie in the two-phase region.

In this chapter we will treat fluid enthalpy as a primary variable, because formulations in terms of energy per unit mass have certain advantages in describing multiphase and near-critical flow. Enthalpy is the sum of the internal energy of a system plus the product of the pressure–volume work done on the system. For processes at a constant pressure, then, enthalpy and internal energy are identical. At relatively low pressures and temperatures less than about 200°C, the enthalpy of liquid water in the non-SI units of cal/g is nearly numerically equivalent to its temperature in degrees Celsius, the two quantities being related by a nearly constant heat capacity of about 1 cal/(g-°C) ($H \sim c_w T$).

In Figures 3.1b and 3.1c it can be seen that the energy required to vaporize a certain mass of pure water decreases towards the *critical point* of water (374°C, 22.06 MPa, 2086.0 kJ/kg). Above the critical point any sharp distinction between liquid water and steam disappears. For multicomponent systems the position of the critical point will (for example) shift toward higher temperatures and pressures with increasing salinity. For saline fluids there is also the possibility of *supercritical phase separation* between a saline brine and a relatively dilute vapor phase (see, e.g., Bischoff and Pitzer, 1989). In this chapter we will generally confine our discussion to pure-water systems for which the critical point is defined by particular, fixed values of temperature, pressure, and enthalpy.

3.1 Governing equations

Here we will develop equations that describe the flow and transport of pure water and heat. A fuller description would consider the effects of solutes and noncondensible gases on fluid properties and include coupled descriptions of solute transport like those developed in the previous chapter. However, systems of equations more complex than those considered here are difficult to solve, even numerically, and the pure-water assumption is reasonable for many applications. The pure-water assumption is discussed further in Chapter 3.1.7.

3.1.1 Choice of dependent variables

Groundwater flow and heat transport can be described by a set of coupled equations expressing mass and energy conservation. At first inspection it might seem reasonable to pose these equations in terms of hydraulic head and temperature, respectively. However, the familiar groundwater flow formulation in terms of head (Chapter 1.5) is predicated on constant fluid density. For moderate- to high-temperature applications we cannot ignore the thermal expansivity of the fluid, or even assume that it is constant. Thus it is necessary to calculate mass fluxes in terms of the forces acting on fluid(s) – pressure and gravity – rather than attempting to define an equivalent hydraulic head. It is also preferable to describe changes in mass storage in terms of changes in fluid density and porosity, rather than hydraulic head, because under some conditions, in situ changes in fluid density can contribute significantly to the mass flux.

Use of temperature in the energy-balance equation is also unsuitable for the general case, because a pressure–temperature combination does not specify the thermodynamic state of a steam and liquid water mixture. For any particular combination of pressure and salinity, steam and liquid coexist at a single temperature; for a pure-water system at 1 atmosphere (0.1 MPa), that temperature is of course 100°C. The single pressure–temperature combination of 1 atmosphere and 100°C thus applies to steam–liquid mixtures ranging from all steam to all liquid and, as a result, to large ranges in fluid enthalpy (418–2,676 kJ/kg) and density (0.6–987 kg/m^3), corresponding to the differences between the properties of steam and liquid water. In contrast, a pressure–enthalpy combination of 1 bar and 418 kJ/kg uniquely specifies a temperature of 100°C, a liquid saturation of 1.0 (all liquid), and a density of 987 kg/m^3.

In addition to posing difficulties in the two-phase region, choosing temperature as a dependent variable greatly complicates equation-of-state descriptions of fluid properties near the *critical point of water*. Near the critical point, maxima in the thermal expansivity and heat capacity of water nearly coincide with a minimum in kinematic viscosity, and these fluid-property extrema may influence transport in hydrothermal systems (Norton and Knight, 1977). In pressure–temperature coordinates the critical point is at the vertex of the vaporization curve (Figure 3.1a) and represents a singularity in equations of state. For example, fluid *heat capacity* ($c_f(P, T)$), which is required in order to convert temperature to energy content, diverges to $+\infty$, and the partial derivatives of fluid density ($\rho(P, T)$) diverge to $\pm\infty$ (Johnson and Norton, 1991). In pressure–enthalpy coordinates, where two-phase conditions are represented as a region rather than a single curve (Figure 3.1b), all relevant properties of liquid water and steam merge smoothly to finite values at the critical point (Figure 3.1c).

In general, if the governing equations are solved in terms of energy content per unit mass, fluid heat capacity does not enter into the formulation, equation-of-state singularities at the critical point are avoided, and fluid properties are uniquely specified under both single- and two-phase conditions. In the development that follows we will emphasize the pressure–enthalpy formulation presented by Faust and Mercer (1979a) and used in the U.S. Geological Survey's HYDROTHERM simulation model (Hayba and Ingebritsen, 1994). An alternative approach, employed in the TOUGH2 simulation model developed at

Lawrence Berkeley Laboratories (Pruess, 1991), is to switch variables, solving for pressure and temperature under single-phase conditions and pressure and saturation under two-phase conditions.

One possible disadvantage of the pressure–enthalpy formulation is that values of enthalpy, unlike temperature, are not readily obtained in a field situation. However, codes such as HYDROTHERM generally allow for input and output in terms of pressure–temperature, converting to enthalpy internally for purposes of calculation.

3.1.2 Statements of mass and energy conservation

For the sake of clarity, we will first write the governing equations in one-dimensional forms, choosing the vertical (z) dimension because, under ordinary conditions, gravity acts in the z direction and most conductive heat transfer also occurs vertically. We will begin with equations that are simply statements of conservation of mass and energy, equating changes in storage with fluxes and any sources or sinks. Conservation equations are commonly developed in terms of a *representative elementary volume* (see Chapter 1.4). Here we will allow for the possibility of two fluid phases and adopt a sign convention such that an increase in mass or energy stored is viewed as a positive quantity.

The law of conservation of mass dictates that the time rate of change in mass stored in a control volume be equal to the net mass flux through the volume plus any internal mass sources (or sinks). That is,

change in mass stored = (mass steam in − mass steam out)
+ (mass liquid in − mass liquid out)
+ mass sources,

or, in differential form,

$$\frac{\partial(n\rho_f)}{\partial t} = -\frac{\partial(\rho_s q_s)}{\partial z} - \frac{\partial(\rho_w q_w)}{\partial z} + R_m, \tag{3.1}$$

where the subscripts s, w, and f denote steam, liquid water, and the single- or two-phase fluid mixture in place, respectively. The time-derivative term $\partial(n\rho_f)/\partial t$ represents changes in the fluid mass ($n\rho_f$) stored within the differential volume. The spatial-derivative terms represent the variation in mass flux of steam ($\rho_s q_s$) and liquid water ($\rho_w q_w$) across the vertical dimension of the control volume, and R_m is a mass source/sink term. In the context of hydrothermal systems, internal sources of mass might include magmatic volatiles, metamorphic reactions that release water, and wells. (See also the discussion of sources and sinks of fluid in Chapter 4.1.)

The q within the spatial derivatives is a *specific discharge* or *Darcy velocity* ($m^3/(s\text{-}m^2)$, or m/s), which when multiplied by density ρ gives a mass flux ($kg/(s\text{-}m^2)$). The mass stored is the product of porosity, n, and the density of the fluid mixture in place, ρ_f, which under two-phase conditions is given by

$$\rho_f = S_s\rho_s + S_w\rho_w, \tag{3.2}$$

where S_s and S_w are volumetric steam and liquid saturations, respectively, and

$$S_s + S_w = 1. \tag{3.3}$$

For a pure-water system, the values assigned to these saturation variables become arbitrary above the critical point.

For energy, allowing for the possibility of two fluid phases and heat transfer by advection and conduction, conservation of energy dictates

change in heat stored
    = (heat advected in by steam $-$ heat advected out by steam)
    $+$ (heat advected in by liquid $-$ heat advected out by liquid)
    $+$ (heat conducted in $-$ heat conducted out) $+$ energy sources.

Or, in differential form,

$$\frac{\partial[n\rho_f H_f + (1-n)\rho_r H_r]}{\partial t} = -\frac{\partial(\rho_s q_s H_s)}{\partial z} - \frac{\partial(\rho_w q_w H_w)}{\partial z} - \frac{\partial q_h}{\partial z} + R_h, \tag{3.4}$$

where the subscript r denotes rock properties, and the enthalpy of the fluid in place, H_f, is

$$\frac{(H_s S_s \rho_s + H_w S_w \rho_w)}{(S_s \rho_s + S_w \rho_w)}. \tag{3.5}$$

Under the time derivative we must account for changes in the amount of energy stored in the rock ($(1-n)\rho_r H_r$) as well as the fluid ($n\rho_f H_f$), because for most reasonable values of porosity (n) the quantity $(1-n)\rho_r H_r \gg n\rho_f H_f$. The spatial-derivative terms are advective-flux terms and are simply equivalent to the mass-flux terms of Eq. (3.1) multiplied by the enthalpy of the appropriate fluid phase. Heat transfer by conduction is represented by q_h, and R_h is an energy source/sink term. In the context of hydrothermal systems, internal sources of energy might include latent heat of crystallization, metamorphic reactions, and wells. For more general geothermal conditions in the Earth's crust, the radioactive decay of certain isotopes is the dominant heat source (Chapter 7.1).

We will eventually arrive at the final form of the governing equations by substituting empirical expressions for q_f (*Darcy's law*) and q_h (*Fourier's law*) into the mass- and energy-balance equations and then generalizing them to three dimensions. It is important to note, though, that we have already introduced some significant assumptions. We have not included a separate term for heat exchange between fluid and rock and thus are implying local water–rock thermal equilibrium; we have also neglected the possibility of heat transfer by *dispersion* and *radiation*. These and other assumptions will be discussed further in Chapter 3.1.7, after the governing equations have been more fully developed.

3.1.3 A form of Darcy's law for two-phase flow of compressible fluids

A simple one-dimensional form of Darcy's law, appropriate for applications in which the density of liquid water can be assumed constant, states that the volumetric flow rate q_w is directly proportional to the gradient in hydraulic head h; the coefficient of proportionality K is called *hydraulic conductivity* (Chapter 1.1). Thus for vertical flow,

$$q_w = -K\left(\frac{dh}{dz}\right). \tag{3.6}$$

For fluids having variable density a more complex form of Darcy's law is required. Flow must be computed directly in terms of the pressure energy (P) and gravitational energy ($\rho_f g z$) of a unit volume of fluid, rather than hydraulic head. In standard derivations of the groundwater flow laws these quantities are divided by the specific weight of water, $\rho_w g$, which is assumed to be a constant, and then summed to obtain h (e.g., Chapter 1.1.2). This simplification is not possible if ρ_w varies significantly. Moreover, hydraulic conductivity depends on fluid density ρ_f and viscosity μ_f as well as rock properties, so that if there is significant variation in these fluid properties K must be expanded to

$$K = \frac{k\rho_f g}{\mu_f} \tag{3.7}$$

(Nutting, 1930), where only the *intrinsic permeability*, k, is a property of the rock. Finally, for systems in which two fluid phases may be present, *relative-permeability* (k_r) variables are introduced to describe the reduction in the flux of one phase due to the interfering presence of the other (Figure 3.2). Relative permeabilities are empirically based functions of volumetric saturation. The functional relationship is generally assumed to be nonhysteretic, and values of k_r vary from 0 to 1.

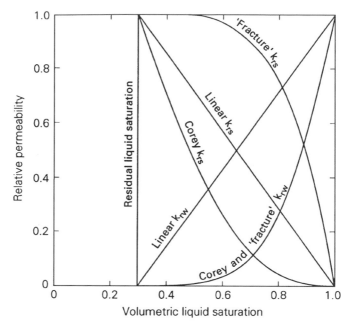

Figure 3.2 Linear, Corey-type (Corey, 1957), and "fracture-flow" (Sorey and others, 1980) *relative-permeability functions*. These functions bracket the range of behavior that has been suggested for steam–liquid water systems. The Corey and "fracture-flow" k_{rw} functions are identical, but their k_{rs} relations are very different, so that whereas the Corey functions give $k_{rs} + k_{rw} \ll 1$ for a large range of saturations, the "fracture-flow" functions give $k_{rs} + k_{rw} = 1$. Values of k_{rs} for the linear functions lie between the Corey and "fracture-flow" values. In these examples, the residual steam and liquid saturations are 0.0 and 0.3, respectively.

Thus, forms of Darcy's law appropriate for describing the vertical flow of variable-density steam and liquid water can be written

$$q_s = -\frac{k_{rs}k}{\mu_s}\frac{\partial(P + \rho_s g z)}{\partial z} \tag{3.8}$$

and

$$q_w = -\frac{k_{rw}k}{\mu_w}\frac{\partial(P + \rho_w g z)}{\partial z}, \tag{3.9}$$

respectively. Such forms of Darcy's law are widely used to describe both miscible and immiscible flow of variable-density fluids; to paraphrase Scheidegger (1974), the relative permeabilities are essentially "fudge factors" that allow Darcy's law to be applied to various empirical data on multiphase flows. There are very limited empirical data from which to infer relative permeabilities for the specific case of steam–liquid water systems. Some laboratory column-scale data for unconsolidated porous media suggest Corey-type (Corey, 1957) relations, with $k_{rs} + k_{rw} \ll 1$ over a fairly wide range of saturations (Piquemal, 1994), whereas some well-test data from fractured media suggest complementary relations with $k_{rs} + k_{rw} \sim 1$ (Sorey and others, 1980). Perhaps there is less opportunity for interference between phases where the intrinsic permeability is dominated by fractures.

It is worth noting here that the fluxes described by these expanded versions of Darcy's law, unlike those described by Eq. (3.6), are not *potential flows*. For a constant-density fluid in an isotropic media, flowlines will be completely dictated by a potential field. That is, they will parallel the gradient vectors of some measure of fluid potential, such as hydraulic head. Such a potential field cannot be defined for variable-density flow, except for the very special case in which lines of equal pressure, temperature, and concentration are parallel. In the general variable-density case, the field of force has a rotational component that tends to cause convective movement. This rotational component can perhaps be regarded as superimposed on the potential field (Hubbert, 1940).

3.1.4 Conductive heat flux

Conductive heat flux is given by *Fourier's law* of heat conduction as

$$q_h = -K_m \left(\frac{dT}{dz} \right), \tag{3.10}$$

where K_m is the thermal conductivity of the medium and T is temperature. Fourier's law is similar to the form of Darcy's law shown as Eq. (3.6); it is also analogous to *Fick's first law of diffusion* (Chapter 2.1.1) and *Ohm's law* for conduction of electricity. Each of these laws describes a linear relation between a flux and a gradient in potential, and such relations give rise to diffusion-type conservation statements such as the standard equation for groundwater flow (Chapter 1.5). Carslaw and Jaeger's (1959) classic *Conduction of Heat in Solids* is a rich source of analytical solutions to diffusion-type equations.

Whereas the *hydraulic conductivity* of crustal materials varies over approximately 16 orders of magnitude (Chapter 1.2), the *thermal conductivity* of the upper crust generally varies by less than a factor of 5. The thermal conductivity of lacustrine clay, one of the least conductive materials, is about 1 W/(m-K); that of granite, a relatively good conductor, is typically about 3 W/(m-K). There is substantial empirical evidence (Sass and others,

1971) to show that *medium thermal conductivity* is well modeled as the geometric mean of the rock conductivity and the pore-fluid conductivity, that is,

$$K_m = K_r^{(1-n)} K_f^n, \tag{3.11}$$

where n is porosity and the subscripts r and f denote rock and fluid properties, respectively. At 25°C, K_f is about 0.6 W/(m-K).

Because we are developing the governing equations in terms of pressure and enthalpy, temperature will not be not calculated directly. It is possible to expand the temperature gradient in Eq. (3.10) to

$$\frac{dT}{dz} = \left[\left(\frac{\partial T}{\partial P} \right)_H \left(\frac{\partial P}{\partial z} \right) + \left(\frac{\partial T}{\partial H} \right)_P \left(\frac{\partial H}{\partial z} \right) \right] \tag{3.12}$$

(Faust and Mercer, 1979a). Alternatively, one can calculate $T(P, H)$ for every z location and then calculate dT/dz as $[T(P, H)_1 - T(P, H)_2]/[z_1 - z_2]$ (Hayba and Ingebritsen, 1994).

3.1.5 One-dimensional forms of the governing equations

One-dimensional forms of the flow and transport equations are obtained by substituting versions of Darcy's law (Eqs. 3.8 and 3.9) and Fourier's law (Eq. 3.10) into the mass- and energy-balance equations (Eqs. 3.1 and 3.4), which results in

$$\frac{\partial(n\rho_f)}{\partial t} - \frac{\partial\left[\frac{\rho_s k_{rs} k}{\mu_s} \frac{\partial(P+\rho_s gz)}{\partial z} \right]}{\partial z} - \frac{\partial\left[\frac{\rho_w k_{rw} k}{\mu_w} \frac{\partial(P+\rho_w gz)}{\partial z} \right]}{\partial z} - R_m = 0 \tag{3.13}$$

and

$$\frac{\partial[n\rho_f H_f + (1-n)\rho_r H_r]}{\partial t} - \frac{\partial\left[\frac{H_s \rho_s k_{rs} k}{\mu_s} \frac{\partial(P+\rho_s gz)}{\partial z} \right]}{\partial z} \tag{3.14}$$
$$- \frac{\partial\left[\frac{H_w \rho_w k_{rw} k}{\mu_w} \frac{\partial(P+\rho_w gz)}{\partial z} \right]}{\partial z} - \frac{\partial\left[K_m \left(\frac{\partial T}{\partial z} \right) \right]}{\partial z} - R_h = 0.$$

These equations are coupled and nonlinear. They are coupled by the appearance of both dependent variables (P and H) in the heat transport equation (Eq. 3.14). They are nonlinear because many of the coefficients (e.g., ρ_s, ρ_w, k_r, μ_s, and μ_w) are functions of the dependent variables.

In the flow equation (Eq. 3.13) the dependent variable P appears only under the second-order spatial derivative; it is thus a diffusion-type equation, like the standard groundwater flow equation (Eq. 1.34). The heat transport equation (Eq. 3.14), which includes a diffusive heat-conduction term, also includes advective terms in which the dependent variables H_s and H_w appear under the first-order spatial derivative. It is an *advection–diffusion* equation, like the solute transport equation (Chapter 2.1).

The coupled pair of Eqs. (3.13) and (3.14) is generally analogous to Eqs. (2.28) and (2.21), developed to describe variable-density groundwater flow and solute transport in Chapter 2. The heat transport equation includes advective-flux terms that are directly analogous to those in the solute transport equation. However, the conductive-flux term of the heat transport equation, $\partial[K_m(\partial T/\partial z)]/\partial z$, is only partially analogous to the dispersive-flux term of the solute transport equation. The heat-conduction term is directly comparable to the diffusive flux that is embedded in the dispersive-flux term for solute transport. The other key element of the dispersive-flux term, *hydrodynamic dispersion* (Chapter 2.1.3), has been omitted in the heat transport case. The omission of *thermal dispersion* has been noted previously in association with Eq. (3.4) and will be discussed more later in this chapter.

In general, in solute transport problems only a single term is used to describe mass flux and advective transport, because only one fluid phase is considered. In the description of solute transport the role of the concentration variable can be regarded as roughly analogous to that of H_s, H_w, and T in the heat transport equation, and the diffusion coefficient as analogous to K_m. Although the conductive heat-flux term in Eq. (3.14) is written in terms of temperature, the temperature gradient ($\partial T/\partial z$) can readily be expanded in terms of the actual independent variables H and P, as shown in Eq. (3.12).

3.1.6 *Extending the governing equations to three dimensions*

We arrive at the final, three-dimensional form of the governing equations by introducing the vector operator ∇ to indicate $\mathbf{i}\partial/\partial x + \mathbf{j}\partial/\partial y + \mathbf{k}\partial/\partial z$ and rearranging slightly, giving

$$\frac{\partial(n\rho_f)}{\partial t} - \nabla \cdot \left[\frac{\rho_s k_{rs}\overline{k}}{\mu_s}\nabla(P + \rho_s gz)\right] \tag{3.15}$$

$$- \nabla \cdot \left[\frac{\rho_w k_{rw}\overline{k}}{\mu_w}\nabla(P + \rho_w gz)\right] - R_m = 0$$

and

$$\frac{\partial[n\rho_f H_f + (1-n)\rho_r H_r]}{\partial t} - \nabla \cdot \left[\frac{\rho_s k_{rs} \overline{k} H_s}{\mu_s} \nabla(P + \rho_s g z)\right] \quad (3.16)$$

$$- \nabla \cdot \left[\frac{\rho_w k_{rw} \overline{k} H_w}{\mu_w} \nabla(P + \rho_w g z)\right] - \nabla \cdot K_m \nabla T - R_h = 0,$$

where the overline indicates that $\overline{k}$ is a second-rank tensor in the general case
(Chapter 1.2.2).

3.1.7 Assumptions

We have already noted several of the assumptions involved in the derivation
of the governing Eqs. (3.13–3.16): that the circulating fluid is pure water; that
water and rock are in thermal equilibrium; that heat transfer by modes other than
advection and conduction is negligible; and that a two-phase form of Darcy's
law applies. The equations also imply that *capillary-pressure* effects (Chapter
6.2.1) are negligible, because a single value of pressure (P) applies to both
fluid phases. Finally, in solving these equations it is generally assumed that the
relative-permeability (k_r) terms can be represented as nonhysteretic functions
of volumetric liquid saturation. In this section we will discuss some of these
assumptions in greater detail (see also Faust and Mercer, 1979a; Hayba and
Ingebritsen, 1994).

Pure water

For many moderate- to high-temperature applications it is reasonable to neglect
the effects of solutes on fluid density and assume that the circulating fluid is pure
water. The density contrasts encountered in hydrothermal systems are often
much larger than those encountered in isothermal solute-transport problems
and are mostly due to pressure/temperature effects, rather than variations in
concentration. At a pressure of 1 atmosphere, the density contrast between steam
and liquid water is $>10^3$, and over the full temperature range of a magmatic–
hydrothermal system (as much as 0–1,200°C) the density of the liquid phase
alone may vary by more that a factor of 3. For comparison, the concentration-
related density difference between freshwater and seawater is only about 3%.
This is not to suggest that the influence of solutes is trivial; they can have major
effects on boiling-point behavior and, since the salinity of magmatic brines may
be greater than 20 weight percent (Fournier, 1987), their effects on density will
sometimes be significant. Furthermore, in saline systems we can expect two
fluid phases to be present even above the *critical point* (220.6 bars, 374°C) of

pure water: a NaCl-rich brine and a dilute vapor (Bischoff and Pitzer, 1989). When applying pure-water models, we must be aware of their limitations.

Local thermal equilibrium

The assumption that fluid and rock are in local thermal equilibirium ($T_f = T_r$) is commonly made in geothermal reservoir engineering applications. It is justified by the relatively high thermal conductivity of typical Earth materials and the generally slow, steady rates of subsurface fluid flow. The assumption of water–rock thermal equilibrium may be inappropriate for descriptions of rapid transients such as those involved in geysering (Chapter 7.6) or phreatic eruptions.

Thermal dispersion

The assumption that thermal dispersion is negligible is also common and justified by the high thermal conductivity of Earth materials. Because heat conduction is relatively efficient, it tends to homogenize local temperature variations, making the influence of hydrodynamic dispersion less important in heat transport than in solute transport (see Chapter 3.4.1).

Radiative heat transfer

A phenomenon similar to black body radiation, radiative heat transfer is nearly universally ignored in heat-transport modeling, but it may become significant at temperatures greater than about 600°C. Because the radiative heat flux, like the conductive flux, is proportional to the temperature gradient, one approach to radiative heat transfer would be to treat K_m as a temperature-dependent conduction–radiation coefficient (see Chapter 7.1.2).

Capillary pressures

It is difficult to estimate appropriate capillary-pressure functions for hydrothermal systems, as relevant data are limited. The surface tension of water decreases with temperature and vanishes at the *critical point*, where the properties of steam and liquid water merge. The formulation adopted here assumes that fluid pressures in the steam and liquid phases are equal, but the relative-permeability functions can incorporate some capillary effects by allowing for residual (immobile) liquid water and steam saturations. For particular analyses, such as those involving commercial development of two-phase geothermal reservoirs, the assumption of negligible capillary-pressure effects may be untenable. For example, geothermal development at The Geysers, California, has reduced reservoir pressures from an initial level of 3.0–3.5 MPa to about 1.0–1.5 MPa; meanwhile, reservoir temperature has remained nearly constant at about 235°C. The initial reservoir $P-T$ condition at The Geysers lay along the vaporization curve that defines the boundary between the steam and liquid-water regions, whereas the current $P-T$ condition lies well into the steam region (Figure 3.1a). Nevertheless, both field and experimental data indicate that substantial reserves of liquid water remain in place. This liquid is being "held" by capillary and adsorptive forces under conditions where we would predict only steam on the basis of purely thermodynamic considerations. Clearly, realistic models of The Geysers system under production stress would have to include such forces.

Relative permeabilities

A wide range of functional relations has been invoked to describe steam–liquid water relative permeabilities (Figure 3.2). Production data from geothermal fields would seem to indicate that there is little interference between phases; that is, k_{rs} and k_{rw} sum to 1 (e.g., Grant, 1977; Horne and Ramey, 1978; Sorey and others, 1980). This apparent lack of interference might perhaps be explained in terms of capillary forces that would tend to make liquid water favor smaller fractures and pore spaces, with steam occupying the larger voids. A single laboratory study in unconsolidated porous media (Piquemal, 1994) suggests Corey-type functions that give $k_{rs} + k_{rw} \ll 1$ over a large saturation range, and many modeling studies invoke such functions. Because relative permeabilities are highly dependent on pore and fracture geometry, the appropriate functional relationship may reasonably be expected to vary within the system of interest. However, there are seldom sufficient data to warrant the use of more than one or two generalized functions. In general, it seems reasonable to assume that the residual steam saturation is near zero on the basis of a thermodynamic analysis by Verma (1986) that suggests that steam bubbles are unlikely to be trapped. In contrast, the retention of liquid water by capillary and adsorptive forces may lead to high residual liquid saturations (e.g., Economides and Miller, 1985), as at The Geysers.

3.1.8 Fluid properties

The governing equations invoked here (Eqs. 3.13–3.16) require fluid densities, viscosities, and temperature to be calculated as functions of pressure and enthalpy. Correlations of experimental data such as those by Haar and others (1984) can be used to obtain these values. The U.S. Geological Survey's HYDROTHERM simulation model uses a large (2 megabyte) look-up table, based in part on the Haar "steam table" equations, to obtain values of fluid properties and their derivatives and cross derivatives (Hayba and Ingebritsen, 1994).

3.1.9 Numerical solution

Analytical solutions to Eqs. (3.13)–(3.16) pertain only to very special cases involving idealized geometries, homogeneous media, and minor variations in fluid properties. In general, these equations must be solved numerically. Though numerical solution techniques are beyond the scope of this book, a few guidelines are worthy of mention.

Numerical solution is complicated by the extreme nonlinearity introduced by the relative-permeability functions and variations in fluid properties with pressure and enthalpy. A standard approach is to treat the nonlinear coefficients by using *Newton–Raphson iteration*, which produces a system of linear equations. Within a Newton–Raphson iteration loop, a simple direct solution technique such as Gaussian elimination affords stability advantages, particularly in the case of multiphase problems. Node-by-node mass and energy balances can be monitored and used to define convergence criteria for the Newton–Raphson iteration.

For further information on numerical solution techniques the interested reader can turn to general monographs such as *Numerical Recipes* (Press and others, 1986) or to the documentation available for specific algorithms, such as HYDROTHERM (Faust and Mercer, 1979b; Hayba and Ingebritsen, 1994) and TOUGH2 (Pruess, 1991).

3.2 Initial and boundary conditions

In order to solve differential equations such as (3.13)–(3.16) one has to specify initial conditions throughout the domain of interest and boundary conditions at the boundaries of the domain (see also Chapter 2.1.6). In this case, *initial conditions* are values of the dependent variables, pressure and enthalpy, at the initial problem time. *Boundary conditions* generally comprise pressure–enthalpy values that are held constant through time and/or specified fluxes of mass and energy. In most cases, near-steady-state solutions to these equations do not show any dependence on initial conditions, but the choice of reasonable boundary conditions is crucial to successful problem solving.

A hydrostatic pressure gradient and a regional geothermal gradient comprise useful initial conditions for many problems, including most studies of regional-scale flow or fluid circulation near magma bodies. Because the governing equations are stepped through time, a faster approach to the steady-state solution can be achieved by estimating initial conditions that roughly approximate the expected steady state. Conversely, choosing arbitrary, constant initial values of pressure and enthalpy will retard the approach to steady state. For regional-scale problems involving 5- to 10-km deep crustal sections, temperatures will

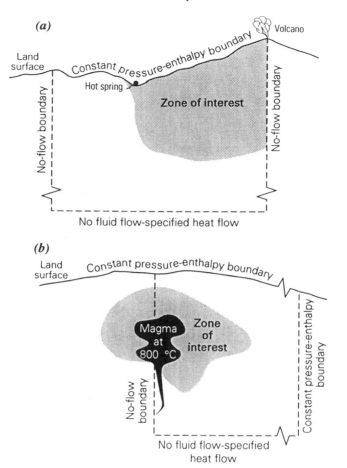

Figure 3.3 Schematic diagrams indicating possible *boundary conditions* for problems involving (a) topographically driven groundwater flow and heat transport and (b) hydrothermal circulation about a cooling pluton.

approach steady-state values over simulation times of 10^5 years or more. Reasonable, quasi-steady initial conditions for analyses of geothermal development can be obtained through long-term simulations of the natural predevelopment state of the system (e.g., Bodvarsson and others, 1984; Ingebritsen and Sorey, 1985).

In vertical sections and three-dimensional models the top boundary, which often represents the land surface (or water table), is commonly treated as a constant pressure–enthalpy boundary (Figure 3.3). The pressure is typically specified as one atmosphere (0.1 MPa) and, if topographic relief is significant, the specified enthalpy might vary with elevation, perhaps according to an *adi-*

abatic lapse rate (~5.5°C per km of elevation). Constant-through-time lateral boundaries are also common, and along such boundaries it is usually assumed that pressure follows a hydrostatic gradient and enthalpy follows a geothermal gradient. Unless such boundaries represent well-documented hydrogeologic features, they should be placed at a large distance from the area of interest so as to minimize their effects on the solution.

For regional and subregional flow problems, the lateral and bottom boundaries are commonly treated as controlled-flux boundaries. The lateral boundaries might be associated with topographic divides that can reasonably be treated as no-flow (symmetry) boundaries (Figure 3.3a) or, for a problem involving cooling of a regularly shaped pluton, symmetry might be invoked on a plane through the center of the pluton (Figure 3.3b). The most common lower boundary for large-scale flow problems is one with no fluid flow and a specified heat flow. The lack of fluid input is justified by the assumption of low permeability at midcrustal depths, and an appropriate heat input can be estimated on the basis of the regional conductive heat flow. It is usually much easier to estimate fluxes at the lower boundary than it is to estimate appropriate pressures and enthalpies, and imposing constant pressure–enthalpy (or –temperature) lower boundaries tends to have an undesireable controlling effect on the solution. A constant heat-flux lower boundary should be separated from any regions of active fluid flow by a substantial thickness of low-permeability material. Like a poorly known lateral boundary, the lower boundary should be placed at a large distance from the actual area of interest.

3.3 Temperature-based formulations

Despite the advantages of a pressure-energy per unit mass formulation for the general hydrothermal case, many heat-transport models are formulated in terms of temperature, as are most analytical solutions to heat-transport problems. Most single-phase heat-transport models are posed in terms of pressure or hydraulic head and temperature, so that they cannot readily be adapted to two-phase or near-critical problems. These include models suitable for a temperature range of about 0–1,000°C that have been used to simulate cooling plutons (e.g., Norton and Knight, 1977; Cathles, 1977) as well as models with linear equation-of-state approaches that are viable only to about 100°C (e.g., Voss, 1984, and Kipp, 1986). Many multiphase models, on the other hand, are posed in terms of pressure and energy per unit mass. Because most of the multiphase models have been designed by reservoir-engineering groups, they are intended to handle the sub-critical point conditions encountered in geothermal-reservoir development, and thus they generally employ equation-of-state descriptions that are adequate only to 300–350°C. However, with extended equations of state, such models can readily be adapted to higher-temperature problems (e.g., Hayba and Ingebritsen, 1994).

Because of the historical importance of temperature-based formulations, and their continued widespread use, we will describe them briefly here, working by analogy with the enthalpy-based formulation (Eqs. 3.15 and 3.16). In the temperature-based formulation, the mass- and advective-flux terms for steam drop out, as do relative permeabilities, and fluid and rock enthalpies are replaced by the product of heat capacity and temperature ($c_w T$ and $c_r T$, respectively). Thus the mass and energy balance equations become

$$\frac{\partial(n\rho_w)}{\partial t} - \nabla \cdot \left[\frac{\overline{k}\rho_w}{\mu_w}(\nabla P - \rho_w g \nabla z) \right] - R_m = 0 \tag{3.17}$$

and

$$\frac{\partial[n\rho_w c_w T + (1-n)\rho_r c_r T]}{\partial t} - \nabla \cdot \left[\frac{\overline{k}\rho_w c_w T}{\mu_w}(\nabla P - \rho_w g \nabla z) \right] \tag{3.18}$$
$$- \nabla \cdot K_m \nabla T - R_h = 0.$$

As noted in Chapter 3.1.1, such formulations can be adapted to two-phase problems by switching dependent variables at the boundary of the two-phase region; adding a third dependent variable is another possibility.

In modeling practice, further simplifications of Eqs. (3.17) and (3.18) are commonly invoked. Perhaps the most significant of these are arrived at through the *Boussinesq approximation*, which assumes that fluid density is constant except insofar as it affects the gravitational forces acting on the fluid. The traditional volume-based *stream function* (Slichter, 1899; de Josselin de Jong, 1969), which can be contoured to represent groundwater flow paths, implicitly invokes the Boussinesq approximation by assuming that fluid volume (rather than fluid mass) is conserved at a steady state and that changes in storage are negligible. The volume-based stream function can be significantly in error for cases of transient, variable-density flow, and some of its limitations have been elucidated by Furlong and others (1991), Hanson (1992), and Evans and Raffensperger (1992). Although it is possible to derive a mass-based version of the stream function that avoids the particular limitations imposed by the Boussinesq approximation (e.g., Evans and Raffensperger, 1992), the volume-based version has been widely applied to variable-density flow in studies involving cooling plutons (e.g., Norton and Knight, 1977; Cathles, 1977) and regional groundwater flow and heat transport (e.g., Garven and others, 1993). The Boussinesq approximation in the context of magmatic–hydrothermal systems will be further discussed in Chapter 7.2.

3.4 One-dimensional groundwater flow

Analytical solutions are available to describe the temperature fields caused by one-dimensional flow of a single fluid phase. Particularly useful examples are the steady-state solution for vertical flow presented by Bredehoeft and Papadopolous (1965) and the transient solutions for flow in a confined aquifer developed by Bodvarsson and others (1982) and Ziagos and Blackwell (1986). It is illuminating to compare the solution for steady vertical flow of groundwater and heat with the analogous solution for steady flow of groundwater and a solute.

3.4.1 Steady vertical flow

The solution described by Bredehoeft and Papadopolous (1965) applies to steady-state, vertical groundwater flow between constant-temperature boundaries. They point out that such conditions might reasonably be expected to apply to flow in a low-permeability *confining layer* between two aquifers. Their solution is

$$\frac{(T_z - T_U)}{(T_L - T_U)} = f\left(\beta, \frac{z}{L}\right), \tag{3.19}$$

where

$$f\left(\beta, \frac{z}{L}\right) = \frac{\left[\exp\left(\frac{\beta z}{L}\right) - 1\right]}{[\exp(\beta) - 1]};$$

$$\beta = \frac{c_w \rho_w q_w L}{K_m};$$

T_U, T_L, and T_z are temperatures at the upper and lower boundaries and at an intermediate depth z; and L is the distance between the constant-temperature boundaries. The value of z is zero at the upper boundary and increases downward, and the dimensionless parameter β is negative for upflow, positive for downflow. The solution can be applied to field data by calculating $(T_z - T_U)/(T_L - T_U)$ from measured temperatures, then plotting these dimensionless temperature values versus z/L at the same scale as type curves of $f(\beta, z/L)$.

Equation 3.19 indicates that if *medium thermal conductivity* (K_m) is constant and there is no groundwater flow ($q_w = 0$), the temperature profile will be linear. Choosing a typical value of 2 W/(m-K) for K_m and plotting temperature solutions for a series of flow rates (Figure 3.4a), we find that the temperature profiles become detectably *non*linear at volumetric flow rates of about 10^{-10} m/s, or about 0.3 cm/year. For the 100–200°C fluid used in our example, a flow rate of 10^{-11} m/s corresponds to an advective heat flux of $\sim$4–7 mW/m^2 ($c_w \rho_w q_w T \sim$ 4–7 mW/m^2) and results in an essentially linear

Heat transport

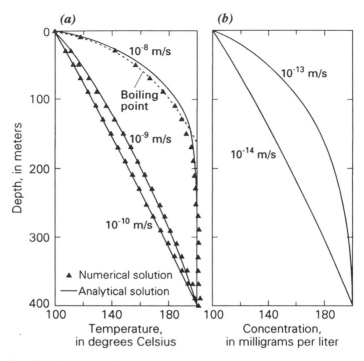

Figure 3.4 Steady-state solutions for one-dimensional flow groundwater with (a) transport of heat and (b) transport of a nonabsorbing solute. Temperatures and concentrations are held fixed at the upper and lower boundaries. Note that the temperature profile remains nearly at flow rates that are sufficient to greatly perturb the concentration profile. In (a), solid triangles indicate numerical solutions obtained with the HYDROTHERM model.

temperature profile (not shown). A flow rate of 10^{-10} m/s corresponds to an advective heat flux of 40–70 mW/m^2 and causes a subtle but detectable curvature; larger flow rates cause pronounced curvature (Figure 3.4a). The mean continental conductive heat flux is about 60 mW/m^2 (Jessop and others, 1976) , so if vertical flow rates are as large as 10^{-9} m/s (3 cm/year), advection is likely the dominant mode of vertical heat transport, even if temperatures are substantially lower than those used in our example. The "upflow" profiles in Figure 3.4a are convex upward; the corresponding "downflow" profiles would be concave upward. Convex profiles are typically encountered in groundwater discharge areas and concave profiles in recharge areas. The Bredehoeft and Papadopolous (1965) solution can be used to estimate vertical flow rates in recharge and discharge areas, although in most such areas the lower boundary condition will not strictly apply.

In Figure 3.4a we compare the temperature profiles obtained with Eq. (3.19) with solutions to the more general equations (3.15–3.16). Solutions to

Eqs. (3.15–3.16) were obtained numerically with the HYDROTHERM model (Hayba and Ingebritsen, 1994). At flow rates of $<10^{-8}$ m/s there is good agreement between the analytical and numerical solutions; the assumption of constant fluid properties for the analytical solution is responsible for the minor differences. For these particular boundary temperatures, the upflowing fluid intercepts the *boiling-point-with-depth curve* (Chapter 7.4.4) at flow rates $\geq 10^{-8}$ m/s, so that the analytical solution is no longer a good approximation. Over the boiling portion of the flow path, the numerical solution agrees well with steam-table data (Haar and others, 1984).

Solutions to the analogous solute transport problem of steady flow between constant-concentration boundaries (Figure 3.4b) can also be obtained analytically (e.g., Finlayson, 1992). Written in the same form as Eq. (3.19) and with the assumption that $C_L > C_U$, the solution for the solute transport problem is

$$\frac{(C_z - C_U)}{(C_L - C_U)} = f\left(\xi, \frac{z}{L}\right) \tag{3.20}$$

where

$$f\left(\xi, \frac{z}{L}\right) = \frac{\left[\exp\left(\frac{\xi z}{L}\right) - 1\right]}{[\exp(\xi) - 1]};$$

$$\xi = \frac{q_w L}{nD};$$

C_U, C_L, and C_z are concentrations at the upper and lower boundaries and at an intermediate depth z; L is the distance between the constant-concentration boundaries; D is a dispersion-diffusion coeffient; n is porosity; and the dimensionless parameter ξ is negative for upflow, positive for downflow. Whereas thermal conductivity can often be treated as a simple constant, the dispersion-diffusion coefficient D is a velocity-dependent parameter that can be expressed as

$$\frac{\alpha_L q_w}{n} + D_m, \tag{3.21}$$

where α_L is known as the *longitudinal dispersivity* (Chapter 2.1.3) and, in porous-media applications, D_m is some fraction of the *coefficient of molecular diffusion* in Fick's first law (Chapter 2.1.1). For common, nonadsorbed dissolved species, values of D_m are generally in the range of 10^{-11}–10^{-9} m²/s (e.g., Chapter 2.1.1; Freeze and Cherry, 1979). The *dispersivity*, α_L, is generally regarded as a scale-dependent parameter (Chapter 2.1.3), and for flow path lengths on the order of 10^2 meters, values of α_L on the order of 1–10 meters might be taken as reasonable (e.g., Chapter 2.1.3; Gelhar and others, 1992).

In the heat transport case, a linear temperature profile indicates that heat transfer occurs predominantly by conduction, and in the solute transport case a linear concentration profile indicates transport predominantly by molecular diffusion. Taking typical values of α_L and D_m to be 5 m and 10^{-10} m^2/s, respectively, and n as 0.1, we note that the effects of advection on the concentration profile become obvious at flow rates that are about 10^4 times smaller than those required to significantly perturb the temperature profile (Figure 3.4b). This difference reflects the relative efficiency of conduction as a heat transport mechanism as compared to molecular diffusion as a solute transport mechanism. The relative efficiency of heat conduction was invoked previously to justify the omission of *thermal dispersion* in the development of the governing Eqs. (3.15–3.16).

It should be noted that for the flow rates depicted in Figure 3.4b (10^{-14} to 10^{-13} m/s) the dispersive solute flux is also negligible ($\alpha q_w / n < 5 \times 10^{-12}$ m^2/s); the curvature of the concentration profiles is controlled only by the competing influence of advection ($v_w C$; Chapter 2.1.2) and molecular diffusion ($D_m = 10^{-10}$ m^2/s). Sanford and Wood (1995) analyzed similarly curved concentration profiles through a clay confining layer in west Texas to determine the relative importance of advection and molecular diffusion.

In nature these one-dimensional solutions for temperature and concentration would both generally apply to steady, vertical flow through a confining layer between two aquifers, with lateral flow rates in the aquifers sufficient to hold temperature or concentration constant at the boundaries of the confining layer. The solutions indicate that, for relatively high vertical flow rates, most variations in temperature and concentration will occur within relatively thin *boundary layers*, which therefore may be the loci of certain temperature- and concentration-dependent chemical reactions.

3.4.2 Flow in a confined aquifer or fault zone

Transient solutions developed by Bodvarsson and others (1982) and Ziagos and Blackwell (1986) apply to lateral groundwater flow in a confined aquifer or fault zone, with heat transfer by conduction into the adjacent rock. The solution derived by Bodvarsson and others (1982) assumes that the temperature is fixed at some distance below the aquifer, whereas the Ziagos and Blackwell (1986) solution assumes a semi-infinite half space below the aquifer. The two solutions converge as the depth to the fixed-temperature boundary becomes large. These are really quasi-analytical solutions; both involve numerical transforms. At any distance from the fluid source, the initial condition is a linear geothermal gradient (e.g., Figure 3.5, "background"), and the final steady-state solution involves two different, linear geothermal gradients: the initial "background" gradient below the aquifer and an elevated gradient above the aquifer (e.g. Figure 3.5b, $t = \infty$). The transient evolution between these end members involves periods

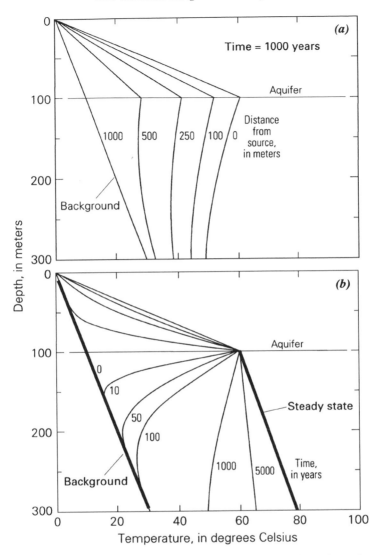

Figure 3.5 Transient solutions for the temperature profiles caused by horizontal ground-water flow. The aquifer is bounded by impermeable layers, so that heat flow into the surrounding rocks is by conduction only. In (a) the solution is shown as a function of distance from the fluid source at a time of 1,000 years. In (b) the solution is shown as a function of time at 0 distance from the source. After Ziagos and Blackwell (1986).

in which the geothermal gradient below the aquifer is reversed (Figure 3.5b, $t = 10$ to 1,000 years) and then nearly isothermal (Figure 3.5b, $t = 1,000$ to 5,000 years). Both types of transient profiles are fairly common in tectonically active areas.

The solutions for flow in a fault zone or confined aquifer allow the distinctive temperature profiles associated with confined lateral flow (Figure 3.5) to be interpreted in terms of such parameters as flow rate, the temperature of and distance to (Figure 3.5a) the fluid source, the "background" conductive heat flow below the aquifer, and the age of the system (Figure 3.5b). However, the data requirements for a unique interpretation are substantial, and they usually are not met in a field situation. On the basis of these solutions, episodes of widespread lateral flow of hydrothermal fluids during the Tertiary have been invoked to explain anomalous, near-constant hydrocarbon maturation profiles in the northwestern United States (Summer and Verosub, 1987). These particular solutions have also been pertinent to a debate about the thermal structure of the Oregon Cascade Range, where the temperature profile from a 2.5-km-deep well can be interpreted as resulting from either a relatively short-lived (10^4 years) flow system with a relatively high (120 mW/m^2) background heat flow (Blackwell and Baker, 1988; Blackwell and others, 1990) or a relatively long-lived (10^5 years) flow system with lower (70 mW/m^2) background heat flow (Ingebritsen and others, 1992). (See Chapter 4.3.4 for further discussion of heat flow in the Oregon Cascades.)

3.5 Dimensionless numbers

Several dimensionless quantities are useful in heat transport theory; those most frequently invoked are the *Nusselt* and *Peclet numbers*, used to describe the relative importance of advection, and the *Rayleigh number* criterion for the onset of free convection. In this section we will briefly describe each of these quantities and their applications.

3.5.1 Nusselt number

The Nusselt number Nu is perhaps the easiest of the dimensionless quantities to grasp; it is the ratio, in a particular dimension, of the total heat transfer to the heat transfer that would be expected in the absence of advection. It thus has a value of unity if $q_w = 0$. For a finite layer, in the notation used previously for steady vertical flow (Eq. 3.19), Nu can be taken as

$$\frac{\left[c_w \rho_w q_w T + \frac{K_m (T_L - T_U)}{L} \right]}{\left[\frac{K_m (T_L - T_U)}{L} \right]}. \tag{3.22}$$

For the particular cases shown in Figure 3.4a, the expected conductive heat flux is 0.5 W/m^2 (2.0 W/(m-K) × 0.25°C/m). At a flow rate of 10^{-10} m/s, taking the average value of T to be 150°C, we get an advective flux of about 0.06 W/m^2

(4,200 J/kg/K $\times$ 920 kg/m^3 $\times$ 10^{-10} m/s $\times$ 150°C). Thus the value of *Nu* is about 1.1 (0.56/0.5). For flow rates of 10^{-9} and 10^{-8} m/s the corresponding *Nu* values are about 2.2 and 15, respectively. These are the *Nu* values that apply across the entire 0- to 400-m depth interval. However, particularly for the higher flow rates, it is obvious that *Nu* will vary substantially depending on the specific depth interval over which it is calculated. Because these are steady flows, the value of the total heat flux, $c_w \rho_w q_w T + K_m(T_L - T_U)/L$, must remain constant over the entire 0- to 400-m interval. But as the temperature gradient increases toward the land surface, the conductive component of heat flux increases, whereas the advective component and the value of *Nu* decrease along with the temperature itself. For example, let us consider the case with a flow rate of 10^{-8} m/s (Figure 3.4a). If we had data only from the 200–400 m depth interval, we would define $T_L = 200$°C and $T_U \sim 195$°C, so that the expected conductive component in the absence of advection would only be about 0.05 W/m^2, and *Nu* would be about 150. Considering the 0–200 m depth interval alone, with $T_L \sim 195$°C and $T_U = 100$°C, the conductive component would be about 1.0 W/m^2, and *Nu* would be about 7. Thus the value of *Nu* varies by a factor of 20 between the 0–200 and 200–400 m depth intervals, and even larger variability in *Nu* would be calculated if the system were further subdivided.

3.5.2 *Peclet number*

The Peclet number, *Pe*, is another measure of the relative importance of advective heat transport. In fact, we have already encountered *Pe* in the guise of the dimensionless parameters β for heat transport problems (Eq. 3.19) and ξ for solute transport problems (Eq. 3.20). Because the entire numerator of *Pe* is multiplied by q_w, its value is zero when transport is entirely by conduction or molecular diffusion. The particular heat transport problems shown in Figure 3.4a involve Peclet numbers ranging from 0.08 to 7, versus Nusselt Numbers of 1.1 to 15. Unlike *Nu*, *Pe* does not vary when evaluated at different depths, because it does not depend directly on the temperature distribution.

The Peclet number is often invoked in the context of numerical analysis (e.g., Chapter 2.2). For that purpose, *L* is taken as some characteristic grid spacing, and the grid spacing is sometimes chosen such that *Pe* will be sufficiently small to minimize numerical errors (e.g., Anderson and Woessner, 1992). A *Pe* criterion can also indicate the threshold at which upstream weighting schemes, rather than simple averaging, might be used to estimate the values of certain flow parameters at internodal interfaces (e.g., Spalding, 1972).

3.5.3 Rayleigh number

The Rayleigh Number Ra indicates the tendency toward *free convection*, that is, flow driven purely by density differences. Derivations of Ra are found in many texts that deal with fluid mechanics and transport theory; a particularly complete and lucid derivation is that of Turcotte and Schubert (1982).

Classic Rayleigh convection theory was developed in the context of an infinite, permeable, horizontal layer bounded at top and bottom by isothermal, impermeable boundaries. The Rayleigh number is based on the ratio of "buoyant" forces, which drive convective fluid flow, to the viscous forces inhibiting fluid movement; it is obtained through linear stability analysis after linearizing an energy-balance equation, such as Eq. (3.18), by making the *Boussinesq approximation*. The value of Ra is given by

$$\frac{[\alpha_w \rho_w^2 c_w g k L (T_L - T_U)]}{[\mu_w K_m]}, \tag{3.23}$$

where α_w is *thermal expansivity* ($1/^\circ C$) and the other notation is as defined earlier. Lapwood (1948) showed that the fluid in an infinite horizontal layer will begin to convect at a critical Ra value (Ra_{crit}) of $4\pi^2$. As noted previously in Chapter 3.1.3, in the general case of variable-density flow there will always be some rotational component (Hubbert, 1940; Criss and Hofmeister, 1991). Nevertheless, Lapwood's criterion provides a useful, approximate indicator of the point at which convection should begin to perturb the thermal regime.

Sorey (1978) noted that if $(T_L - T_U)$ is large enough, the values of the relevant fluid properties ($\alpha_w, \rho_w, c_w, \mu_w$) as evaluated at T_L and T_U will be significantly different. He defined a "cold-side" Rayleigh number, Ra_U, based on fluid properties evaluated at T_U and a "mean" Rayleigh number, Ra_M, based on fluid properties evaluated at $(T_L + T_U)/2$. For large temperature differences the actual Ra_{crit} values were found to be as small as 2 on the basis of Ra_U and as large as 60 on the basis of Ra_M. In the region near the critical point of the fluid, very small temperature differences can cause large differences in fluid properties and trigger vigorous convection (Ingebritsen and Hayba, 1994). Near-critical convection is further discussed in Chapter 7.3.

In convecting systems, the value of the Nusselt number Nu is usually found to be positively correlated with Ra. Combarnous and Bories (1975) suggested that a general relation $Nu \sim 0.218 Ra^{0.5}$ is applicable to the case of an infinite horizontal layer for Ra in the range of 60–4,000.

3.6 Buoyancy-driven flow

An estimate of the steady, vertical flow rate in an upflow plume caused by a local, intense thermal anomaly can be derived directly from Darcy's law (e.g.,

Norton and Cathles, 1979; Turcotte and Schubert, 1982). Taking Eq. (3.9) as a starting point, and assuming a single fluid phase with a density that is a linear function of temperature ($\rho_w = \rho_0(1 - \alpha_w(T - T_0))$), we obtain

$$q_w = -\frac{k}{\mu_w}\frac{\partial[P - \rho_0(1 - \alpha_w(T - T_0))gz]}{\partial z}, \tag{3.24}$$

where ρ_0 is a reference density and T_0 is a reference temperature. If we further assume that the pressure gradient is approximately equal to the hydrostatic gradient at the surrounding temperature T_0 (i.e., $\partial P/\partial z \sim \partial(\rho_0 gz)/\partial z$), and rearrange slightly, we can obtain

$$q_w = -\frac{[k\rho_0 g\alpha_w(T - T_0)]}{\mu_w}, \tag{3.25}$$

which expresses a linear relation between q_w and k, given a particular value of $(T - T_0)$ with an associated set of fluid properties. Taking k as 10^{-16} m^2 and $(T - T_0)$ as 300°C, a value reasonably appropriate to magmatic–hydrothermal systems, we find $q_w > 10^{-9}$ m/s, a flow rate sufficient to noticeably perturb the thermal regime (see Figure 3.4a).

It is important to recognize that Eq. (3.25) is usually a very incomplete description of hydrothermal circulation. It was derived assuming that only buoyancy drives fluid flow, whereas in the general case the effects of topography, thermal pressurization, and fluid sources might also be significant. Furthermore, the assumption of constant μ_w and α_w is problematic for large values of $(T - T_0)$, and the approximation ($P \sim \rho_0 gz$) is sometimes grossly in error, as we shall see in the following section.

3.7 Heat pipes

Some two-phase hydrothermal systems include *vapor-dominated zones* that function like *heat pipes* (White and others, 1971). Within vapor-dominated zones, steam is the pressure-controlling phase, so that the vertical pressure gradient is usually close to vaporstatic, rather than hydrostatic. At a given pressure, steam and liquid water coexist at only one temperature (Figure 3.1a); thus, the small vertical pressure gradient corresponds to a small temperature gradient, and conductive heat flux through the vapor-dominated zone is very small. Heat transport through the the vapor-dominated zone occurs by means of a countercurrent flow of steam and liquid water that is analogous to the industrial heat-transfer process known as the heat pipe. A large fraction of the heat advected upward by rising steam is released by *condensation* at the top of the vapor-dominated zone. The relatively low-enthalpy liquid condensate trickles back down through the vapor-dominated zone. The upward extent of the vapor-dominated zone is usually limited by a low-permeability *caprock*, and vertical heat transport through the caprock to the land surface may be largely by conduction. Because the vapor-dominated zone is generally underpressured with respect to local hydrostatic pressure, it must also be isolated from surrounding flow systems by low-permeability barriers.

A simple analytical approach reveals some characteristics of a closed heat pipe, that is, a heat pipe with no net mass flux. If we assume that the system is at steady state, and that the lateral pressure and temperature gradients within the heat pipe are negligible, then the energy balance can be described simply by

$$
\left[\frac{H_s \rho_s k_{rs} k}{\mu_s} \frac{\partial(P - \rho_s g z)}{\partial z} \right]
$$

$$
+ \left[\frac{H_w \rho_w k_{rw} k}{\mu_w} \frac{\partial(P - \rho_w g z)}{\partial z} \right] + q_{cond} = 0, \tag{3.26}
$$

where in this case q_{cond} represents the net heat flux through the system, expressed as pure conduction above or below the closed heat pipe. Because there is no net mass flux through the system, the mass fluxes of the two phases are equal and opposite in sign ($\rho_s q_s = -\rho_w q_w$), so that Eq. (3.26) can be rearranged to obtain, for example,

$$
q_s = \frac{q_{cond}}{\rho_s(H_s - H_w)} \tag{3.27}
$$

and

$$
-q_w = \frac{q_{cond}}{\rho_w(H_s - H_w)}, \tag{3.28}
$$

where q_s and q_w are the *volumetric flow rates* of Eqs. (3.8) and (3.9).

The conductive heat flux above the vapor-dominated zone at The Geysers, California, is about 0.3–0.5 W/m^2 (Walters and Combs, 1992), and by assigning these values to q_{cond} and evaluating fluid properties at 240°C, a typical vapor-dominated zone temperature, we obtain $q_s = 1.0$–1.7×10^{-8} m/s and $-q_w = 2.1$–3.5×10^{-10} m/s. The one-dimensional balance equation (3.26) can also be rearranged to show that a near-vaporstatic vertical pressure gradient ($\partial P/\partial z \sim \rho_s g$) implies $k_{rs} \gg k_{rw}$.

One should not be unduly seduced by the analytical tractability of the one-dimensional, closed heat pipe; many vapor-dominated zones have significant throughflow of mass. Some areas at Lassen Volcanic National Park, California are believed to be underlain by vapor-dominated zones at about 300 m depth, and have total near-surface heat flows on the order of several hundred W/m^2 (Sorey and Colvard, 1994). Such high heat flow values can only be explained in terms of substantial steam discharge from the underlying vapor-dominated zone. Vertical and lateral exchange of mass is also a critical component of the long-term, natural evolution of vapor-dominated zones (e.g., Ingebritsen and Sorey, 1988). Vapor-dominated zones are further discussed in Chapter 7.4.2.

Problems

3.1 Use the pressure–enthalpy diagram of Figure 3.1c to estimate the mass fraction of steam in a fluid mixture that has decompressed *adiabatically* (isenthalpically) from (a) 200 bars, 200°C to 1 bar, 100°C and (b) 200 bars, 300°C to 1 bar, 100°C.

3.2 Use the pressure–enthalpy diagram of Figure 3.1c to estimate the *latent heat of vaporization* and the densities (ρ_s, ρ_w) and viscosities (μ_s, μ_w) of the coexisting steam and liquid water phases at (a) 100°C, (b) 200°C, and (c) 300°C.

3.3 Use the following data set and Eqs. (3.10) and (3.11) to determine *conductive heat flow*. Note that the appropriate value of K_m to apply to the entire depth interval is the *harmonic mean* of the individual K_m values (Chapter 1.2.2).

Depth	T	K_r	n
100 m	16.7°C	1.37 W/(m-K)	0.11
110	17.3	1.46	0.15
120	18.0	1.48	0.19
130	18.5	1.49	0.14
140	19.2	1.34	0.09
150	20.0	1.40	0.16

3.4 Use the method of Bredehoeft and Papadopolous (Eq. 3.19) to calculate the temperature profiles associated with upflow at a rate of 10^{-8} m/s, for $L = 100$ m, $P_U = 100$ bars, and (a) $T_L = 10°C$ and $T_U = 0°C$ and (b) $T_L = 510°C$ and $T_U = 500°C$. Use a steam-table reference such as Haar and others (1984) to determine appropriate values of fluid density and heat capacity, and assume $K_m = 2.0$ W/(m-K). Do the shapes of the two curves differ? If so, how and why?

3.5 Consider a 1-km-thick layer at 2–3 km depth, with $T_L = 300°C$ and $T_U = 200°C$. For what values of k should we expect *free convection* to take place? Use the Rayleigh number approach (Eq. 3.23) and assume $Ra_{crit} \sim$ 40. Do the calculation twice, using fluid properties evaluated at **(a)** T_U and **(b)** $(T_L + T_U)/2$. Assume $K_m = 2$ W/(m-K), and use a steam-table reference such as Haar and others (1984) to determine fluid properties.

3.6 Working from Eq. (3.26), calculate k_{rs}/k_{rw} for a closed, 250°C heat pipe, given that the vertical pressure gradient is twice vaporstatic and assuming that q_{cond} is 1.0 W/m^2.

4

Regional-scale flow and transport

This chapter represents a transition between the basic theory of previous chapters and the particular applications of Chapters 5 to 10. Here we will apply the theory of groundwater flow and transport in fairly generalized geologic contexts, whereas in later chapters we focus on applications of these ideas to more particular geologic processes and environments. Specific topics to be addressed in this chapter include sources and sinks of fluids in geologic environments (Chapter 4.1.1), anomalous fluid pressure generation (Chapter 4.1.2), natural hydraulic fracturing (Chapter 4.1.3), regional-scale solute transport (Chapter 4.2), and regional-scale heat transfer (Chapter 4.3).

Regional-scale groundwater flow can frequently be regarded as being driven by hydraulic head (Chapter 1.1.2) gradients between high-elevation recharge areas and low-elevation discharge areas. The layered heterogeneity of many geologic media, and of sedimentary basins in particular (e.g., Figure 4.1), leads to anisotropic permeabilities (Chapter 1.2.2), with horizontal permeabilities (roughly parallel to layering) typically much larger than vertical permeabilities. Such permeability structures facilitate the regional-scale lateral flow of groundwater that is described in Chapters 4.2 and 4.3.

In real systems the processes of groundwater flow and mechanical deformation, solute transport, and heat transport are fully coupled (e.g., Person and others, 1996), and various mechanisms combine to drive groundwater flow in particular localities. For example, in the Gulf Coast basin of the south-central United States and eastern Mexico (Chapter 4.1.4) there is shallow topographically driven flow, thermohaline convection near salt domes (Chapter 9.3.1), and compaction-driven flow. However, in this chapter, as in Chapters 1 to 3, we will generally simplify the analyses by treating groundwater flow and deformation, solute transport, and heat transport as if they were independent processes. In later chapters we will introduce a few of the intercouplings and other complexities associated with particular geologic processes.

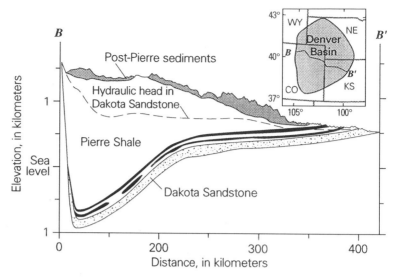

Figure 4.1 Vertical cross section through the Denver basin of the North American midcontinent, showing how fluid pressures in the basal Dakota sandstone depart from hydrostatic. At this vertical scale the water table is coincident with the land surface. The Dakota aquifer is isolated from high-elevation recharge areas to the west by the lowpermeability Pierre Shale and directly connected to low-elevation outcrops in Nebraska and Kansas to the east. From Neuzil (1995).

4.1 Sources and sinks of fluid

In topographically driven groundwater flow systems, fluid circulation occurs because hydraulic head (Chapter 1.1.2) decreases from high-elevation recharge areas (highest head) to low-elevation discharge areas (lowest head) (Toth, 1963; Freeze and Witherspoon, 1966). Patterns of flow may be much more complex where chemically (Chapters 2.3 and 9.3.1) and/or thermally induced (Chapters 3.5.3–3.6, 7.2–7.3, and 10.2.2) variations in fluid density also act to drive flow. In general, fluid pressures in most topographically and density-driven systems are close to the *hydrostatic* gradient (perhaps within 10% of hydrostatic). In crystalline rocks, near-hydrostatic conditions may persist to >10 km depth in the crust (see, e.g., the data summarized by Zoback and Zoback, 1997). Even in purely topographically driven systems, however, fluid pressures sometimes depart substantially from hydrostatic. For example, in the North American midcontinent, hydraulic heads in deep aquifers of the Denver basin are more than 25% below the hydrostatic level because they are isolated from high-elevation recharge areas in the Rocky Mountains by the low-permeability Pierre Shale and are relatively well-connected to low-elevation discharge areas in Kansas (Figure 4.1; Belitz and Bredehoeft, 1988). Substantial departures from hydrostatic conditions in regional flow systems might also be related to chemically driven flow or *osmosis* (see Chapters 1.1.2 and 4.2.3).

Departures from hydrostatic conditions related to heterogeneous permeability (e.g., the Denver basin), heterogeneous pore fluids (e.g., oil and gas), or chemical osmosis can be treated as equilibrium phenomena; that is, the hydrodynamic regime, as represented by the fluid-pressure distribution, can be regarded as being in equilibrium with the geologic framework. In contrast, departures from hydrostatic conditions related to ongoing geologic processes such as sediment compaction, hydrocarbon generation, production and consumption of water by metamorphic reactions, and degassing of magma are disequilibrium phenomena, in the sense that the fluid-pressure regime has not adjusted to the evolving geologic framework. Neuzil (1995) termed such processes *geologic forcings* with respect to the groundwater system and showed that they can induce significant fluid flow in some geologic environments. In the context of standard equations of groundwater flow (e.g., Eq. 1.31), geologic forcings may be treated as sources or sinks of fluid (Bredehoeft and Hanshaw, 1968), that is, for example,

$$s_s \frac{\partial h}{\partial t} = \nabla \cdot \overline{K} \nabla h + \Gamma, \qquad (4.1)$$

where Γ is a source term ($L^3/L^3/T$ or $1/T$ in this case). For such forms of the groundwater flow equation, in fact, any processes that act to change fluid density or porosity, other than than the uniaxial (vertical) compression invoked in deriving the specific storage s_s (Chapter 1.5.2), must be treated as spatially distributed sources or sinks of fluid. Thus, for example, variations in fluid density due to thermal expansion must be regarded as sources or sinks. For more general forms of the groundwater flow equation, such as

$$\frac{(\partial n \rho_f)}{\partial t} = \nabla \cdot \left[\frac{\overline{k} \rho_f}{\mu_s} (\nabla P + \rho_f g \nabla z) \right] + \Gamma_m, \qquad (4.2)$$

where Γ_m is a source term ($M/L^3/T$ in this case), changes in porosity n (e.g., due to compaction) and fluid density ρ_f (e.g., due to thermal expansion) can be accommodated in the storage (left-hand side) term. In this case the source term can be reserved for processes that truly introduce "new" fluid to the system – for example, hydrocarbon generation.

We will proceed by describing the effective source/sink rates associated with various geologic processes, using the concept of geologic forcing (Neuzil, 1995) to treat diverse sources and sinks of fluid in a unified conceptual framework. We will then evaluate the conditions under which sources and sinks of fluid can lead to anomalous fluid pressures and even hydraulic fracturing ("hydrofracturing") of the geologic medium. Particular geologic environments that we will consider

Table 4.1 *Sources and sinks of fluid ((L^3/T)/L^3, or $1/T$) in various geologic settings. "Virtual" fluid sources are those that act by changing porosity and/or fluid density.*[a]

Source	Magnitude (1/seconds)	Type
Devolatilization in a contact-metamorphic setting	3×10^{-13}	actual
Heating in a contact-metamorphic setting	3×10^{-13}	virtual
Petroleum generation	1×10^{-14}	actual
Compaction in accretionary prisms	10^{-15} to 10^{-13}	virtual
Pressure solution of quartz	10^{-16} to 10^{-14}	virtual
Compaction and heating in subsiding sedimentary basins	$<7 \times 10^{-15}$	virtual
Decompaction and cooling in uplifting sedimentary basins	$<7 \times 10^{-15}$	virtual
Dewatering of smectite in subsiding sedimentary basins	$<3 \times 10^{-15}$	actual
Devolatilization in a regional metamorphic setting	$<3 \times 10^{-15}$	actual
Deformation in the vicinity of a transform fault	$<2 \times 10^{-15}$	virtual
Deformation in a stable intraplate setting	10^{-23} to 10^{-20}	virtual

[a] After Neuzil (1995).

include the Gulf Coast of North America and the wedges of sediment (accretionary prisms) that overlie subducting oceanic plates.

4.1.1 Geologic forcing

Approximate values of the source-term strength (Γ in Eq. 4.1) that might be associated with various geologic processes can be obtained from the literature. In Table 4.1, various of these *geologic forcings* are ranked according to the approximate magnitude of Γ and, again employing Neuzil's (1995) terminology, classified as either actual or virtual forcings. The virtual forcings are those that involve changes in porosity or fluid density ($\partial(n\rho_f)/\partial t$ in Eq. 4.2) rather than an actual addition of fluid to the system (Γ_m in Eq. 4.2).

Inspection of Table 4.1 reveals a surprisingly narrow range of actual and virtual geologic forcings (about 10^{-16} to 10^{-14}/second) to be characteristic of active sedimentary basins and other rapidly deforming environments. Somewhat larger forcings ($\sim 3 \times 10^{-13}$/second) might be expected in contact metamorphic environments, whereas substantially smaller forcings ($<10^{-20}$/second) are expected in stable intraplate environments.

4.1.2 Anomalous fluid pressures

Equation 4.1 and the rate data compiled in Table 4.1 provide us with a basis for estimating when geologic forcing is likely to have a significant effect on the fluid-pressure distribution. For a homogeneous, isotropic hydraulic-conductivity field (Chapter 1.2.2), Eq. (4.1) can be written in dimensionless form as

$$\frac{\partial h_d}{\partial t_d} = \nabla^2 h_d + \Gamma_d, \qquad (4.3)$$

where the operator ∇^2 and hydraulic head h_d have been nondimensionalized with a representative length L, dimensionless time t_d is given by $Kt/s_s L^2$, and the dimensionless source/sink (geologic forcing) Γ_d is given by $\Gamma L/K$. Neuzil (1995) assumed conservatively that hydraulic head gradients greater than unity indicate a significant level of geologic forcing, and he noted that such gradients would be expected for a variety of domain geometries if

$$K < |\Gamma|L. \qquad (4.4)$$

The potential for significant effects on the fluid-pressure regime thus depends only on the strength of the forcing, the size of the domain, and the hydraulic conductivity.

We can now easily estimate the conditions under which certain geologic processes are likely to lead to anomalous pore-fluid pressures. For example, *devolatilization* of a 1-km-thick layer of rock in a *regional metamorphic* setting (Table 4.1: $\Gamma < 3 \times 10^{-15}$/second) might lead to anomalous fluid pressures if the hydraulic conductivity of the layer, K, is less than 3×10^{-12} m/s (or, alternatively, if the layer were bounded above and below by comparably thick and poorly conductive layers). At a temperature of about 150°C this limiting value of hydraulic conductivity would translate to an intrinsic permeability, k, of $\sim 1 \times 10^{-19}$ m^2. In a 100-m-thick *contact-metamorphic* zone adjacent to an igneous intrusion the combined effects of heating ($\Gamma \sim 3 \times 10^{-13}$/second) and devolatilization ($\Gamma \sim 3 \times 10^{-13}$/second) might lead to anomalous pressures for wallrock $K < 6 \times 10^{-11}$ m/s or, at $T \sim 300$°C, for wallrock $k < 1 \times 10^{-18}$ m^2 (see also Hanson, 1992, in this context). The widespread occurrence of reticulate fracture networks ("stockworks") in the wallrock adjacent to intrusions is, in fact, commonly attributed to fluid pressures that temporarily exceeded the stress required for failure, leading to hydraulic fracturing of the rock mass (e.g., Titley, 1990).

4.1.3 Hydraulic fracturing

The term *hydraulic fracturing* (also "hydrofracturing" or "hydrofracting") derives from the petroleum industry. It has been used since the late 1940s to describe the common oil-well stimulation practice of pumping fluid into a well at high pressures in order to fracture the formation and increase its permeability.

The mechanics of this artificial process have become quite well understood (e.g., Hubbert and Willis, 1957; Bredehoeft and others, 1976; Raleigh and others, 1976), and it seems reasonable to assume that the mechanics of natural hydraulic fracturing are to some extent analogous. Here we will describe the criteria for hydraulic fracturing in general, qualitative terms. In Chapter 8, failure criteria are described more quantitatively in terms of Mohr–Coulomb theory.

Hydraulic fracturing will occur when the fluid pressure exceeds the sum of the least principal stress and any tensile strength of the rock mass. Empirical observations indicate that this can occur for fluid pressures as low as ~ 1.5 times *hydrostatic* (hydrostatic pressure $P_h = \rho_f g d \sim 100$ bars/km depth) or, equivalently, ~ 0.6 times *lithostatic* (lithostatic pressure $P_l = \rho_r g d \sim 250$ bars/km depth) (Bredehoeft and others, 1976). The record of induced hydrofracturing also suggests that prospects for frictional failure in the upper crust are much enhanced if fluid pressures reach 1.25 times hydrostatic, or 0.5 times lithostatic (Rojstaczer and Bredehoeft, 1988). A convenient upper bound for failure is the lithostatic pressure or load. When and if the fluid pressure at any depth exceeds the lithostatic load, it is capable of lifting the entire overburden, creating (sub)horizontal fractures – for example, partings along bedding planes.

Commonly, however, fluid-induced failure occurs at fluid pressures substantially less than the total overburden pressure, and hydraulically induced fractures are oriented (sub)vertically rather than horizontally. This hydrofracture orientation can be understood in terms of the normal state of stress in the Earth. The upper crust of the Earth is often thought of as an elastic solid, so that the state of stress can be described in terms of three mutually orthogonal *principal stresses*, σ_1, σ_2, and σ_3. In general, one of the principal stresses will be nearly vertical, at least in regions with gentle topography. The other two principal stresses must thus be nearly horizontal. In an extensional (normal-faulting) environment the *maximum principal stress*, σ_1, is vertical, and fracturing will occur in vertical planes that are orthogonal to the *least principal stress* σ_3 (Figure 4.2). In compressional (reverse- or thrust-faulting) environments, σ_1 is horizontal, and (sub)horizontal fracturing will indeed occur at fluid pressures close to the total overburden pressure (e.g., Hubbert and Willis, 1957).

4.1.4 The Gulf Coast

The Gulf Coast of the south-central United States and eastern Mexico is underlain by a gulfward thickening wedge of unconsolidated to semiconsolidated sediments with relatively low hydraulic conductivity. Oil wells drilled into this thick sedimentary wedge encounter pressures substantially in excess of hydrostatic (Figure 4.3) at depths ranging from less than 2 km in parts of coastal

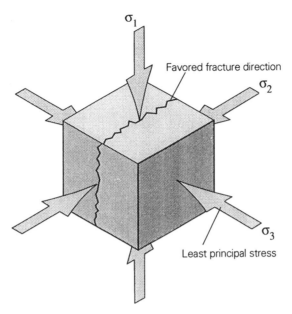

Figure 4.2 Stress element showing preferred plane of fracture orthogonal to the least principal stress σ_3. After Hubbert and Willis (1957).

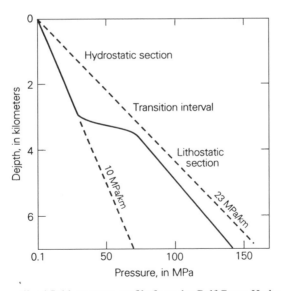

Figure 4.3 Generalized fluid-pressure profile from the Gulf Coast. Hydrostatically pressured section overlies an overpressured section in which fluid pressures approach the lithostatic gradient. After Bethke (1986).

Texas to more than 5 km in parts of coastal Louisiana (Wallace and others, 1981). The origin of the so-called geopressured zone at depth has been attributed to rapid deposition, subsidence, and compaction of sediments (Dickinson, 1953); to dehydration of smectite- (montmorillonite-) group minerals (Powers, 1967); and/or to thermal expansion of pore fluids (Barker, 1972). The transition from hydrostatic to geopressured conditions frequently corresponds to a transition from fluvial (coarse gravel) deposits to delta foreslope (fine gravel) deposits. Some credence has been lent to the smectite dehydration hypothesis by observations that the top of the geopressured zone sometimes coincides with the zone of smectite dehydration.

Bredehoeft and Hanshaw (1968) were the first to recognize that generation of anomalous pressures in subsiding basins could be treated as a transient groundwater flow problem and that the requirement for pressures significantly in excess of hydrostatic is essentially that the rate of fluid flow be slow relative to the rate of sedimentation. They used closed-form analytical solutions to show that for a sedimentation rate of 5 mm/year, appropriate to the Gulf Coast, near-lithostatic pressures might be attained in a sedimentary column with a uniform hydraulic conductivity of $<10^{-10}$ m/s, which would equate to an intrinsic permeability of 10^{-17} m^2 at 15°C or $>3 \times 10^{-18}$ m^2 at the $>100°$C temperatures typical of the overpressured zone.

Bethke (1986) later calculated the permeability profiles required to maintain lithostatic fluid pressures for various assumed subsidence rates and depth–porosity relations. He found that compaction alone might explain overpressure in a Gulf Coast environment; that thermal expansion of pore fluids had only a minor effect; and that the effect of smectite dehydration, while possibly important, was difficult to quantify. Bethke (1986) speculated that the effect of smectite dehydration on permeability might actually be more important than its role as a fluid source.

The areal extent of overpressures in the Gulf Coast region might lead one to expect that the compaction-driven component of flow is approximately one dimensional (vertical). However, differential subsidence and variations in basin thickness may induce focusing of compaction-driven flow toward the basin margins (Harrison and Summa, 1991). Similarly, lateral permeability variations in rift-style sedimentary facies may induce lateral compaction-driven flow from playas toward alluvial fans (Person and Garven, 1994).

4.1.5 Accretionary prisms

Accretionary prisms are wedges of unconsolidated to semiconsolidated sediment that form above subduction zones as seafloor sediments are scraped off

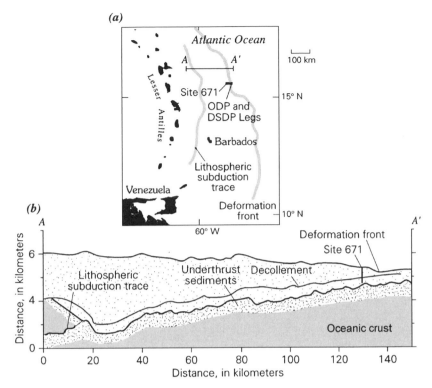

Figure 4.4 (a) Map of the eastern Caribbean showing the location of the Barbados Ridge accretionary prism, which extends approximately from the deformation front to the lithospheric subduction trace. (b) Vertical cross section of the Barbados Ridge accretionary prism at 16° 12′ N. From Bekins and others (1995).

the downgoing plate and accreted to the upper plate (Figure 4.4). The detachment surface separating the accreted and subducted sediments is known as the decollement. Particularly well-studied accretionary prisms include those along the Barbados Ridge of the eastern Caribbean (e.g., Moore and others, 1988; Henry and Le Pichon, 1991; Henry and Wang, 1991; Ferguson and others, 1993; Bekins and others, 1994, 1995), the Cascadia margin of the northwestern United States and southeastern Canada (e.g., Henry and Wang, 1991; Hyndman and others, 1993; Carson and others, 1994; Cochrane and others, 1994; Wang, 1994; Wang and others, 1994; Yuan and others, 1994; Screaton and others, 1995), the Nankai margin along the southeast coast of Japan (e.g., Morgan and others, 1994; Hyndman and others, 1995; Morgan and Karig, 1995), and the Makran prism in the Gulf of Oman (Minshull and White, 1989). Although the accretion process begins deep in the oceanic trench, the accretionary complex may eventually extend above sea level, as on Barbados Island and in the coastal mountain ranges of Oregon, Washington, and Vancouver Island.

Elevated fluid pressures appear to dominate the tectonic evolution of accretionary prisms, and fluid circulation profoundly affects the distribution of heat and solutes (e.g., Moore and others, 1988). Elevated fluid pressures are inferred to be pervasive in the lower parts of most prisms. In the Barbados Ridge (Bangs and others, 1990, 1996; Bangs and Westbrook, 1991) and Cascadia Margin (Cochrane and others, 1994), seismic imaging suggests that large patches near the decollement are dilated by near-lithostatic pore-fluid pressure. Such pressures presumably facilitate the imbricate thrust faulting within the prisms that has been observed both through drilling and seismic imagery.

As on the Gulf Coast (Chapter 4.1.4), both sediment compaction and smectite dehydration likely contribute to the high fluid pressures. Sediment compaction is probably the dominant fluid source, as the extraordinarily high accretion and consolidation rates on the seaward side of the prism can create virtual fluid sources as large as 10^{-13}/second (Screaton and others, 1990), about 15 times larger than the maximum rate suggested for subsiding sedimentary basins (Table 4.1). The clay dehydration process, although probably less important from a tectonic standpoint, may be required to explain the chemistry of relatively low-chloride (20% less than seawater) fluids sampled near the decollement (e.g., Bekins and others, 1995).

In modeling fluid flow and associated transport in this dynamic environment, one must consider the movement of the solid matrix as well as the fluids (Screaton and others, 1990; Bekins and others, 1995). Convergence and deformation rates (perhaps 2 cm per year in the Barbados case) can actually exceed the fluid flow rates relative to the sediments that are calculated using Darcy's law (Chapter 1.1), invalidating some of the key assumptions invoked in deriving the groundwater flow equation (Chapter 1.5.3). For the fixed-in-space (*Eulerian*) model reference frame invoked in Chapter 1, one must consider the movement of sediments (carrying pore water) through the reference frame, as well as the fluid velocities relative to the sediments obtained from Darcy's law. The velocities that appear in the transport equations (Chapters 2.1 and 3.1) must incorporate both transport due to sediment motion v_r and the Darcian flow nv_w. For example, the quantity $v_r + v_w$ might be used in the advective solute transport term (Chapter 2.1.2), whereas the relative velocity v_w might be used to compute the mechanical dispersion tensor (Chapter 2.1.3) (Bekins and others, 1995). However, most researchers concerned with flow in such continuously deforming environments actually employ a *Lagrangian* reference frame in which equations are derived with respect to a control volume that moves through space (e.g., Person and others, 1996).

4.2 Regional-scale solute transport

This section deals with regional-scale ($\geq 10^4$ m) transfer of solutes via flowing groundwater. We describe various methods of groundwater age determination, hydrodynamic dispersion in regional-scale flow, and the general evolution of groundwater chemistry between regional recharge and discharge areas.

4.2.1 Groundwater age

Groundwater chemistry and the chemical composition of rocks are dynamically coupled by mineral–water equilibria and by the rate and mechanisms of groundwater transport. The transport of solutes in regional groundwater systems can operate on many different time scales. Travel times for solutes in shallow, local flow systems can be on the order of several years. Geologic processes often occur on regional scales where travel times can be thousands to even millions of years or longer. Flow rates or travel times for paleohydrologic systems associated with geologic processes of the distant past can usually only be roughly estimated using Darcy's law and estimates of past permeabilities and hydraulic gradients. Currently active flow systems, however, can serve as analogs to past systems, and groundwater ages can be measured directly by using certain radioactive isotopes. Initial concentrations of these isotopes are usually established at the land surface through interaction with the atmosphere. They then decay over different time scales at known rates as they are transported through the subsurface, and their concentration at any given point along a flow path then yields the time since they were isolated from the atmosphere. Measurement of such ages helps to reduce the level of uncertainty in estimates of flow rates and travel times.

Groundwater travel times of less than fifty years can be measured using certain anthropogenic sources as tracers (Plummer and others, 1993). The most widely used anthropogenic age tracers are based on the radioactive fallout from above-ground nuclear testing that occurred mainly in the late 1950s and early 1960s and on the introduction of chlorofluorocarbons into the atmosphere over the past fifty years. The atmospheric nuclear testing produced elevated levels of *tritium, carbon-14,* and *chlorine-36* that can now often be observed in the shallow subsurface. The depth of the radionuclide "peak" can be used to estimate an integrated recharge flux over the past few decades. Tritium (^{3}H), not really a solute but a minor component of the water itself, decays to helium-3, and measurement of both isotopes can yield a groundwater age. The recent appearance of anthropogenic tritium limits its use to waters less than about 40 years old; its short half-life of 12.5 years means that it will not be useful for waters more than about 60 years old. Thus, unless further above-ground nuclear blasts occur, tritium will not be a useful tracer after about the year 2030 A.D. Chlorofluorocarbons began to appear in the atmosphere in measurable amounts in about the mid 1940s, and thus they, too, can only be used to detect water that is less than about 50 years old. Direct measurement of groundwater travel times in regional systems requires the use of radionuclides with longer half-lives.

The most common age-determination technique used in regional groundwater systems is based on the carbon-14 isotope. With a half-life of about 5,700 years, ^{14}C is ideal for measuring travel times in the range of about 10^3 to 3×10^4 years. Carbon-14 is produced naturally in the upper atmosphere from atoms of carbon-12 by gamma-ray bombardment. A relatively constant gamma-ray influx produces a relatively constant concentration of ^{14}C in the atmosphere; past fluctuations of about 10% have been estimated from tree-ring data. The atmospheric ^{14}C is initially tied up in carbon dioxide but dissolves into rainwater as bicarbonate ions, which can then enter into groundwater systems. Isolation from the atmosphere precludes further production of ^{14}C, and decay is then the only mechanism for removing ^{14}C atoms. However, many other factors complicate the measurement of ^{14}C in groundwater and its use in determining groundwater age.

The most significant factor complicating carbon-14 analysis is the addition of "dead" carbon to the flow system. This often occurs due to dissolution of carbonate minerals that exist within the aquifer framework; the problem is thus most pronounced in carbonate rocks and may be negligible in sandstone aquifers. The carbonate minerals are usually of an age similar to that of the aquifer itself (perhaps millions of years) and thus contain virtually no ^{14}C atoms. Addition of such ^{14}C-free carbon to the groundwater dilutes the overall ^{14}C concentration and results in an apparent age that is too old. Several methods have been developed to attempt to correct for the presence of this dead carbon. Since much of the dead carbon is often acquired very near the recharge area, one approach is simply to estimate the typical value of ^{14}C after recharge occurs. Another popular method is to use the isotopic ratios of carbon-13 to carbon-12 to try to distinguish the different sources of carbon. Carbonate minerals in aquifers are usually enriched in ^{13}C relative to atmospheric carbon. The ^{13}C concentration along a flow path can usually be divided into an atmospheric fraction and a dissolved carbonate fraction based on ^{13}C values assumed or measured for those sources. Carbon-14 corrections can then be made based on the equivalent weighting between atmospheric and dead carbon. For more complex systems a full chemical-mass-balance approach can be used to try to account for all of the potential carbon sources (Plummer and others, 1994).

One classic study in which carbon-14 was used to estimate groundwater travel times and flow rates involved the Madison aquifer system of the north-central

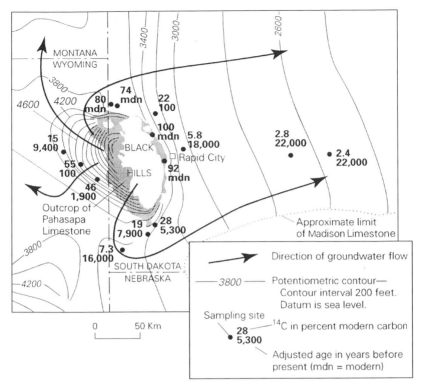

Figure 4.5 Potentiometric map with diagrammatic groundwater flowlines in the Pahasapa Limestone, showing down-gradient increase in carbon-14–based ages. Shaded area shows the outcrop of the Pahasapa Limestone. Modified from Back and others (1983).

United States (Back and others, 1983). Groundwater recharges around the Black Hills of South Dakota (Figure 4.5) and then moves through the Pahasapa Limestone (a local equivalent of the Madison) to depths of well over 1 km. Some water recharged on the western flank of the Black Hills flows westward and eventually to the Powder River basin, whereas water recharged on the northern, southern, and eastern flanks of the Black Hills eventually flows eastward. Carbon-14 ages vary from essentially modern near the recharge area to over 20,000 years at a distance of 200 km down the flow path. The ages calculated by Back and others (1983) agree well with hydraulic estimates of flow velocities of about 10 m/yr (Cooley and others, 1986). Konikow (1985) later extended these interpretations and estimated the regional-scale *effective porosity* of the carbonate aquifer to be between 3.5 and 7.5 percent.

Determining the carbon-14 ages of the Madison aquifer required adjustments that included accounting for chemical reactions along the flow path.

Dedolomitization, a primary reaction within the aquifer system, affects the ^{14}C concentrations (Back and others, 1983). Dedolomitization of dolomite aquifers occurs when small amounts of gypsum in the aquifer dissolve, thereby increasing the calcium concentration. Once the calcium to magnesium ratio exceeds a certain value, calcite becomes a more stable phase than dolomite. Dissolution of dolomite and precipitation of calcite then proceed congruently. The dissolution of the dolomite adds "dead" carbon to the groundwater, causing the uncorrected ^{14}C-based ages to be too old. Corrected ages in the Madison aquifer were several thousand years younger than the uncorrected ages. More extensive geochemical modeling of the entire aquifer system was done by Plummer and others (1990). They found dedolomitization to be important throughout the system and sulfate reduction, cation exchange, and halite dissolution to be locally important, especially in central Montana.

Groundwater travel times longer than 3 to 4×10^4 years require even longer-lived isotopes. Two such isotopes are chlorine-36 and iodine-129, which have only recently begun to be used for groundwater dating. Chlorine-36 has a half-life of about 3×10^5 years and a useful dating range of about 5×10^4 to 2×10^6 years. Iodine-129 has a half-life of about 1.7×10^7 years and a useful dating range of about 3×10^6 to 10^8 years. Chlorine-36 has been used to date groundwaters of the Great Artesian basin in Australia (Bentley and others, 1986) and the Milk River basin in Canada (Phillips and others, 1986). In the Great Artesian basin case, flow paths are as long as 1,200 km, flow rates are about 1 m/yr, and groundwater residence times are about 10^6 years. Iodine-129 dating has yet to be clearly demonstrated as a viable method but has potential to distinguish between deep meteoric and magmatic waters and between meteoric and connate waters of thick shale units where molecular diffusion is the dominant transport process. In utilizing either of these long-lived isotopic methods one must assume a constant source signal over millions of years – an assumption that may be valid but is very difficult to prove. Both isotopes can also be produced in situ by uranium, a potential conflicting source of the isotope for certain rock types.

4.2.2 Large-scale dispersion

The role of *hydrodynamic dispersion* in groundwater was discussed in Chapter 2.1.3. The conceptual model represented by the *advection–dispersion equation* for groundwater systems (e.g., Eq. 2.21) produces a smoothly varying concentration profile in space whose gradient is proportional to the dispersivity parameters. At the scale of heterogeneous regional-flow systems, no smoothly varying concentration gradient actually exists, and such a concentration profile

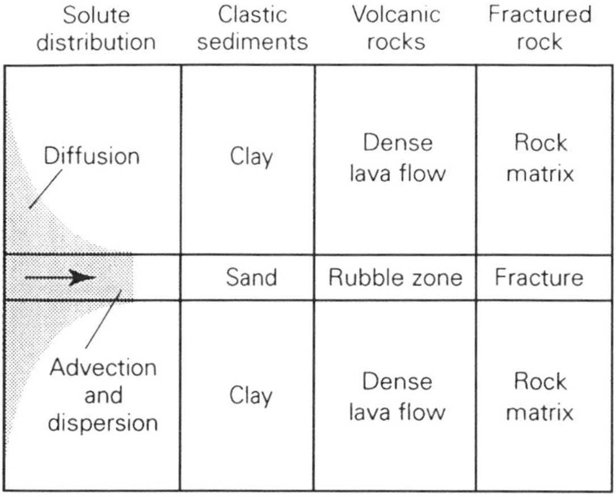

Solute distribution	Clastic sediments	Volcanic rocks	Fractured rock
Diffusion	Clay	Dense lava flow	Rock matrix
	Sand	Rubble zone	Fracture
Advection and dispersion	Clay	Dense lava flow	Rock matrix

Figure 4.6 Schematic diagram illustrating the role of diffusion in regional solute transport in heterogeneous porous media.

only represents the likelihood of finding a concentration of a particular value at a particular point in space. At the regional scale, solutes are transported predominantly by *advection* through narrow permeable zones, and *diffusion* often operates to transport the solute into adjacent, relatively stagnant low-permeability zones (Figure 4.6). If the solute input to the system has varied in time, *matrix diffusion* into these stagnant zones will slow the arrival of the solute at points downstream. If the solute input is constant, stagnant zones will eventually reach equilibrium solute concentrations and the arrival time of the solute will not be affected at later times. If the solute is undergoing radioactive decay, it is likely that a steady-state concentration gradient will develop into the stagnant zone, such that the solute will be continuously lost from the flow zone by a combination of diffusion into and decay within the stagnant zone. Such heterogeneities can exist in a variety of geologic materials (Figure 4.6).

The extent to which matrix diffusion will cause loss of a decaying solute can be calculated (Sanford, 1997). Assume that the advection–dispersion equation (Eq. 2.21) operates within the flow zone with decay but no retardation ($R_f = 1$). Then the steady state transport equation can be written for a flow zone of width a and porosity n_f to include loss by diffusion:

$$\nabla \cdot (\overline{D} \nabla C) - \nabla \cdot (vC) - \lambda C - \frac{2q_d}{n_f a} = 0. \tag{4.5}$$

The diffusive loss term $2q_d/n_f a$ can in turn be derived from *Fick's first law* (Eq. 2.1) as a function of the diffusion coefficient, the decay constant, and the width of the stagnant zone, b, such that

$$\nabla \cdot (\overline{D} \nabla C) - \nabla \cdot (vC) - \lambda C - \lambda' C = 0, \tag{4.6}$$

where λ' is a constant relating the loss due to diffusion and is defined as

$$\lambda' = \frac{\sqrt{\lambda n_s D_m}}{n_f a} \tanh \left(\frac{b}{2} \sqrt{\frac{\lambda n_s}{D_m}} \right), \tag{4.7}$$

where n_s is the porosity of the stagnant zone.

The loss due to diffusion to the stagnant zone, like decay, is a linear function of the concentration. The effect that this has on a carbon-14–based groundwater age, for example, can be expressed as

$$t_c = t_u \left(\frac{\lambda}{\lambda + \lambda'} \right), \tag{4.8}$$

where t_c is the age corrected for diffusion and t_u is the uncorrected age. For many hydrogeologic systems, loss of ^{14}C by diffusion can be not only significant but dominant relative to decay. In a study of groundwater flow in the Bangkok basin of Thailand, for example, it was estimated that half of all the ^{14}C was being lost by diffusion (Sanford and Buapeng, 1996).

4.2.3 Evolution of regional groundwater chemistry

The chemical composition of groundwater evolves during regional flow due to in situ reactions. This evolution can be generalized by considering the water types that are typically found in different zones of groundwater flow systems. Weathering, ion exchange, salt dissolution, and/or seawater intrusion tend to cause water to evolve from a dilute calcium bicarbonate type in recharge areas toward a more concentrated sodium chloride or calcium chloride type (Figure 4.7).

In recharge areas the weathering of carbonate and silicate minerals provides the dominant control on concentrations of the major ions in solution. Calcite and dolomite dissolution add calcium, magnesium, and bicarbonate to solution, whereas feldspar dissolution typically adds calcium, sodium, and potassium. Bicarbonate enters solution through the input of atmospheric carbon dioxide to the soil zone, which can be greatly enhanced by the respiration of bacteria acting upon decaying organic matter. The dilute carbonic acid that results reacts to dissolve many minerals.

As flow through the system progresses, ion exchange between groundwater and clay minerals tends to play a more significant role in controlling the overall solute chemistry. An equilibrium distribution between the different ions is

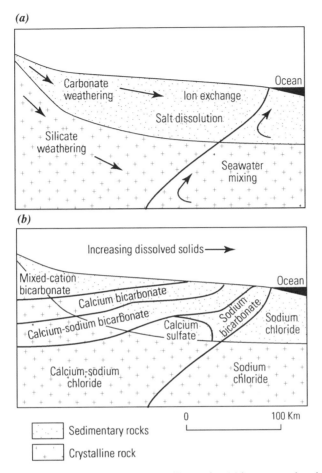

Figure 4.7 Hypothetical regional flow system illustrating (a) important chemical interactions and (b) different water types associated with the long-term evolution of groundwater chemistry. Modified from Back and others (1993).

maintained as a function of the clay mineralogy and ionic strength of the solution. (Reactions of this type were described quantitatively in Chapter 2.5.2.) The exchange sites of clays that were deposited in a marine environment are originally dominated by sodium, due to the high sodium content of seawater. Subsequent gradual flushing by meteoric water under equilibrium conditions usually results in the groundwater becoming enriched in sodium and depleted in calcium, because meteoric water tends to have a higher calcium to sodium ratio than seawater. Because the exchange sites may greatly outnumber the dissolved ions in the groundwater, one flushing of the pore spaces may cause only a slight shift in the sorbed concentrations. Clays must often be flushed by hundreds of pore volumes of meteoric water before their exchange sites reach a final equilibrium with respect to the meteoric water.

The existence of soluble-salt deposits within a regional aquifer system can have a profound effect on groundwater chemistry. Gypsum or anhydrite beds are major sources of calcium and sulfate. Groundwater usually reaches equilibrium with respect to calcite and/or dolomite relatively quickly, and subsequent contact with gypsum will increase the dissolved calcium concentrations even further. Such an increase will cause a simultaneous oversaturation and precipitation of calcite, with the result that the bicarbonate levels will drop to relatively low values. If dolomite is present, this can also cause the replacement of dolomite by calcite, a process known as *dedolomitization*. (See Chapter 4.2.1 for discussion of dedolomitization of the Madison Limestone.) Dissolution of bedded halite can lead to large increases in sodium and chloride. A brine in equilibrium with respect to halite will have a dissolved solids concentration of about 300 g/liter (about ten times that of seawater). It may often be difficult to distinguish the origin of a sodium chloride brine; the salt source may be dissolved halite, evaporated seawater, or even magmatic fluids. The ratio of chloride to bromide (two conservative halides) can be useful in distinguishing between the first two possibilities – most bromide will remain in the brine during evaporation and halite precipitation. (The evolution of highly saline brines and the associated precipitation of evaporite minerals is discussed in detail in Chapter 9.)

The mixing of meteoric groundwater with seawater can also be significant to the chemical evolution. Due to the greater density of seawater with respect to fresh groundwater, seawater will exist as a coastal wedge beneath a lighter freshwater lens. In a static system the depth to the transition zone should be about 40 times greater than the height of the freshwater water table above sea level (Du Commun, 1828). In reality, dynamic movement of the water, changes in sea level, and heterogeneous permeabilities can cause large deviations from the theoretical static position. Mixing occurs directly in the transition zone between the freshwater and seawater owing to mechanical forcing by tidal and other sea level changes, or mixing may occur more slowly over time as a result of diffusion of solutes into or out of stagnant layers after a sea level change has caused a lateral shift in the transition zone. In either case, such mixing can theoretically lead to unique geochemical conditions of under- or over-saturation (Runnels, 1969) and subsequent mineral transformations such as calcite dissolution (Back and others, 1979), dolomitization (Hanshaw and others, 1971), manganese deposition (Force and others, 1986), or silica diagenesis (Knauth, 1979).

Another process that may affect regional groundwater chemistry is *chemical osmosis* (see also Chapter 1.1.2). Shales may act as *semipermeable membranes* that transmit water more readily than solutes. This could lead to anomalous osmotic pressures and water being driven by solute concentration gradients in order to establish equilibrium pressures and concentrations. Clays are known to behave as semipermeable membranes in the laboratory (Olsen, 1969), and in theory one can use parameters from laboratory measurements to predict field-scale occurrences (Greenberg and others, 1973). It has been postulated that deep shale layers should act as significant barriers to salt movement, but paucity and uncertainty of data have prevented this from being shown, except in isolated cases. In fact, in the case of the Gulf Coast basin it has been shown that membrane diffusion is probably insignificant (Hanor, 1994a, 1994b). The Pierre Shale of the North American midcontinent has exhibited in situ osmotic properties in ambient tests using shallow wells (C. E. Neuzil, U.S. Geological Survey, written communication, 1995), but the Kiamichi Shale of western Texas exhibits no such properties under similar tests (W. E. Sanford, unpublished data). The former has been more deeply buried and compacted (30% porosity) than the latter (40% porosity), suggesting that the degree of compaction may be a critical factor controlling the osmotic efficiency of shales.

Another process that can affect regional groundwater chemistry is the direct input of solutes from hydrothermal sources. For example, magmatic HF was released at Hicks dome at the southern end of the Illinois basin during Permian time (Plumlee and others, 1995), resulting in the Illinois–Kentucky fluorspar district. The Upper Mississippi Valley district at the north end of the Illinois basin is also dated as Permian and may be related to contemporaneous northward migration of brine; the thermal history of the basin is consistent with such a flow pattern (Rowan and Goldhaber, 1996). Magmatic fluorine released at the southern end of the basin may have traveled northward by way of the Mt. Simon and St. Peter Sandstones. Geochemical data from well cores and cuttings in the basin support this hypothesis (Goldhaber and others, 1994), as bulk fluorine concentrations in composited rock samples decrease from south to north across the basin and generally increase with vertical proximity to the St. Peter Sandstone.

4.3 Regional-scale heat transfer

This section deals with regional-scale transfer of heat via flowing groundwater. We describe the thermal regime of sedimentary basins, regional-scale groundwater flow and heat transfer in volcanic terrain, and the role of hydrogeology in the "stress–heat flow paradox" of the San Andreas fault.

4.3.1 The conductive regime in sedimentary basins

Vertical temperature gradients in sedimentary basins are commonly larger than the global average value of about 25°C/km. By inspection of *Fourier's law* (e.g., Eq. 3.10), we can see that the relatively low thermal conductivity of many sediments and consolidated sedimentary rocks implies higher than average temperature gradients in regions of normal heat flow. Aquifers in sedimentary basins such as the Paris basin, France, and the Pannonian basin, Hungary, can be exploited for low-enthalpy geothermal resources because they are insulated by thick blankets of low-conductivity sediments.

If rates of groundwater flow and uplift or subsidence are sufficiently slow, heat transfer in basins is conduction-dominated and can be completely characterized by forms of Fourier's law such as Eqs. (3.10) and (7.1). Under such conditions, the major source of uncertainty in conductive heat-flow determinations is the appropriate value of *medium thermal conductivity*, K_m. In practice, the in situ value of K_m is usually estimated by measuring the thermal conductivity of rock fragments (cuttings) from a drill hole, estimating the in situ porosity, and then using a geometric-mean model such as Eq. (3.11) to approximate the thermal conductivity of the undisturbed rock–water medium. In plutonic environments (for example) the porosity correction is often negligibly small, but in sedimentary environments, larger and variable porosities commonly create both a systematic depth dependence and significant uncertainty in K_m. For example, in the Great Artesian basin of Australia the mean medium thermal conductivity increases from about 2.0 W/(m-K) at 1 km depth to about 2.5 W/(m-K) at 4 km depth, owing to loss of porosity. At any given depth in the basin the thermal conductivity varies over a range of about 0.4 W/(m-K), owing to variations in porosity and lithology (Toupin and others, 1997). A further source of uncertainty in many instances is the large thermal anisotropy of shale, the most common sedimentary rock. The thermal anisotropy of shale has been attributed to the preferred (bedding-parallel) orientation of sheet silicates, which conduct heat quite well along the sheet structures but have much lower thermal conductivities orthogonal to the sheets (Diment and Pratt, 1988). This anisotropy can be directly measured in oriented drill cores but is lost in the randomly oriented cuttings. Most values reported in the literature rely on measurements made on cuttings, and Blackwell and Steele (1989) suggested that these values may commonly be in error by as much as 100% due to oversampling of the bedding-parallel orientation. However, some of the anisotropy measured in shale cores may be due to unloading and changes in fluid saturation prior to measurement. Sample-volume changes due to unloading might be partly accommodated by bedding-parallel partings, which would then be filled by less conductive fluids (air or water). It is worth noting that a core from the flat-lying Pierre Shale of the central United States that was processed with special attention to maintaining the in situ moisture content showed relatively minor anisotropy (horizontal/vertical $\sim$1.16) (Sass and Galanis, 1983).

Other factors besides fluid movement that can affect basin heat flow include rapid uplift or subsidence. Episodes of uplift/erosion tend to increase near-surface temperatures and heat flow, whereas subsidence/sedimentation has the opposite effect. The effects of uplift or subsidence at a particular rate are crudely similar to the effects of vertical groundwater flow (Chapter 3.4.1) at the same rate, because the volumetric heat capacities of water and rock differ only by about 50% ($c_w \rho_w \sim 4.2$ kJ/(kg-K) $\times$ 1,000 kg/m^3 = 4,200 kJ/(m^3-K) versus $c_r \rho_r \sim 1.0$ kJ/(kg-K) $\times$ 2,600 kg/m^3 = 2,600 kJ/(m^3-K)). We can assume that the land surface is a constant-temperature boundary and use a solution for one-dimensional groundwater flow (Figure 3.4a) to infer that uplift or subsidence at rates on the order of 1 millimeter per year can detectably perturb the thermal field. Such rates are not uncommon in tectonically active areas; in fact, the entire Brahmaputra basin of southern Asia appears to be experiencing denudation at a rate of nearly 1 mm/yr (Summerfield and Hulton, 1994). On a more local scale, the vicinity of Cajon Pass, central California, appears to have experienced uplift and erosion at about 1 mm/yr over the past 1 Ma ($\sim$1 km total), and the thermal effects of this uplift cause heat flow measured in a deep drill hole there to vary from about 70 mW/m^2 at 3.5-km depth to nearly 100 mW/m^2 in the near surface (Lachenbruch and Sass, 1992; Sass and others, 1992; see also Chapter 4.3.5). Carslaw and Jaeger (1959, p. 388, Eq. 7) pose an analytical solution that is applicable to steady erosion or sedimentation in terms of heat conduction in an infinite half-space moving through a constant-temperature boundary.

4.3.2 Thermal effects of groundwater flow in sedimentary basins

Intermediate- to regional-scale groundwater flow, if sufficiently rapid, influences the thermal structure of basins by depressing heat flow in groundwater recharge areas and increasing heat flow in discharge areas (see Chapters 3.4.1 and 6.1.2). Vigorous regional-scale flow has been invoked to explain the anomalously high emplacement temperatures associated with some ore deposits; the Mississippi Valley–type lead–zinc deposits of the central United States are probably the best-known example (see Chapter 5.1).

Whether or not the thermal structure of a basin is affected by groundwater flow depends upon the permeability of the basin sediments and the geometry of the flow system, including the regional slope (water table configuration) and vertical/horizontal aspect ratio. Large permeabilities, slopes, and aspect ratios act to enhance advective perturbations. Domenico and Palciauskas (1973) suggested a modified *Peclet number* for regional groundwater flow that is applicable to quasi-rectangular geometries and monotonic water table slopes. Taking their modified Peclet number and substituting ($k\rho_w g / \mu_w$) for hydraulic conductivity,

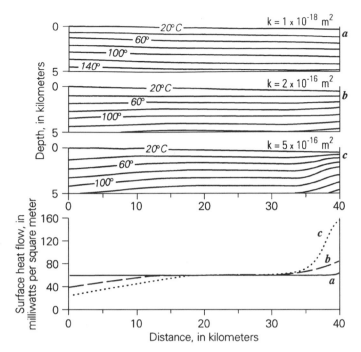

Figure 4.8 Results from numerical simulation of a 40-km-long by 5-km-deep sedimentary basin for three different values of homogeneous, isotropic permeability. A total of 500 m of water table relief (a linear slope of 1:80) drives groundwater flow from left to right. Heat flow at the base of the system is 60 mW/m², and the medium thermal conductivity is approximately 2.5 W/(m-K). From Smith and Chapman (1983).

as per Eq. (3.7), we obtain the dimensionless ratio

$$\frac{\left(\rho_w^2 c_w k g B \Delta z\right)}{2(\mu_w K_m L)},$$

(4.9)

where B is basin thickness, Δz is the water table relief, and L is basin length. (See Chapter 3.5.2 for a discussion of the thermal Peclet number, Chapter 3.4.1 for a Peclet number–based approach to determining rates of vertical groundwater flow, and Chapter 6.1.2 for application of the Peclet number to the problem of hydrocarbon maturation.)

We can use the Peclet number approach (Eq. 4.9) to analyze some numerical-simulation results that also serve to illustrate some basic aspects of basin-scale heat transfer. The particular numerical results shown in Figure 4.8 are from Smith and Chapman (1983). They apply to a 5-km-deep by 40-km-long basin and a linearly sloping water table with 500 m of total relief. For a homogeneous, isotropic permeability of 1×10^{-18} m², advective effects are negligible;

the surface heat flow is uniform and equal to the heat flow specified at the bottom of the model grid (60 mW/m^2). With a permeability of 2×10^{-16} m^2, advective effects reduce surface heat flow to about 40 mW/m^2 at the high-elevation (recharge) margin of the basin and increase heat flow to about 80 mW/m^2 at the low-elevation (discharge) margin. With a permeability of 5×10^{-16} m^2, advective effects dominate the thermal structure: Heat flow varies from 20 mW/m^2 at the recharge margin to 160 mW/m^2 at the discharge margin, and temperatures at similar depths in the recharge and discharge areas differ by as much as 50°C. By evaluating fluid properties (ρ_w, c_w, μ_w) at the average basin temperature of about 90°C, we can use Eq. (4.9) to translate the permeabilities of Figure 4.8 to modified Peclet numbers of 0.002, 0.3, and 0.8. The threshold Peclet number for significant advective perturbation appears to be on the order of 1, as originally suggested by Domenico and Palciauskas (1973).

The relatively large vertical/horizontal aspect ratio and water table slope of Figure 4.8 might apply, for example, to some of the intermontane basins of the western United States; large, midcontinental sedimentary basins would typically have smaller aspect ratios and slopes. The permeabilities of Figure 4.8 are in the lower range of the values expected in sedimentary strata (e.g., Freeze and Cherry, 1979), and in many large midcontinental basins the presence of higher-permeability strata ($k \geq 10^{-14}$ m^2) might compensate for geometries that are less favorable for advective heat transport.

Other numerical results of Smith and Chapman (1983) demonstrate that the thermal effects of basin-scale groundwater flow will often be subtle and difficult to detect. Even in basins where regional-scale heat transfer is significant, temperature profiles can be quasi-linear to substantial depth, and there may be large areas with little lateral variation in heat flow (see especially their Figures 5 and 7 and the associated discussion).

Although modeling exercises using reasonable parameter values demonstrate the feasibility of basin-scale heat transfer by groundwater, applications to real-world basins are often ambiguous, because direct measurement of groundwater velocities (or permeabilities) on the appropriate scale is impossible, and any spatial variations in heat flow are often subject to alternative explanations. We will now proceed to consider some particular examples. In Chapter 7 we consider other geothermal problems, such as groundwater, flow near cooling igneous plutons and the occurrence of hot springs and geysers.

4.3.3 Some case studies of sedimentary basins

A vigorous debate about heat flow in the Western Canada (Alberta) sedimentary basin illustrates the difficulty of proving or disproving groundwater effects.

Majorowicz and others (1984) observed that average heat flow values increase from about 40 mW/m^2 in southern and southwestern Alberta near the Rocky Mountains to 70 mW/m^2 in central Alberta, and they argued for a close relationship between temperature gradients, topography, and (by inference) groundwater flow. A series of related studies generally agreed that heat flow in much of western Canada could be explained in terms of regional-scale patterns of upflow and downflow of groundwater (e.g., Majorowicz and Jessop, 1981; Hitchon, 1984; Jones and others, 1985; Majorowicz and others, 1989). In hydrogeologic terms, all of these studies were qualitative in nature, and Bachu (1988) criticized them for ignoring relevant permeability and hydraulic head data. He used permeabilities obtained from drillstem tests and hydraulic gradient data from wellfields to infer that, in general, rates of lateral groundwater flow are too small to affect the thermal field. As an alternative, he suggested that variations in heat flow within the basin might reflect conditions in the underlying crystalline basement. Garven (1989) contributed to the debate about geothermal conditions in the Western Canada basin by developing a quantitative, regional-scale model of groundwater flow and associated heat transport. His results suggested that, during the late Tertiary, regional-scale groundwater flow was rapid enough to perturb the thermal field and account for petroleum accumulations near the eastern margins of the basin (see Chapter 6.2.3) but that subsequent erosion had disrupted regional-scale groundwater flow (Figure 4.9). A modeling study by Corbet and Bethke (1992) agreed that current rates of groundwater flow are too slow to transport significant heat.

In the Western Canada basin, the general correspondence between areas of low temperature gradients and areas of probable downflow of groundwater and between high gradients and upflow of groundwater comprise strong circumstantial evidence for the influence of groundwater flow, although other explanations for the heat flow pattern cannot be ruled out. The ultimate area of disagreement between those who favor the groundwater explanation and those who would discard it is the appropriate value(s) of large-scale permeability; other relevant parameters are much more amenable to direct measurement. In this regard it is worth noting that Garven (1989) assigned permeabilities of 6×10^{-13} m^2 to the important Paleozoic carbonate aquifers, whereas Bachu (1988) cited drillstem-test values that are generally 10 to 100 times lower. Because of the complex scale-dependence of permeability (Chapter 1.2.3), it is impossible to say which value is more appropriate.

A few site-specific studies of basin-scale advective heat transport have integrated heat flow data and hydraulic measurements in a useful manner; studies of the Uinta basin, Utah, and the North Slope basin, Alaska, are particularly good examples. The Uinta basin group used numerous measurements of bottom-hole

Late Paleozoic

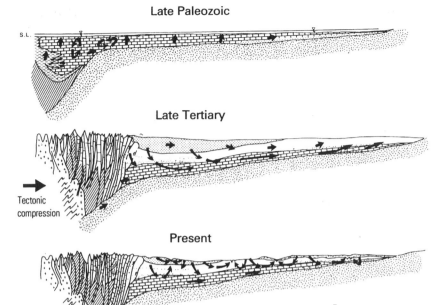

Late Tertiary

Present

Figure 4.9 Conceptual models of fluid flow in the evolving Western Canada sedimentary basin. In the Late Paleozoic, topographic relief was absent, and groundwater flow was driven only by compaction and convection. In the Late Tertiary, after the Laramide orogeny, a throughgoing, topographically driven regional groundwater flow system transported significant amounts of heat and hydrocarbons toward the discharge margins of the basin. Subsequent erosion disrupted the regional-scale flow and resulted in the development of more active local flow systems. Such systems might still transport significant amounts of heat, but not on the spatial scale invoked by Majorowicz and others (1984). From Garven (1989), reprinted by permission of American Journal of Science.

temperature and thermal conductivity to characterize the thermal structure of the basin. They found that temperature gradients are depressed near the high-elevation margin of the basin and enhanced in the central part of the basin (Chapman and others, 1984). By simulating groundwater flow and heat transport, they determined that the thermal observations can be reproduced if the two upper sedimentary units (a total thickness of 1–3 km) are assigned a permeability of about 5×10^{-15} m^2 (Willet and Chapman, 1987). This value is, rather surprisingly, less than the mean values of permeability determined by core- and well-test data from the same rocks (about 10^{-14} m^2 and 10^{-12} m^2, respectively). (See Chapter 6.5.1 for further discussion of groundwater movement, heat flow, and hydrocarbon occurrences in the Uinta basin.) In the case of the North Slope basin, heat flow is depressed in the foothills of the Brooks Range and enhanced near the Arctic Ocean in the vicinity of the Prudhoe Bay oil fields (Deming and others, 1992). Deming (1993) found that the thermal observations were well matched by assigning horizontal permeabilities (k_h) equivalent to the arithmetic mean of drillstem-test values for a particular unit and vertical permeabilities (k_v) equivalent to the harmonic mean of the drillstem-test values. However, the thermal observations were matched equally well by assigning single values of k_h (in the range of 2.5×10^{-14} to 2.5×10^{-13} m^2) and k_v (1.0 to 5.0×10^{-16} m^2) to the entire basin.

Discrepancies between permeabilities determined from hydraulic data and those required to match thermal data might be caused by sample bias, as suggested by Willett and Chapman (1987); hydraulic data are perhaps more likely to be obtained from high-permeability horizons, whereas the lower-permeability horizons would tend to control regional-scale flow. An alternative explanation for some cases in which well-test permeabilities greatly exceed the "thermally reasonable" values is that the well tests sample permeable fracture networks that are discontinuous on a regional scale. (See also Chapter 1.2.3.)

A combination of thermal data and transport modeling can be used to test the "reasonableness" of a permeability structure estimated from independent hydraulic data, and thermal data can also be inverted in order to estimate permeabilities where hydraulic data are unavailable (e.g., Woodbury and Smith, 1988). However, permeability structures obtained by inversion of thermal data should not be too complex: As Deming (1993) demonstrated for the North Slope, one or two values of "basin-scale" permeability are as likely to match the thermal data as a more complex permeability structure.

4.3.4 An example from volcanic terrane

As we have discussed in the context of sedimentary basins, groundwater flow has the general effect of decreasing heat flow in recharge areas and increasing heat flow in discharge areas (e.g., Smith and Chapman, 1983). This effect tends to be enhanced in volcanic highlands due to large topographic gradients and the relatively high permeability of most unaltered volcanic rocks.

A map of near-surface conductive heat flow in the north-central Oregon Cascades (Figure 4.10) shows zones of elevated heat flow in older volcanic rocks flanking a zone of depressed heat flow in the younger volcanic rocks. The younger rocks create a broad ridge that receives heavy snowfall and is a regional groundwater recharge area. Topographically driven groundwater flow from this area feeds thermal springs that discharge to the east and west, in deeply eroded

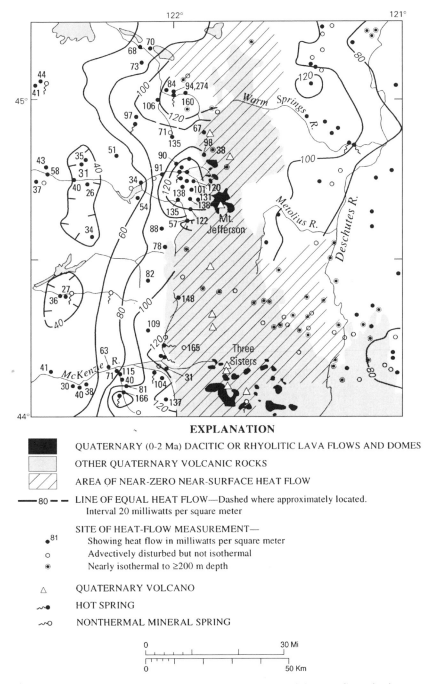

EXPLANATION

QUATERNARY (0-2 Ma) DACITIC OR RHYOLITIC LAVA FLOWS AND DOMES

OTHER QUATERNARY VOLCANIC ROCKS

AREA OF NEAR-ZERO NEAR-SURFACE HEAT FLOW

— 80 — — LINE OF EQUAL HEAT FLOW—Dashed where approximately located. Interval 20 milliwatts per square meter

SITE OF HEAT-FLOW MEASUREMENT—

• 81   Showing heat flow in milliwatts per square meter

○   Advectively disturbed but not isothermal

◉   Nearly isothermal to ≥200 m depth

△   QUATERNARY VOLCANO

～●   HOT SPRING

～○   NONTHERMAL MINERAL SPRING

Figure 4.10 Conductive heat flow map of the north-central Oregon Cascades between 44° and 45° 15′ N latitude. From Ingebritsen and others (1993).

terrains underlain by older, less permeable volcanic and volcaniclastic rocks. Heat flow observations in the northern California Cascades (Mase and others, 1982) define a similar pattern that results from the same process.

In general, the depth to which groundwater recharge affects conductive heat flow measurements in a particular area is poorly known and presumably highly variable. In north-central Oregon, the large area of near-zero conductive heat flow occurs in young volcanic rocks of the High Cascades (Figure 4.10), an area where groundwater recharge is approximately 1 m/yr (Ingebritsen and others, 1994). There, the zone of low-to-zero near-surface conductive heat flow may only be a few hundred meters thick beneath topographic lows. It presumably thickens beneath topographic highs; Swanberg and others (1988) described two drill holes on the flanks of Newberry volcano, Oregon, that are isothermal at mean annual air temperature to depths of 900–1,000 m. In the summit and east-rift zones of Kilauea volcano, Hawaii, groundwater recharge depresses temperatures to depths of 500–1,000 m (Kauahikaua, 1993), effectively masking the existence of high-temperature hydrothermal systems at greater depths.

Assuming steady-state conditions, conservation of energy dictates that depressed heat flow in groundwater recharge areas must be balanced by enhanced heat flow in discharge areas. In north-central Oregon, the anomalously low near-surface heat flow in areas where volcanic rocks younger than about 7 Ma are exposed (e.g., Figure 4.10) is balanced by anomalously high advective and conductive heat discharge in older rock at lower elevations. For a 135-km-long segment of the Cascade Range volcanic arc between 44° and 45° 15' N latitude, the heat deficit represented by near-zero conductive heat discharge in <7 Ma rocks (470 MW) is more than sufficient to balance the anomalous advective and conductive heat flow observed in >7 Ma rocks (315 MW) (Ingebritsen and others, 1989, 1994). Temperature–depth data from the deepest drill hole in the >7 Ma rocks (Figure 4.11) indicate that the high conductive heat flow measured there may in fact be controlled by groundwater flow. Seventeen shallower (<500 m) drill holes in the immediate vicinity had high gradients that corresponded to conductive heat flows >110 mW/m^2. However, a similar gradient in the upper part of the deepest hole changed abruptly below a zone of thermal-fluid circulation at 800-m depth. That such a change was observed in the deepest drill hole suggests that the high gradients in the shallower holes are also controlled by groundwater flow.

4.3.5 The stress–heat flow paradox of the San Andreas fault

The issue of heat transfer via topographically (gravity) driven regional groundwater flow is also pertinent to the so-called *stress–heat flow paradox of the San*

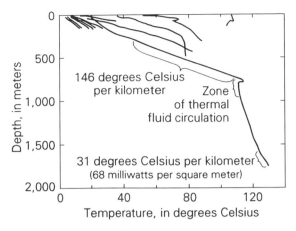

Figure 4.11 Temperature–depth profiles from drill holes collared in rocks older than 7 Ma in the Breitenbush Hot Springs area, north-central Oregon (near long. 121°58′32″ W, lat. 44°46′52″ N). From Ingebritsen and others (1994).

Andreas fault. The San Andreas is a major right-lateral strike-slip fault that defines the boundary between the Pacific and North American plates in California and northwestern Mexico. As first pointed out by Brune and others (1969), displacement rates along the San Andreas are large enough that frictional heat generation should cause a pronounced, local heat flow anomaly unless the fault is "weak," that is, unless it has low frictional resistance. Because no such heat flow anomaly had yet been observed, Brune and others (1969) concluded that the fault is indeed weak. A subsequent compilation of about 100 high-quality heat flow data by Lachenbruch and Sass (1980) confirmed that there is no local conductive heat flow anomaly associated with the fault (Figure 4.12). They agreed with Brune and others (1969) that this implies a frictional resistance of less than 10 MPa. If this rather low value is correct, then the stress drops caused by earthquakes, typically in the range of 1–10 MPa (Kanamori and Anderson, 1975), may represent nearly complete relief of shear stress along the fault.

The inference of low frictional resistance along the San Andreas system is inconsistent with most in situ stress measurements in the upper few kilometers of the crust and with the results of laboratory rock strength studies. The in situ measurements suggest that in intraplate regions the maximum horizontal shear stress generally increases with depth at a rate of 7–8 MPa/km, so that the shear stress at seismogenic depths would be projected as 50–100 MPa. Attempts to extend laboratory studies (e.g., Byerlee, 1968) to midcrustal conditions predict even somewhat larger shear-stress values of 100–200 MPa at 10–15-km depth (Brace and Kohlstedt, 1980). This apparent conflict between the stress and heat

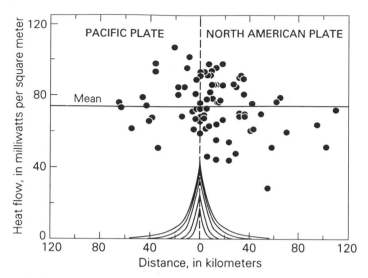

Figure 4.12 Heat flow as a function of distance from the main trace of the San Andreas fault. Dots are measured values; curves represent the conductive heat flow anomaly expected for an average fault friction of 50 MPa and a slip velocity of 3 cm per year at various times. The uppermost curve is the infinite-time (steady-state) solution. From Lachenbruch and Sass (1980).

flow observations became known as the stress–heat flow paradox of the San Andreas fault (e.g., Zoback and Lachenbruch, 1992).

Lachenbruch and Sass (1980) considered and rejected the possibility that groundwater flow is masking frictional heating effects by removing significant quantities of heat from the fault zone. Thermal springs are scarce along the length of the San Andreas, and the total hydrothermal heat discharge is negligible relative to the heat represented by the "missing" conductive heat flow anomaly. Although they could not categorically rule out substantial undetected discharge at temperatures slightly above ambient, they saw no obvious driving mechanism for such flow.

Williams and Narasimhan (1989) later recognized the possibility of topographically driven groundwater flow away from the San Andreas fault. They demonstrated that cross-fault topographic profiles throughout central and southern California show significant relief associated with the fault zone (Figure 4.13). The fault zone is generally elevated and might represent a regional recharge area in which heat flow is depressed. Williams and Narasimhan (1989) found that topographically driven flow could completely suppress a fault-centered conductive heat flow anomaly if the average upper crustal permeability is at least 10^{-16} m^2 (Figure 4.13).

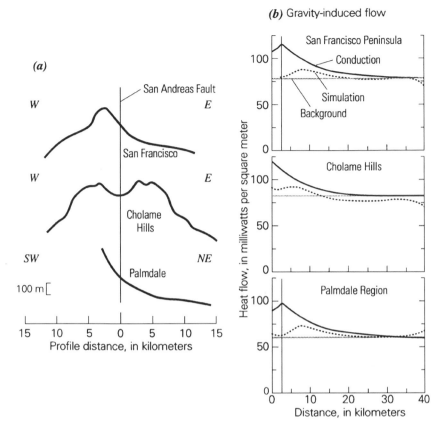

Figure 4.13 (a) Three averaged topographic profiles across the San Andreas fault (0 km on the horizontal axis). Averaging was done for 45-km (San Francisco), 70-km (Cholame Hills), and 60-km (Palmdale) segments of the fault. (b) Effect of topographically driven groundwater flow on near-surface heat flow. For a permeability–depth relation, $k = 10^{-(0.2z+15)}$ m^2, where z is depth in kilometers, the conductive heat flow anomaly caused by frictional heating is totally dissipated. From Williams and Narasimhan (1989), with kind permission of Elsevier Science – NL, Amsterdam.

In the late 1980s a 3.5-km-deep research borehole was drilled in Cajon Pass, southern California, in order to better determine the thermal, mechanical, and hydrologic conditions along the San Andreas fault. A relatively high conductive heat flow value of about 90 mW/m^2 had been measured in a nearby 1.7-km-deep oil well, tending to support the idea of a "strong" fault. However, heat flow in the Cajon Pass borehole itself showed a systematic decrease from about 100 mW/m^2 at less than 400-m depth to 75 mW/m^2 in the lowermost 300 m (Sass and others, 1992). The shape of the Cajon Pass heat flow profile is consistent with uplift and erosion at an estimated rate of about 1 mm/yr over the past

1 Ma (Chapter 4.3.1). The elevated heat flow at shallow depths is likely a transient feature related to this rapid uplift. The deeper heat-flow value more likely represents the thermal steady state and is consistent with the average value obtained from relatively shallow (<300 m) drillholes along the length of the San Andreas (Figure 4.12).

In situ stress measurements in the Cajon Pass borehole revealed that the frictional strength of the crust near the fault is actually quite high, consistent with laboratory values. However, surprisingly, there was no evidence of right-lateral shear on planes parallel to the San Andreas fault. Instead, the stresses measured in the borehole are consistent with the active normal faults in the vicinity and with *left*-lateral displacement on a fault that locally runs parallel to the San Andreas. This suggests that the San Andreas fault itself moves at low shear stresses and in fact is extremely weak relative to the surrounding crust (Zoback and Healy, 1992). The fault seems to operate in the context of roughly fault-normal compression (Zoback and others, 1987), with the surrounding crust failing under more typical "high-stress" frictional faulting conditions.

The permeability of core samples obtained from 2,100- to 3,500-m depth in the Cajon Pass borehole ranged from 10^{-22} to 10^{-19} m^2 (Morrow and Byerlee, 1992), much lower than the threshold value for heat dissipation via topographically driven flow determined by Williams and Narasimhan (1989). However, the cores are from crystalline rocks in which fluid movement occurs dominantly in major fracture zones. It is well known that core measurements on such rocks tend to underestimate the larger-scale in situ permeability by a factor of about 10^3 (e.g., Chapter 1.2.3; Brace, 1980), mainly because the cores fail to sample the important fractures. Thus the permeability measurements do not exclude the possibility of effective advection of heat away from the fault.

At this point, arguments about the strength or weakness of the San Andreas ultimately depend upon the occurrence and movement of pore fluids. The current consensus is that the fault is weak; if so, the most likely explanation is that pore fluids within the fault zone are at near-lithostatic pressures, so that the effective stress is very low. One possible explanation for elevated pore-fluid pressures is some sort of fluid source at depth (Chapter 4.1). If the fault is strong (which now seems less likely), the associated frictional heating must be effectively obscured by groundwater movement. Various models for the fluid-pressure regime of the San Andreas fault are discussed in Chapter 8.4.

Problems

4.1 Use Eq. (4.4) to estimate the value of hydraulic conductivity required to allow **(a)** thermal pressurization or **(b)** compaction to induce anomalous fluid pressures at depth in a subsiding basin. Assume a 5-km-thick basin which receives sediment and subsides at a steady rate of 5 mm/year and has a 25°C/km thermal gradient. Use a value of $10^{-3}/°C$ for the thermal expansion of the pore fluid and assume that porosity varies from 0.40 at the land surface to 0.05 at 5-km depth. **(c)** Are the results for (a) and (b) in accord with Bethke's (1986) assessment of the relative importance of thermal pressurization versus compaction (Chapter 4.1.4)? Why or why not? **(d)** Is the result for (a) in reasonable agreement with the limiting value suggested by Bredehoeft and Hanshaw (1968)?

4.2 Assume that groundwater is flowing through a heterogeneous aquifer system composed of alternating layers of sands and clays. Given that the mean thicknesses of both the sand and clay lenses are about 10 meters, how will diffusion of carbon-14 from the sand layers into the stagnant clay layers affect an apparent ^{14}C- based age in the sand sections of the aquifer system? Assume typical values for any transport parameters.

4.3 Consider a 500-km-long by 4-km-deep sedimentary basin with a total of 300 m of water table relief and an average vertical temperature gradient of 25°C/km. Further assume that the bulk permeability of the basin sediments is about 1×10^{-14} m^2 and that the medium thermal conductivity is about 2.0 W/(m-K), and use Eq. (4.9) to assess whether the thermal structure of the basin is likely to be affected by groundwater flow.

4.4 Using the map of the central Oregon Cascades given as Figure 4.14, draw a rough topographic section from Sweet Home to Black Butte. Project the drillhole locations to section, and indicate their positions and depths, using depth information from Figure 4.14b. On the basis of topography (Figure 4.14a) and the temperature profiles (Figure 4.14b), draw arrows showing the approximate magnitudes and directions of groundwater flow. How does permeability vary along section and with depth?

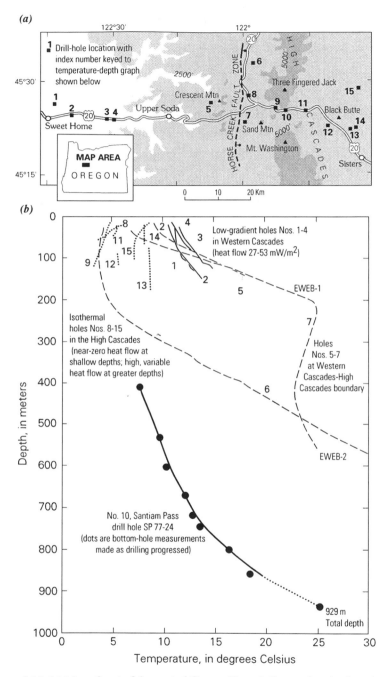

Figure 4.14 (a) Map of part of the central Oregon Cascade Range showing locations of drill holes where temperature–depth profiles of (b) were obtained. In the High Cascades, Pliocene and Quaternary volcanic rocks are exposed, and the topography is constructional and relatively smooth. This section of the High Cascades is bounded to the west by normal faults of the Horse Creek fault zone, which exhibit about 1 km of vertical displacement. Older, deeply incised volcanic rocks are exposed west of the Horse Creek fault zone.

5

Ore deposits

Most economically significant ore deposits exist because of the advective transport of solutes and heat by flowing groundwater. Mobilization, transport, and deposition of chemical species are all linked to fluid flow. The thermal effects of flowing groundwater can explain the emplacement of some ore deposits at anomalously shallow depths. Although many ore deposits are associated with magmatic–hydrothermal systems or metamorphic environments, in this chapter we focus on the important subset of ore deposits formed in sedimentary basins with no associated igneous rocks. In particular, we will emphasize Mississippi Valley–type (MVT) stratabound lead–zinc deposits and sediment-hosted tabular uranium deposits. Both types of deposits are economically significant, were apparently controlled by regional patterns of groundwater flow, and have been the subject of many "paleohydrologic" reconstructions over the past two decades. They are thus of particular interest to hydrogeologists. We will also consider supergene enrichment of copper porphyry deposits and, finally, the Colombian emerald deposits, as examples of economic mineralization caused by essentially in situ diagenetic processes. The specific material in this chapter is complemented by the generalized discussions of chemical and heat transport in Chapters 2 and 3 and, more directly, by the discussion of basin-scale flow and transport in Chapter 4. The more numerous ore deposits associated with magmatic–hydrothermal systems are considered briefly in Chapter 7.8 and in considerable detail in other texts (e.g., Barnes, 1979).

5.1 Mississippi Valley–type deposits

Mississippi Valley–type ore deposits (MVTs) are stratabound lead–zinc deposits that are found in porous carbonate rocks or (more rarely) sandstones. Most occur near basin margins or basement highs in tectonically stable continental interiors. The common ore minerals include sphalerite (ZnS), galena

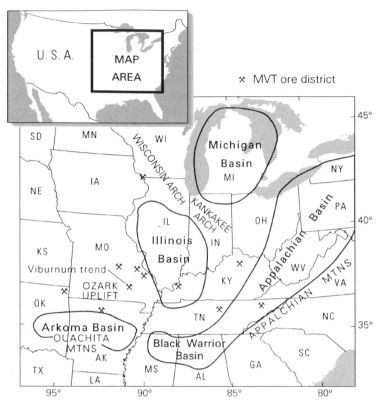

Figure 5.1 Map of the central United States showing location of geologic structures and major Mississippi Valley–type ore districts. After Garven and others (1993).

(PbS), barite ($BaSO_4$), and flourite (CaF_2). MVTs occur worldwide, but those of the United States midcontinent have historically been the most productive and, considered as a group, are among the largest known economic lead–zinc accumulations in the world. As of the mid-1970s a single ore district in southeast Missouri, the Viburnam Trend (Figure 5.1), accounted for 15% of world lead production and 85% of United States production (Vineyard, 1977).

It is now generally accepted that the United States MVTs were formed by brines that migrated laterally for substantial distances. However, there is continued debate regarding the driving mechanism(s) for fluid flow and the source of the highly saline brines.

5.1.1 Evidence for regional-scale brine migration

The most important argument for the role of regional fluid flow in MVT genesis is that the deposits were emplaced at moderately high temperatures (80 to 150°C) and at relatively shallow depths (<2 km). This temperature–

depth relation implies paleotemperature gradients that would be implausibly high in the context of a conduction-only system and seems to require advective heat transfer via flowing groundwater.

Ore-emplacement temperatures and salinities have been inferred primarily from analyses of microscopic *fluid inclusions* in crystals of sphalerite and other minerals. In these analyses, thin slices of rock are heated or cooled while under observation through a microscope. The temperature at which vapor bubbles within the inclusions disappear during heating is called the *homogenization temperature* and represents a lower limit on the formation temperature of the crystal. *Pressure corrections* can be made to account for formation in the single-phase liquid field above the two-phase curve (Figure 3.1a), but these tend to be minor in MVT deposits. Because these deposits form at moderate to shallow depths and therefore low pressures, the pressure corrections are generally small and the homogenization temperatures are close to the true temperatures of formation. The salinity of the included fluid can be inferred from a related analysis in which the inclusion is frozen; the depression of the freezing point (relative to pure water) provides an index of salinity. (For a detailed discussion of fluid-inclusion techniques, see, e.g., Roedder, 1984.) Ore emplacement depths have been inferred primarily from stratigraphic reconstruction, which is relatively straightforward, because most deposits occur in undeformed Paleozoic (250 to 570 Ma) strata.

The Ozark region of Missouri (Figure 5.1) affords a particularly well-proven example of elevated ore emplacement temperatures at shallow depths. There, Leach (1979) found that most fluid inclusion temperatures from sphalerite lie within the narrow range of 83 to 101°C, and, on the basis of nearby, nearly complete stratigraphic sections, that the maximum depth of burial was less than 800 m. Leach (1979) further noted that it is essentially impossible to define the boundaries of the individual Ozark ore districts, because minor sphalerite and galena occurrences are ubiquitous in particular strata throughout northern Arkansas and southern Missouri. He proposed a regional-scale ore-deposition model involving northward migration of brines from the Arkoma basin through permeable Paleozoic carbonate rocks toward the Ozark uplift (see Figures 5.1 and 5.2a for locations).

In a later paper, Leach and Rowan (1986) compiled fluid inclusion temperatures that are compatible with a gradual cooling with distance from the Ouachita front along this proposed flow path. Homogenization temperatures from fluid inclusions in the four main Ozark MVT deposits decrease very gradually northward, by about 0.09°C/km or 27°C over a distance of 300 kilometers. Figure 5.2b shows a composite of data from different stages of sphalerite growth with overlapping temperature ranges. The age of each sphalerite band hosting inclusions could not always be identified, and the scatter inherent in these

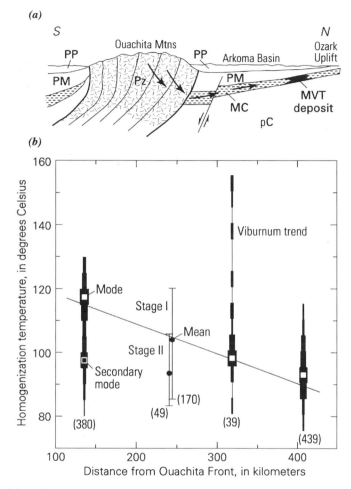

Figure 5.2 (a) Schematic north–south geologic cross section from Ouachita Mountains through Ozark uplift in Late Pennsylvanian–Early Permian (320–260 Ma) time. PP, Late Pennsylvanian to Early Permian shales and sandstones; PM, Carboniferous flysch and molasse; MC, Cambrian to Mississippian formations, largely dolomitized platform carbonates and sandstones; Pz, thrust-faulted rocks of Ouachita foldbelt; pC, Precambrian basement. Arrows show proposed direction of fluid flow. (b) Fluid-inclusion homogenization temperature data for sphalerite versus distance north of the Ouachita Mountains. After Leach and Rowan (1986).

data made it difficult or impossible to detect small temperature differences that would be expected on the scale of tens of kilometers across a single deposit or district. However, the means and modes of fluid-inclusion homogenization temperatures show a northward decreasing pattern. At the scale of hundreds of kilometers, a gentle temperature gradient is apparent.

5.1.2 The salt problem

Fluid inclusion data indicate that the mineralizing fluids associated with MVTs were highly saline Na–Ca–Cl brines (generally ≥ 10 weight % NaCl). Na–Ca–Cl brines are defined by

$$2m_{Ca} > 2m_{SO_4} + m_{HCO_3} + 2m_{CO_3}, \tag{5.1}$$

where m is in molal units (moles solute/kg water) (Hardie, 1983). That is, some of the calcium (Ca^{2+}) present in solution is electrically balanced by chloride (Cl^-). Na–Ca–Cl brines may evolve from Na–Cl brines by processes such as albitization of plagioclase:

$$\begin{array}{cc} \text{anorthite} & \text{albite} \\ CaAl_2Si_2O_8 + 2Na^+ + 4H_4SiO_4 = 2NaAlSi_3O_8 + Ca^{2+} + 8H_2O, \end{array} \tag{5.2}$$

where anorthite represents the calcium-rich end member of plagioclase. Oil field brines commonly have a $CaCl_2$ component, and $CaCl_2$ waters have also been produced from deep wells in igneous and metamorphic rocks of the Canadian Shield (Hardie, 1983), but salt concentrations as high as 10 weight percent are rare except in evaporitic environments (Chapter 9). Unfortunately for proponents of regional brine migration, evaporite beds are essentially absent from the sedimentary strata that have been invoked as source areas for the MVT brines. One must either assume that any evaporitic strata have been completely dissolved or call upon another process such as shale-membrane filtration to explain the elevated salinities. Neither of these explanations is completely satisfying; it seems unlikely that evaporite sections could disappear without a trace, and osmotic processes have not yet been clearly demonstrated to generate high salinities (Chapter 1.1.2).

5.1.3 Controls on ore deposition

As noted previously, most economic MVT deposits occur in porous carbonate rocks near basin margins or basement highs. In either of these environments the carbonates are particularly likely to have significant secondary porosity as the result of carbonate dissolution and/or dolomitization. The deposits are probably further localized in areas where organic material containing reduced sulfur was available to induce precipitation of metals by reactions such as

$$\begin{array}{c} \text{sphalerite} \\ Zn^{2+} + H_2S = ZnS + 2H^+ \end{array} \tag{5.3a}$$

and

galena

$$Pb^{2+} + H_2S = PbS + 2H^+, \tag{5.3b}$$

where Zn^{2+} and Pb^{2+} are metals in solution. Alternatively, the sulfur source might have been nonlocal, and local organic matter may simply have provided reductants (e.g., organic carbon) to reduce sulfate in the brine, ultimately precipitating metal sulfides by the same reactions. In fact, association of the large majority of MVTs with subeconomic hydrocarbon occurrences argues for the importance of organically reduced sulfur. The relatively uniform fluid inclusion temperatures suggest that cooling of the brines was not an important control on ore deposition, as does the paucity of silica in the deposits (the solubility of silica is strongly temperature dependent, so that significant cooling is often accompanied by deposition of abundant silica). Because they generate acids, the reactions shown as Eqs. (5.3a) and (5.3b) may themselves enhance porosity through dissolution or even brecciation of the carbonate host rock.

5.1.4 Driving forces for fluid flow

Any acceptable fluid flow model for the genesis of MVT deposits must satisfy the observations summarized above. In the case of the United States' MVTs, most fluid flow models that have been suggested fall into two main categories: those that emphasize topographically driven flow and those that emphasize compaction-driven flow. In the compaction-driven-flow models it is assumed that the mineralizing brines originated as pore waters and then were driven upward and toward basin margins by vertical loading and/or tectonic compression. The tectonic compression model was expressed most fluently by Oliver (1986) as the *tectonic squeegee*. In recent years the pure compaction models have generally been dismissed, mainly because they rely on a finite and apparently insufficient source of fluids: the pore waters stored in the sediments. The rates of flow expected in steadily compacting basins would probably be too small to influence the thermal regime significantly (Cathles and Smith, 1983; Bethke, 1985). In a purely compaction-driven regime, flow rates would be sufficient only to raise near-surface temperatures in areally restricted discharge areas, whereas the fluid inclusion data indicate anomalous temperatures on a very broad scale. It also appears that, given the relatively low solubilities of lead and zinc, the total volume of pore fluid would be insufficient to transport the huge mass of metals now stored in the MVTs.

In the case of the topographically driven flow models the availability of fluid is limited only by the plausible recharge rates and aquifer permeabilities and by the longevity of the flow system. As in the systems outlined in Chapters 4.3.2 and 4.3.3, the widespread elevated temperatures documented by the ore deposits are caused by the upward and lateral flow of fluids heated at depth. Several numerical-modeling studies of topographically driven flow systems indicate that such flow could indeed explain the widespread thermal anomalies associated with MVTs (e.g., Garven and Freeze, 1984a, 1984b; Bethke and Marshak, 1990; Garven and others, 1993).

In the topographically driven flow models, vigorous regional-scale flow is initiated by major orogenic events. Groundwater recharged in uplifted orogenic belts sweeps through adjacent sedimentary basins, driving saline fluids that have evolved deep in the basins into relatively undeformed, shallow strata. In the particular case of the Ozarks MVT deposits, the Late Pennsylvanian–Early Permian (320–260 Ma) Ouachita orogeny may have initiated a flow system that drove hot, saline fluids from the Arkoma basin toward the Ozark uplift (Figure 5.2a). The small lateral temperature gradients and elevated temperatures at distances of several hundred kilometers from the orogenic front (Figure 5.2b) indicate that lateral groundwater flow must have been rapid; Garven and others (1993) invoke Darcy velocities of 1–5 m/yr.

Vigorous, topographically driven, regional flow systems might easily have lasted for 10^6 years or more, waning as erosion reduced the overall relief and dissected the topography. Numerical simulations (e.g., Garven and others, 1993; Figure 5.3) suggest that maximum temperatures in the mineral districts would have occurred 10^5 to 10^6 years after the onset of gravity-driven flow and that the thermal anomalies there could have lasted for up to 5×10^5 years. In the topographically driven flow models the mineralizing fluids are assumed to have acquired metals and salt, in addition to heat, within the sedimentary strata. Some of the numerical-modeling studies include solute transport, and these generally assume as an initial condition that a saline brine is present deep within the basin. In the absence of any other source of salt, mass-balance calculations suggest that such brines would be depleted by about 10^6 years of vigorous circulation (e.g., Deming and Nunn, 1991).

Deming and Nunn (1991) used a numerical model that coupled groundwater flow, heat transport, and solute transport to assess the viability of the topographically driven flow models. They were able to reproduce the required ore deposit temperatures only by invoking an implausibly high regional heat flow of 100 mW/m^2. Moreover, when simulated temperatures in the mineralized (discharge) areas were sufficiently high, temperatures and heat flow in the elevated recharge areas were depressed well below the values observed today in active orogenic belts. They concluded that the topographically driven flow model is inadequate. This motivated Deming (1992) to propose a model involving

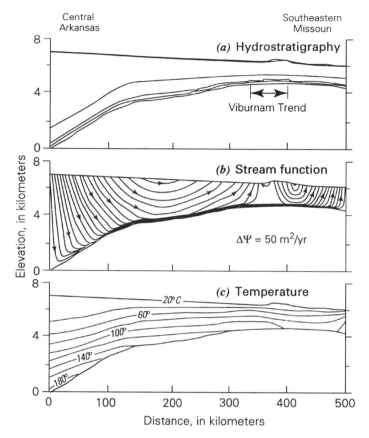

Figure 5.3 Simulation results after Garven and others (1993) showing the (a) hydro-stratigraphy, (b) streamlines, and (c) temperature field under topographically driven flow from south to north in the mid-Mississippi Valley area.

convection in the crystalline basement rock underlying the sedimentary strata. Enhanced basement permeability induced by the orogenic event itself allows convection cells to develop; heat and metals mined by the convection cells feed into an overlying topographically driven flow system (Figure 5.4). Convection in the basement was assumed to be an inherently episodic process that might account for the "banded" nature of many of the MVT deposits. From a thermal standpoint it could have the same effect as a very large increase in regional heat flow.

Although the combined topographic/convective flow model has attractive features, Garven and others (1993) questioned Deming's (1992) premise that the pure topographically driven flow model is thermally implausible. Their numerical simulations invoked reasonable regional heat flow values of

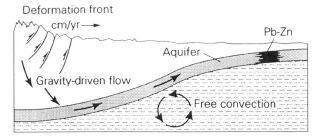

Figure 5.4 Conceptual model for MVT genesis that combines topographically driven flow in the sedimentary section with deformation-induced convection in the underlying crystalline basement. From Deming (1992), reprinted by permission of Geological Society of America.

70–80 mW/m^2, and they stated that "...our calculations replicate thermal conditions of ore formation... without imposing unrealistic thermal constraints."

However, it appears to us that the issues raised by Deming and Nunn (1991) and Deming (1992) have not been fully resolved. Whereas Garven and others (1993) were marginally successful in reproducing the thermal anomalies associated with ore deposition, they employed two approximations that would tend to overestimate the maximum temperatures achieved. First, they assumed that the regional-flow system was initiated by "instantaneous" uplift. Thus hot brines from deep within the basin, with initial temperatures controlled by a purely conductive regime, were flushed out at the maximum possible rates. A more realistic model of gradual uplift, with increasingly vigorous lateral flow, could cause temperatures at depth in the basin to be reduced by advection before the maximum topographic relief is attained. In this more realistic case the maximum flow rates would not be coincident with maximum temperatures at depth in the basin. Second, and perhaps more importantly, Garven and others (1993) invoked a constant heat flux at the base of the sedimentary section. Under transient conditions it is unrealistic to impose a constant heat-flux boundary at the base of an active circulation system; such boundaries need to be placed at a large distance from the area of interest (see, e.g., Figure 3.3). In fact, a laterally flowing thermal aquifer at a depth of about 1 km would be conducting heat downward, rather than accepting a regional heat flux, for the first 10^4 to 10^5 years of its existence (see Figure 3.5b and associated discussion).

5.1.5 The Irish MVTs

Significant stratabound lead–zinc deposits in Ireland also seem to have been formed by hot, highly saline brines at fairly shallow depths. Despite many similarities to United States MVTs, workers who have studied these Irish deposits tend to favor ore-genesis models involving compaction-driven flow and/or relatively local, deep convection through crystalline basement rock; rarely do they invoke topographically driven regional flow (see various papers in Andrew and others, 1986). Lead–isotope data suggest that lead was derived from relatively local basement rock (LeHuray and others, 1987), a hypothesis that is compatible with the deep-convection model.

There is an apparent disagreement between the most widely accepted genetic interpretations of United States MVTs (regional brine migration) and similar Irish MVTs (deep convection). However, the tectonic settings during mineralization were fundamentally different, and there may in fact be real genetic differences between the two sets of MVTs. At the time of mineralization ($\sim$350 Ma), Ireland was probably undergoing extension (e.g., Russell, 1986; Williams and Brown, 1986). The associated block faulting may have created deep, subvertical permeable conduits for convective circulation. In contrast, the orogenic events associated with the United States MVTs resulted from collision and suturing of continental plates, so that the stress regime may have been dominantly compressive. Compression is compatible with compaction- or topographically driven brine migration, but perhaps not with deep convective circulation.

5.2 Sediment-hosted uranium

Uranium occurs in most igneous rocks, but generally at such low concentrations that it is not of economic value. The original source of uranium for most sediment-hosted deposits is probably silicic magmatism (see Chapter 7.2.1 for a brief discussion of silicic magmatism). Because the uranium ion has a large charge, it is unlikely to substitute for similarly sized ions in the structures of major early-crystallizing minerals and tends to be concentrated in the late (silicic) differentiates of igneous melts (e.g., Levinson, 1980). Thus silicic plutons, volcanic rocks, and volcaniclastic sediments are relatively uranium-rich, containing approximately 5 ppm uranium. Circulating groundwater can leach dispersed uranium from such rocks and concentrate it in hydrologically and geochemically favorable areas.

5.2.1 Redox control of uranium solubility

Dissolution, transport, and deposition of uranium are controlled by its redox chemistry. The solubility of uranium at a +6 valence state is about 10^4 times higher than that of uranium at a +4 valence state; uranic (U^{6+}) or uranyl (UO_2^{2+}) ions form the relatively soluble hydroxide UO_2OH^+, whereas the uranous (U^{4+})

ion reacts with bases to form the insoluble hydroxide $U(OH)_4$ (e.g., Krauskopf, 1979). In neutral-pH groundwaters the solubility of U^{6+} is increased approximately ten fold by the formation of carbonate complexes (Krauskopf, 1979), and the effective solubility contrast between the $+6$ and $+4$ valence states increases to about 10^5. Thus, uranium tends to be dissolved and transported as a uranyl complex in environments where sufficient dissolved oxygen is available and gets deposited in reducing environments. A genetic relationship between tabular uranium deposits and organic reducing agents is clearly demonstrated by their intimate association with pore-filling "humate" (Turner-Peterson and Hodges, 1986) and by the concentration of uranium minerals around buried logs, bone fragments, and lenses of dark shale (Krauskopf, 1979). (We use "humate" in the sense of Swanson and Palacas (1965) to denote dark, amorphous, soluble organic material.)

5.2.2 Tabular uranium deposits

Tabular uranium ore deposits generally occur in feldspathic or tuffaceous sandstones. The common ore minerals are microcrystalline uranium oxides generically known as pitchblende, which vary in actual composition between UO_2 (uraninite) and U_3O_8, and uranium silicates such as coffinite ($U(SiO_4)_{1-x}(OH)_{4x}$, with $0 \leq x \leq 1$). Uranium for atomic fission has been economically and strategically important since World War II, and tabular-type uranium deposits, mostly on the Colorado Plateau, account for about 65% of United States production and reserves (Chenoweth and McLemore, 1989).

Like the MVT lead–zinc deposits of the central United States, tabular uranium deposits can be placed in the context of regional groundwater flow systems that may have extended laterally for hundreds of kilometers. In the case of the famous deposits of the Colorado Plateau, Hansley and Spirakis (1992) and Sanford (1990, 1992) developed a unified ore-genesis model. At the time of ore deposition, groundwater flow was driven largely by topography (gravity), and the general direction of flow was northeastward from the Mogollon and Elko highlands toward low-lying groundwater discharge areas (Figure 5.5a). The dilute gravity-driven flow system was underlain by a relatively stagnant basinal brine (Figure 5.5b). The gravity-driven system dissolved rhyolitic ash in tuffaceous sandstones of the Morrison Formation, resulting in a uranium-rich, slightly alkaline solution. The enhanced alkalinity favored the dissolution of scattered organic material in further transit through the Morrison Formation. Organic material (as humate) and uranium were eventually codeposited and concentrated along the freshwater–brine interface, near the discharge areas of the gravity-driven flow system (Figure 5.5b).

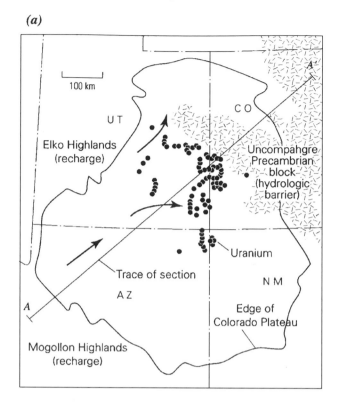

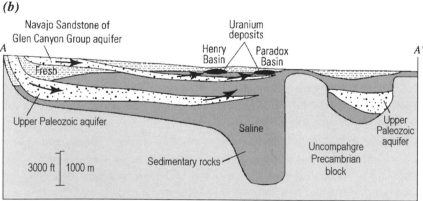

Figure 5.5 (a) Map of the Colorado Plateau area showing inferred groundwater recharge areas, general direction of groundwater flow (arrows), and location of uranium deposits. (b) Southwest–northeast cross section across the Colorado Plateau showing general direction of groundwater flow and location of uranium deposits relative to inferred position of the freshwater–brine interface. In (b) the Morrison Formation overlies the Navajo Sandstone. Hydrogeologic conditions are reconstructed for Late Jurassic (163–144 Ma) time. After Sanford (1992).

Field observations at many localities along the Gulf Coast of Florida (Swanson and Palacas, 1965) and laboratory experiments (Ortiz and others, 1980) both confirm the tendency of humate to deposit at the interface between two solutions. Codeposition of humate and uranium at a stable interface between two density-stratified fluids can explain both the tabular shape of the deposits (up to thousands of meters in breadth by a few meters thick) and their tendency to dip away from inferred areas of groundwater discharge. The fairly common occurrence of subparallel sets of tabular deposits can be explained in terms of shifts in the position of the freshwater–brine interface, which could be due to changes in recharge rates and/or the waxing and waning of playa lakes.

5.2.3 Unconformity-type uranium deposits

A second important class of sediment-hosted uranium deposits is the unconformity-type (Marmont, 1987). Most of these deposits are associated with Precambrian rocks within the McArthur basin, Northern Territory, Australia and the Athabasca basin, Saskatchewan, Canada. Deposits in the McArthur basin have been described by Needham and Stuart-Smith (1976), Ypma and Fusikawa (1980), Gustafson and Curtis (1983), Ludwig and others (1987), and Wilde and Wall (1987), whereas deposits in Athabasca basin have been described by Dahlkamp (1978), Hoeve and Sibbald (1978), Bell (1981), Tremblay (1982), and Mellinger and others (1987). These deposits commonly are located near Precambrian unconformities at the base of sedimentary basins, are overlain by thick sandstone sequences, are associated with graphite-rich, faulted basement structures surrounded by intense hydrothermal-alteration halos, and are associated with warm brines (on the basis of fluid inclusion data). Theories on the genesis of these deposits have included supergene enrichment (Langford, 1974), hydrothermal deposition (Ryan, 1979), and focusing by thermally driven convection cells (Darnley, 1981; Hoeve and Quirt, 1984, 1987).

Detailed numerical experiments by Raffensperger and Garven (1995a, 1995b) tested the feasibility of the thermally driven convection-cell model. They solved coupled equations of groundwater flow, heat transport, solute transport, and geochemical reactions for an idealized system that was representative of the conditions present during the formation of the deposits in the Athabasca basin. As depicted in Figure 5.6, a small fraction of the groundwater circulating in the thermally driven convection cells within the basement sediments penetrates into the less-permeable basement rock, where it is then focused upward through a graphite-rich fracture zone. A high contrast in redox conditions between oxygenated waters in the sandstones and reduced waters in the graphite zone provides a major control on ore deposition. An alteration halo also surrounds the

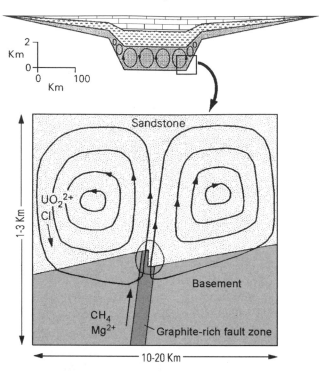

Figure 5.6 Conceptual hydrothermal model for the unconformity-type uranium deposit simulated by Raffensperger and Garven (1995a).

deposit (Figure 5.7). These halos are typified by outward progressing zones of hematite, clay alteration, sandstone alteration, and quartz cementation.

Raffensperger and Garven (1995b) solved a set of coupled partial differential and algebraic equations similar to those presented in Chapters 2 and 3 for heat transport, solute transport, and chemical equilibria over a hypothetical section through a basin 6-km deep by 10-km wide. The simulated section included one half of a convection cell in a sandstone layer 2.5-km thick. The geochemical description of the system included 12 solute components, 56 aqueous species, and up to 21 minerals. To solve this system required 12 solute transport equations in the form of Eq. (2.34), 56 aqueous-speciation equations in the form of Eq. (2.45), and up to 21 mass-action equations in the form of Eq. (2.47). All equilibrium constants were functions of temperature, fluid density was a function of temperature and solute fraction, and viscosity was a function of temperature. Although the calculated mineralogical changes would significantly affect porosity and permeability, the groundwater flow field was held constant over time. It is only with the recent advent of high-performance computational workstations that such comprehensive reactive-transport simulations

CIGAR LAKE DEPOSIT

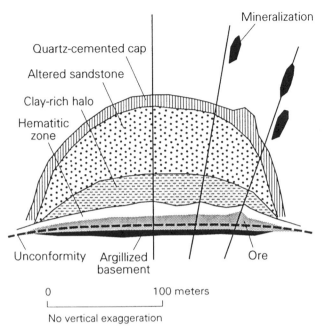

Figure 5.7 Cross section through the Cigar Lake uranium deposit, Saskatchewan. After Raffensperger and Garven (1995a).

have been achievable for multidimensional problems without the use of vector processing on supercomputers. The simulations successfully reproduced the main features of the ore deposits and alteration halos (compare Figures 5.7 and 5.8).

Some of the main conclusions of Raffensperger and Garven (1995a, 1995b) were that (1) a basin thickness of at least 3 km was required to sustain a vigorous free-convection system, (2) the upward flowing limbs of paired convection cells must coincide with the graphite-rich fault zone for an ore deposit to form, (3) concentrations of dissolved uranium of less than 30 ppm in the basin fluid are sufficient to generate the uranium ore in less than one million years, (4) fracturing and faulting of the graphitic basement rocks are essential to focus water into a geochemically reducing environment where uranium can precipitate, (5) the alteration zones are caused mainly by the transport of chemical mass through temperature gradients and across compositional boundaries (see Figure 2.10a), and finally (6) the thermally driven convection cell model for the generation of unconformity-type uranium deposits is consistent with the chemistry, mineralogy, and physics of the geologic system as it is believed to have existed during the Proterozoic.

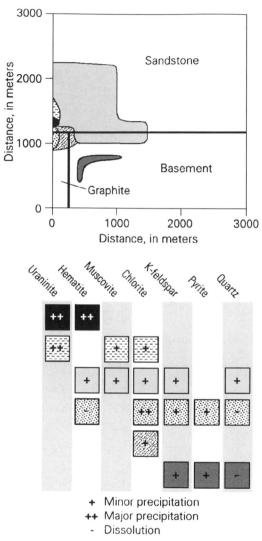

Figure 5.8 Summary of the zones of alteration in a simulated unconformity-type uranium deposit. The ore zone (solid black and horizontal hatching) is characterized by major precipitation of uraninite and hematite and minor precipitation of muscovite and chlorite. Immediately below the ore zone (dotted pattern) there is dissolution of hematite and quartz and precipitation of chlorite, potassium feldspar, and pyrite. In an outer alteration halo (gray tone), hematite, muscovite, chlorite, and quartz form at the expense of potassium feldspar. This calculated pattern of alteration is generally consistent with field observations (Figure 5.7). After Raffensperger and Garven (1995b).

5.3 Supergene enrichment of porphyry copper

In this section we retreat temporarily from the deep sedimentary basin to an important ore-forming process controlled by groundwater nearer the land surface. *Porphyry copper* deposits form at depth in magmatic hydrothermal systems (Chapter 7.8), but many are economically viable only because they have later undergone *supergene enrichment* due to weathering and percolation of groundwaters near the water table. Classic examples of such enriched deposits exist in Chile (Alpers and Brimhall, 1988; 1989), Papua New Guinea (Bamford, 1972; Titley, 1978), and Butte, Montana (Brimhall, 1979; 1980). For a general overview of copper porphyry deposits we refer the reader to Titley and Beane (1981) and Beane and Titley (1981). Supergene enrichment in these systems occurs as copper in a proto-ore, or *protore*, deposit of low grade is leached out in an oxidizing environment above the water table and then reprecipitated in a *blanket zone* after it is transported downward to a reducing environment below the water table. The process continues to concentrate the copper through time as the water table elevation falls concomitantly with the erosion of the land surface. An idealized cross section through such a deposit, illustrated in Figure 5.9, depicts the typical leached, blanket, and protore zones. As many of the minerals are copper sulfides, the mobilization of sulfur plays an important role in the enrichment process. Fully coupled models of groundwater transport and geochemistry describing this process have been developed by Ague and Brimhall (1989) and Lichtner and Biino (1992).

Ague and Brimhall (1989) published the first model that included the coupling of geochemistry and transport through all three zones of an enriched copper deposit. They used the model TRUST (Narasimhan and Witherspoon, 1977) to calculate flow through the variably saturated porous material and the models EQ3NR and EQ6 (Wolery, 1979) to calculate the distribution of aqueous species and extent of water–rock interactions, respectively. The computation time required by the coupled model restricted them to a spatial discretization of only five volume elements: two in the leached zone, one in the blanket zone, and two in the protore zone. Each volume element was 30-m long, and the recharge rate was assumed to be 20 cm/yr. The initial rock composition was chosen to approximate the protore lithology of the Butte, Montana deposit: approximately 20% K-Feldspar, 34% plagioclase (An_{30}), 21% quartz, 10% biotite, 6% muscovite, 4% pyrite, 2% chalcopyrite ($CuFeS_2$), and 2% magnetite by volume. Oxygen fugacity in the leached-zone elements was fixed at 10^{-35} (a value that was measured at the Ok Tedi site in Papua New Guinea, where active weathering is occurring). The rate of mineral dissolution in the leached zone was calculated using linear rate laws with rate constants from the literature.

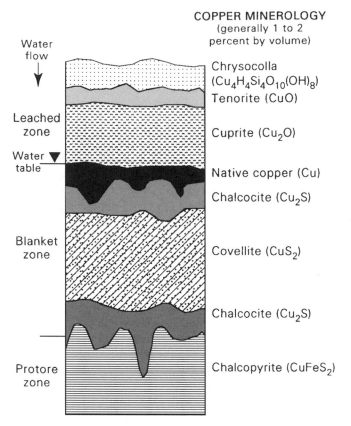

Figure 5.9 Idealized section through an enriched porphyry-copper deposit showing weathering zones and typical copper mineralogy. Modified from Lichtner and Biino (1992).

The results showed pyrite and chalcopyrite weathering to form hematite, covellite (Cu_2S), and an acid drainage. The dissolution in the leached zone generally follows the overall reactions:

$$\frac{15}{2}O_2 + 4H_2O + 2FeS_2 \rightarrow Fe_2O_3 + 4SO_4^{-2} + 8H^+ \tag{5.4}$$

and

$$\frac{17}{2}O_2 + 2H_2O + 2CuFeS_2 \rightarrow 2Cu^{+2} + Fe_2O_3 + 4SO_4^{-2} + 4H^+. \tag{5.5}$$

Precipitation in the blanket zone generally follows the overall reaction:

$$2H_2O + 3Cu^{+2} + CuFeS_2 \rightarrow 2Cu_2S + Fe^{+2} + 4H^+ + O_2. \tag{5.6}$$

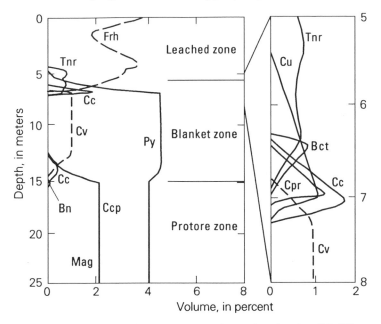

Figure 5.10 Distribution of primary and secondary minerals after 200,000 years of supergene enrichment of a porphyry host rock initially consisting of 4% pyrite (Py), 2% magnetite (Mag), and 2% chalcopyrite (Ccp). Alteration minerals include bornite (Bn), brachantite (Bct), chalcocite (Cc), covellite (Cv), copper (Cu), cuprite (Cpr), ferrihydrite (Frh), and tenorite (Tnr). After Lichtner and Biino (1992).

Conclusions from their study were that the amount of oxygen present provides a key control on the system and that the dominant source of sulfur for precipitation of the ore minerals was the primary minerals in the protore zone, not sulfate migrating from the leached zone.

Lichtner and Biino (1992) studied the process in greater detail using the code MPATH (Lichtner, 1992) to simulate moving reaction fronts on a spatial grid containing hundreds of grid points. They invoked many of the same assumptions as Ague and Brimhall (1989), however, including a constant concentration of dissolved oxygen above the water table, no dispersion of the solutes, and constant porosity and permeability. Unlike Ague and Brimhall (1989), they did not include *gangue* minerals (nonvaluable minerals in the ore) in their calculations. Results from Lichtner and Biino's (1992) simulations allowed tracking of reaction fronts that bracket two types of zones. What they called "ghost" zones do not change in width, only in position, once stationary-state conditions are reached (e.g., the bornite zone). The other type of zone increases in width with time. An example of the latter is the blanket zone itself, which exhibits upper, middle, and lower zones (Figure 5.10). The upper, or "enrichment," zone and lower, or "nascent," zone are typical features observed in these deposits and are often characterized by the presence of chalcocite. Lichtner and Biino (1992) concluded that their results are in qualitative agreement with field observations and that two end-member types of enrichments can be recognized. The first type represents a deep-water-table environment where rapid oxidation causes a sizable blanket to develop rather quickly (in tens of thousands of years). The second type represents a shallow-water-table environment with slow oxidation and enrichment that may take place over hundreds of thousands to a million years. The deposits in northern Chile may be an example of the former end member, whereas the deposits in Papua New Guinea may be an example of the latter.

5.4 Colombian emeralds

Emerald is a clear green variety of beryl ($Be_3Al_2Si_6O_{18}$); its distinctive coloration is due to the presence of trace amounts of chromium and vanadium. Most beryl occurs in granite pegmatites, but for centuries the world's most prized gem-quality emeralds have been mined from organic-rich black shales in parts of Colombia that are remote from any evidence of igneous activity. Ottaway and others (1994) proposed that the Colombian emerald deposits result from diagenetic processes involving basinal brines. The emeralds are associated with highly altered, ash-colored *cenicero zones* of the host shales that are depleted in organic matter. (The Spanish word *cenicero* translates as "ash tray.") The Ottaway et al. model involves basinal brines (40 weight % salt) carrying evaporite sulfate. The brines move through the shales along faults and permeable stratigraphic horizons, and the transported sulfate is reduced to native sulfur and pyrite where it encounters concentrations of organically derived H_2S. There, exothermic reactions such as

$$SO_4^{2-} + 3H_2S = 4S + 2H_2O + 2OH^-$$

and   (5.7)

$$4S + 1.33(CH_2) + 2.66H_2O = 4H_2S + 1.33CO_2$$

eventually proceed to consume all available organic matter, resulting in the cenicero zones. These reactions produce CO_2, and the presence of a fluid source within the generally low-permeability shale environment causes repeated episodes of hydraulic fracturing (Chapter 4.1.3) that are reflected in brecciation of the host rock and the repeated fracturing and healing of individual emerald crystals. The emeralds themselves occur in the distal parts of calcite–albite–pyrite veins that radiate from the cenicero zones. The depletion of Be, Cr, and

V in the cenicero zones, relative to the unaltered shales, appears to be sufficient to account for the observed quantity of emerald. Thus the Colombian emeralds may result from an essentially in situ process.

Though the elemental constituents of the Colombian emeralds can be accounted for locally, the source of heat is somewhat more problematic. Fluid-inclusion temperatures and quartz–calcite oxygen-isotope equilibria both indicate that the emerald deposits formed at temperatures of 350 to 400°C, and the pressure of formation has been estimated to be about 1,000 bars (see Ottaway and others, 1994, and references cited therein). This temperature–pressure relation would appear to be anomalous under a lithostatic pressure regime but fairly normal under a hydrostatic regime. Under near-lithostatic conditions, a pressure of 1,000 bars would correspond to depths of 4 to 5 kilometers, and the implied paleotemperature gradient of 70 to 100°C/km would lie above the normal range of conductive gradients, so that hidden igneous heat sources might be suspected. Under near-hydrostatic conditions, a pressure of 1,000 bars would correspond to depths in excess of 10 kilometers, and the implied paleotemperature gradient of <40°C/km would seem feasible in a normal sedimentary environment (see Chapter 4.3). However, 10 km is an unusually large depth for a sedimentary basin, and the fact that the shales are now within 1 km of the surface, and thus accessible to mining, would imply an unusual amount of uplift as well. In this instance the estimates of formation temperature are probably more accurate than the pressure estimates; perhaps the emeralds actually formed at lower pressures (lesser depths), and the hot brines, unlike the key elemental constituents of the emeralds, were somehow derived from a deeper source.

Problems

5.1 Assume that, during the period of MVT mineralization, temperatures at 1-km depth in the Ozark uplift (Figure 5.1) averaged 100°C, that land-surface temperatures were about 20°C, and that the average medium thermal conductivity at 0- to 1-km depth was about 2 W/(m-K). Under these assumptions the average conductive heat loss was 160 mW/m^2, or about 100 mW/m^2 above the "normal" value. Thus the total rate of anomalous heat loss over the ~10^4 km^2 area of the Ozark uplift was about 1,000 MW (1 MW = 1×10^6 J/s). For approximately how long could such heat loss be balanced by the arrival of >100°C fluids derived from the Arkoma basin deep? Explain your reasoning. (*Hint:* A useful first step is to estimate the amount of heat stored in the Arkoma basin at temperatures ≥100°C. Use the map and cross section of Figures 5.1 and 5.2a as guides.)

5.2 Deep-basin brines are thought to be the source waters for many of the metals associated with MVT ore deposits, but the dense brines must first be pushed up and out of the center of the basin by less dense freshwater. Hubbert (1953) stated that $dz/dx = (\rho_w/(\rho_w - \rho_o))(dh/dx)$, where dz/dx is the slope of the oil–water interface, dh/dx is the slope of the potentiometric surface, and ρ_w and ρ_o are the densities of liquid water and oil, respectively. Hubbert's equation can also be applied to a freshwater–brine interface if one assumes that mixing along the transition zone can be neglected. Given that the underlying basement contact in the center of a basin is 5-km deep 500 km from the edge of the basin and ρ_w and ρ_{brine} are 1,000 and 1,200 kg/m^3, respectively, what head gradient would be required to allow the brines to escape from the basin? Given that the hydraulic conductivity of the formation in which the interface resides is 1×10^{-6} m/s, what is the required groundwater flow rate? (See Figure 6.6 and associated discussion.)

6

Hydrocarbons

The origin, migration, and entrapment of hydrocarbons are all intimately linked
to groundwater flow. The thermal effects of groundwater movement may influence the position of the temperature "window" that favors the initial maturation
of organic matter into hydrocarbons. The poorly understood process of expulsion or *primary migration* of hydrocarbons from shaley source beds into more
permeable rocks is likely influenced by the capillary and relative-permeability
relations between aqueous fluids and hydrocarbons. The required expulsive
force may be provided by elevated fluid pressures (Chapter 4.1.2) related to hydrocarbon maturation. Hydrodynamic forces influence patterns of *secondary
migration* through more permeable rocks toward reservoirs or "traps" where
hydrocarbons are concentrated in economically significant quantities. The actual location and configuration of hydrocarbon traps are influenced both by
regional-scale hydrodynamic forces and by more local capillary effects. Finally,
groundwater flow with bacterial action can change a mobile oil to a viscous,
relatively immovable oil by biodegradation. The movement of hydrocarbons
can be described by a set of three coupled governing equations for groundwater, oil, and gas. Each of these expressions is analogous in most respects to the
equation for variable-density groundwater flow that was derived in Chapter 1.5
and invoked in Chapters 2 and 3. Some quantitative analyses of petroleum systems (e.g., the Uinta basin case study described here) solve a complete system
of governing equations; others (e.g., the Los Angeles basin case) solve only
for groundwater flow, and then use the groundwater flow field as a basis for
inferring hydrocarbon movement.

6.1 Maturation

Economically significant quantities of hydrocarbons are generated from
terrigenous, lacustrine, or marine organic matter deposited and preserved in

low-energy marine or lacustrine sedimentary environments. In higher-energy sedimentary environments, organic matter is winnowed out or oxidized so that it does not accumulate in significant quantities. Thus most hydrocarbon source rocks are shales, rather than sandstones or carbonates.

Fossilized, insoluble organic matter is termed *kerogen*. It consists of a mixture of indistinguishable degradation products and *macerals*, that is, remains of various types of plant and animal matter that can be distinguished by their chemistry, morphology, and reflectance. Hydrocarbons are generated through a thermally driven partial decomposition of the insoluble kerogen to *bitumen*, which is soluble in an organic solvent, followed by another partial decomposition of bitumen to oil and gas.

6.1.1 The oil window

Kerogen is converted to bitumen and then to oil and gas as increasing temperature breaks carbon–carbon bonds in the organic material, resulting in the formation of lighter hydrocarbons (e.g., Lewan, 1994). Kerogen is often classified by type (Tissot and others, 1974), ranging from type I, which tends to generate mostly oil, to type III, which tends to generate mostly gas. In petroleum geology the breaking of carbon–carbon bonds is called *cracking*, and the conversion of kerogen to oil is known as thermal *maturation*. The cracking process begins at a temperature of about 70°C, and temperatures of 80 to 130°C are regarded as optimal for generation of oil. This optimal temperature range for oil generation is often referred to as the *oil window*. At temperatures above about 150°C, all of the hydrocarbon chains are cracked, leaving only gas, mainly methane (CH_4). As hydrocarbon maturation progresses, the optical reflectance of some organic macerals also increases. The reflectance of one coal maceral, *vitrinite*, is often used as an independent measure of the thermal maturity of source rocks (e.g., Burnham and Sweeney, 1989).

Under an average geothermal gradient of about 25°C/km, oil-window temperatures would be expected to occur at depths of 3 to 5 kilometers. Neglecting the effects of groundwater flow for the moment, we can consider probable source-rock temperatures in a sedimentary basin as a function of the rate of sedimentation and subsidence. If we take normal sediment deposition rates to be less than 1 km/Ma, rocks in the oil window would have to be at least 3 Ma old. Sedimentation rates on the order of 1 km/Ma are in fact unusually fast and would apply, for example, to the tectonically active Miocene and Pliocene (<24 Ma) basins of California and the Taranaki basin in New Zealand. In relatively stable environments, rocks in the oil window are more likely to be on the order of 30 Ma or more. In some rare instances (e.g., Railroad Valley,

Nevada), hydrocarbons are produced from relatively young sediments where hydrothermal systems have locally accelerated hydrocarbon maturation.

The temperature dependence of oil generation from kerogen can be modeled by the first-order exponential relation

$$\frac{dV}{dt} = -aV, \tag{6.1}$$

where $a = A \exp(-e/RT)$, V is the volume of kerogen present, t is time, A is an empirical constant, e is the activation energy of the reaction, R is the gas constant, and T is temperature in Kelvins (e.g., Sweeney and others, 1987). The activation energy e is about 200 kJ/mole, and Campbell and others (1978) determined the rate constant a (in 1/sec) for Green River oil shale to be $\sim 2.8 \times 10^{-13} \exp(-26,390/T)$. For this value of a, dV/dt will increase tenfold with a 12°C temperature increase. Thus both the position of the oil window and rates of oil generation will be sensitive to any factor that influences the thermal history of the source rock.

6.1.2 Groundwater flow and the thermal regime

In Chapter 4.3 we discussed the influence of groundwater on the temperature regime of sedimentary basins. We noted that vigorous topographically driven groundwater flow causes cooling in recharge areas and warming in discharge areas. In fact, for the same depths, temperature differences of up to 50°C might be expected between areas of recharge and discharge. Regional groundwater flow could therefore have a profound effect on the position of the oil window. Person and Garven (1992) demonstrated this effect in a study of petroleum generation within the Rhine Graben.

The Rhine Graben is located in northeastern France and southwestern Germany (Figure 6.1) and has been exploited for salt, coal, and petroleum resources. Person and Garven (1992) simulated the temperature history of the basin over the past 30 Ma by simultaneously solving the equations of groundwater flow, sediment compaction, and heat flow (Figure 6.2). Continuous accretion of sediment was simulated by addition of elements along the top of the model mesh at rates that approximated the subsidence history. The computed temperature patterns for 20 Ma ago reflect conduction-dominated heat flow (Figure 6.2a). Under present-day conditions a stronger advective component of heat flow exists (Figure 6.2b) because of a tectonic uplift that occurred approximately 15 Ma ago. As a result, the present-day oil window is shifted significantly upward and toward the current areas of groundwater discharge (Figure 6.2, lower panels). The oil window presently occurs 1,400-m deeper

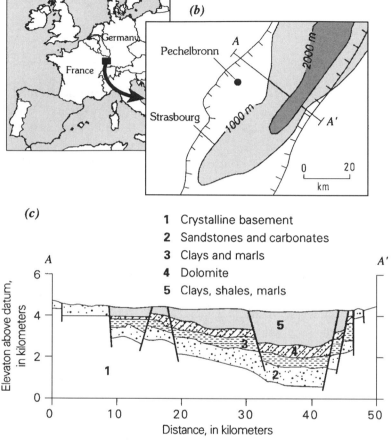

Figure 6.1 Central part of the Rhine Graben showing (a) location, (b) thickness of Cenozoic sediments within the graben, in meters, and (c) the geologic cross section (Doebl and others, 1974) simulated by Person and Garven (1992).

in groundwater recharge areas than in discharge areas. Although the position of the oil window in the simulations was estimated simply by the position of the pertinent isotherms, it corresponds to vitrinite reflectance levels observed within the source and overburden rocks.

The effect of strictly vertical groundwater movement on temperature profiles and petroleum generation was examined by Person and others (1995). Models were devised to represent fluid flow within subsiding columns of fine-grained sediment while holding the ratio of fluid-advected heat flow to conductive

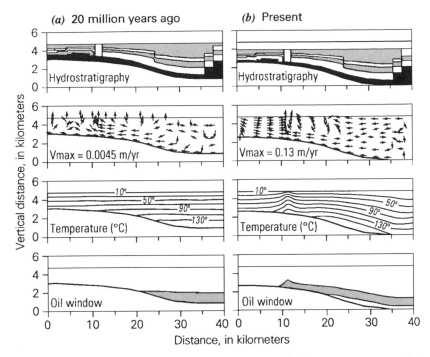

Figure 6.2 Hydrostratigraphy, computed groundwater velocities, temperatures, and the position of the oil window for section A–A′ across the Rhine Graben for conditions (a) 20 million years ago and (b) at present. After Person and Garven (1992).

heat flow (the Peclet number) constant (for more discussion of thermal Pe see Chapter 3.5.2). As the simulated column of sediment lengthened, the groundwater flow rate was reduced to maintain a constant value of Pe. The temperature histories calculated for various values of Pe were used to compute hydrocarbon production profiles, vitrinite-reflectance curves, and apatite *fission-track annealing* patterns and apparent ages. Hydrocarbon production was computed using the kinetic parameters of Tissot and others (1987) and Burnham and Sweeney (1989) for kerogens of type II and III, respectively. The location and extent of hydrocarbon maturation for type II kerogen for several values of Pe are shown in Figure 6.3. The range of Peclet numbers depicted involves groundwater flow rates (*Darcy velocities, q*) of 0.1 to 9 mm/yr. Even when the advective heat flow is only one third as large as the conductive heat flow ($Pe = 0.3$, $q = 0.4$ to 4 mm/yr) the implications for the position of peak oil and gas generation are dramatic. Vertical groundwater flow at the simulated rates should also have a large influence on observed histograms of fission-track annealings, creating distinct distributions that may persist for more

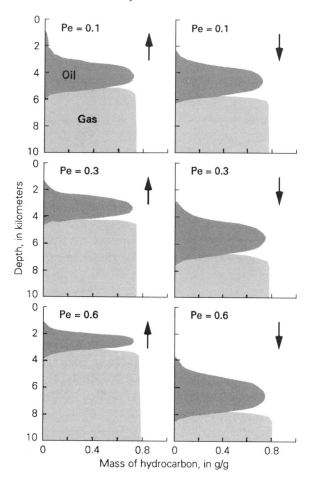

Figure 6.3 Computed distribution of hydrocarbons generated from type II kerogen in groundwater discharge (left-hand panels) and recharge areas (right-hand panels) for selected values of the thermal Peclet number, *Pe*. Vertical arrows indicate the direction of groundwater flow. After Person and others (1995).

than 10^8 years. (The main source of *fission tracks* in geologic media is the spontaneous fission of the uranium isotope ^{238}U. Because the fission rate of ^{238}U is known, it is possible to calculate the age of a mineral on the basis of its ^{238}U content and the number of fission tracks it contains. However, fission tracks fade and disappear when a mineral is heated sufficiently. The healing or *annealing* rate is temperature dependent, and complete annealing requires about 10^5 years at 105°C and about 10^8 years at 150°C (Naeser and others, 1989)).

6.2 Migration

A few hydrocarbon source rocks, such as the Monterey (shale) Formation of California and the Green River Formation of Utah, are sufficiently fractured to act as reservoirs. However, most commercial production of oil and gas requires that hydrocarbons move from low-permeability shales into more permeable rock. The mean lateral migration distance for oil is about 10 kilometers, with a significant number of cases exceeding 80 kilometers (Slujik and Nederlof, 1984). This movement requires *primary migration* from source beds into *carrier beds* (aquifers) and *secondary migration* within the carrier beds.

6.2.1 Capillary effects

Capillary–pressure relations between various fluid phases affect primary migration, as well as patterns of secondary migration and entrapment. *Capillary effects* arise in multiphase systems due to competing molecular forces acting upon the interfaces between two or more fluids. All molecules exert attractive forces upon one another, and molecules within the body of a liquid experience roughly equal attractions from every side. In contrast, liquid molecules at an interface with air, gas, or the liquid's vapor are attracted dominantly toward the body of the liquid, which results in a tendency to reduce the interfacial area to a minimum. This effect is known as *surface tension*. Where two immiscible liquids and/or a liquid and solid are in phase contact, there is an additional attraction across the interface. In this situation, capillary effects can be regarded as resulting from a balance between forces acting in the various interfacial planes.

We will consider a water–oil–rock system as our type example. By convention, a contact angle α is measured in the denser fluid (generally water). The equilibrium force balance is given by

$$\sigma_{wo} \cos \alpha + \sigma_{wr} = \sigma_{or}, \tag{6.2}$$

where σ_{wo}, σ_{wr}, and σ_{or} are surface tensions acting in the planes of the water–oil, water–rock, and oil–rock interfaces, respectively (Figure 6.4). If $\alpha < 90°$, the denser fluid is said to wet the rock, and it is referred to as the *wetting phase*. In water–oil–gas–rock systems, water is usually the wetting phase, and oil and gas are usually *nonwetting phases*.

The existence of a contact angle α causes the interfaces between fluids to be curved, and the resulting interfacial forces are balanced by normal pressure forces P_o and P_w such that

$$P_o - P_w \equiv P_{cow} = \frac{2\sigma_{wo}}{r^*}, \tag{6.3}$$

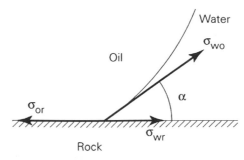

Figure 6.4 Schematic diagram showing balance of interfacial forces (σ) in a water–oil–rock system.

where P_{cow} is the oil–water *capillary pressure* and r^* is the mean radius of curvature of the interface. This expression, known as the Laplace formula, indicates that P_{c} will increase with decreasing capillary radius (or pore size). In fact, for water–oil systems, P_{c} is generally less than 0.1 bar in porous sandstones but may be on the order of tens of bars in source rocks composed of clay-sized material. Thus a shale–sandstone (source rock-carrier bed) interface may allow ready passage of oil from shale to sandstone while amounting to a major fluid-potential barrier with respect to any oil (or gas) present in the sand. In effect, this interface becomes a one-way valve. In the general case where grain-size diameter is variable, oil (or gas) will be acted upon by a net capillary force tending to impel it in the direction of increasing grain size. (For a more complete account of capillary phenomena in the context of water-hydrocarbon systems, see Hubbert (1953), Levorsen (1967), or, for an overview of more recent work, Wendebourg (1994).)

6.2.2 Primary migration

Although capillary effects doubtless influence the movement of hydrocarbons from source rocks into carrier beds (aquifers), several textbooks on petroleum geology suggest that, in general, this *primary migration* is the most poorly understood aspect of hydrocarbon accumulation. Because the volume of water in newly deposited clay-rich sediments is extremely high (generally >50%), one might speculate that progressive dewatering due to compaction and diagenesis plays a role. However, most loss of porosity probably occurs at depths (and temperatures) well above the oil window (Chapter 1.2.5). Also, because the solubility of various types of oil in water is quite low (1–1,000 ppm), the quantity of oil that could be expelled in solution is negligibly small. Expulsion of oil from *bitumen* and/or migration of oil as a separate phase along preferentially oil-saturated pathways are thought to be more important. Either of these processes implies some overpressuring of fluids in the source rock relative to surrounding rock. The kerogen–bitumen conversion itself may generate these overpressures (Chapter 4.1.2) by acting as an in situ source of fluids. Low source-rock permeabilities and/or relatively high temperatures (e.g., Eq. 6.1) would be conducive to such a process, which might ultimately cause hydraulic fracturing (Chapter 4.1.3).

The processes involved in primary migration may be analogous to those that cause oil to be expelled from source-rock samples in laboratory *pyrolysis* (e.g., Lewan, 1994). In *hydrous pyrolysis*, a sample of source rock is heated in a closed reactor containing liquid water. The oil generated when the reactor is heated accumulates on the water surface and is compositionally similar to natural crude oils. In laboratory experiments the development of a continuous bitumen network within the sample seems to be a prerequisite for oil expulsion. The added force needed for the expulsion is believed to be provided by a net volume increase of the organic components within the confining rock matrix. Post-pyrolysis examination of samples may reveal *en echelon* partings that are parallel to the bedding fabric and filled with oil (Lewan, 1987).

The concept of *relative permeabilities* to various fluid phases may be useful in the context of primary migration, but there is very little empirical data upon which to base relative-permeability curves for source rocks. Oil–water relative permeabilities in reservoir rocks are generally described by Corey-like relative-permeability functions (Figure 3.2) with oil immobile at oil saturations of less than perhaps 30 percent. It has been suggested that the oil–water curves for source rocks may be quite different, perhaps with very large residual water saturations ($\sim$80 percent) and oil mobile at relatively low oil saturations (e.g., Waples, 1994). The possibility of transient microfracturing of the source rocks would further complicate the description of relative permeabilities.

6.2.3 Secondary migration

In some instances the reservoir rocks in which hydrocarbons have accumulated are closely associated with the active source rocks. We have mentioned the Monterey Formation of California, which comprises both source rock and reservoir; some other oil reservoirs occur in lenticular sand bodies within massive oil shales (e.g., in the Cherokee Shale, Kansas). In such instances there is little or no secondary migration of hydrocarbons through carrier beds or aquifers. However, most viable hydrocarbon reservoirs occur in porous, permeable sandstones and carbonates at some distance from the probable source rocks. The existence of such reservoirs requires significant secondary migration through permeable rocks. Furthermore, in most instances the hydrocarbons must be concentrated from an originally dispersed state in the active source rock, either in transit or due to conditions at the reservoir site itself. Huge oil reservoirs such as those associated with petroleum systems in Saudi Arabia, Venezuela, and the Western Canada (Alberta) basin have no unusually rich source rocks in the immediate vicinity. In these instances, oil has apparently been collected from a very large volume of ordinary source rocks (thousands of km^3) and traveled as much as several hundreds of kilometers to reach its present position (Dickey, 1979; Garven, 1989).

Secondary migration can occur as *miscible* flow (hydrocarbons in solution) or *immiscible* flow (hydrocarbons migrating as a separate phase). Most workers agree that miscible flow is relatively unimportant, partly because of the low solubility of oil but also because solubility-dependent fractionation of oil between source beds and reservoirs seems to be quite rare. That is, the abundance of various types of oil seems to be fairly consistent between reservoir and likely source rocks and independent of their solubility in water. However, Garven (1989) showed that the giant oil fields near the eastern margin of the Western Canada basin could have been created by miscible transport in a topographically (gravity-) driven groundwater flow system. As we noted previously in the context of Mississippi Valley–type ore deposits (Chapter 5.1), the maximum flux of water through a gravity-driven flow system is limited only by the available recharge, aquifer permeability, and the longevity of the system. Garven (1989) estimated that the Western Canada flow system may have existed for 10 to 50 Ma and, over such time scales, oil concentrations of 1 to 10 ppm are sufficient to account for the oil volume in the deposits. In terms of total oil volume the Western Canada deposits ($400 \ km^3$) are among the world's largest; for example, they are perhaps five times larger than those in the Persian Gulf (see Garven, 1989, and references cited therein). Thus, at least on a mass-balance basis, miscible flow would appear to be a viable transport mechanism in some long-lived, gravity-driven flow systems.

Under miscible conditions, secondary migration of hydrocarbons can be described by the usual set of single-fluid, nonreactive solute transport equations (e.g., Eqs. 2.11 and 2.28). Under immiscible conditions, each of the fluids present is subject to a different set of buoyancy and capillary forces, so that patterns of hydrocarbon migration can be quite different than those predicted for single-fluid conditions. In general, oil will still tend to migrate in the direction of groundwater flow, with an oblique upward drift due to its lower density. Both oil and gas can be trapped by fluid-potential barriers in structures such as anticlines, or where water flows from coarser into finer sediments.

Hubbert (1953) was the first to quantify the hydrodynamic factors governing immiscible secondary migration, and his concept of *impelling force* provides a useful way of visualizing the net forces acting on each fluid. We also use the impelling-force concept to illustrate the mechanism of steam–water *phase separation* in Chapter 7.4.1. There, the impelling force is defined formally (Eq. 7.7); here, we will simply note that it is the negative of the gradient in fluid potential, so that it is a vector quantity that defines the direction in which an element of fluid will tend to migrate and that its derivation omits capillary effects, assuming that these are approximately isotropic except near lithologic contacts. Figure 6.5 shows the impelling forces acting on elements of water, oil, and gas in a hypothetical hydrodynamic environment. Downdip flow of groundwater in a confined aquifer follows the dip-parallel vector E_w. The less dense oil and gas both migrate to the top of the aquifer, but the relative angles of E_o and E_g are such that oil migrates downdip, whereas gas migrates updip.

6.3 Entrapment

Hubbert (1953) also showed that, under hydrodynamic conditions, large-scale (noncapillary) oil–water interfaces will dip in the direction of groundwater flow, according to

$$\frac{dz}{dx} \propto \left(\frac{\rho_w}{\rho_w - \rho_o} \right) \left(\frac{dP}{dx} \right), \tag{6.4}$$

where dz/dx is the slope of the interface and dP/dx is the horizontal fluid-pressure gradient. The significance of this "hydrodynamic tilt" can readily be illustrated in the context of anticlinal structures (Figure 6.6), which constitute some of the more obvious oil and gas traps under both hydrostatic and hydro-dynamic conditions. If the horizontal pressure gradient is sufficiently small, each of the impelling-force vectors (E_w, E_o, and E_g) will be nearly vertical, and the equilibrium distribution of fluids within an anticline will be approximately governed by vertical density segregation (Figure 6.6a). As the lateral pressure gradient increases, so does the tilt of the oil–water interface (Figures 6.6a–c). The anticlinal structure can serve as an effective oil trap only where its own dip exceeds the tilt of the interface. A number of examples of hydrodynamically tilted oil and gas fields are discussed by Dahlberg (1994).

Under immiscible conditions, oil is also likely to be trapped anywhere that groundwater is forced to flow upward into shaley confining units, for example, due to the pinchout of an underlying aquifer. Oil will be left behind at the contact unless the propulsive force of the water is sufficient to overcome a major capillary-pressure barrier. The giant oil deposits of the Western Canada basin occur where sandstone and carbonate aquifers subcrop beneath shale aquitards,

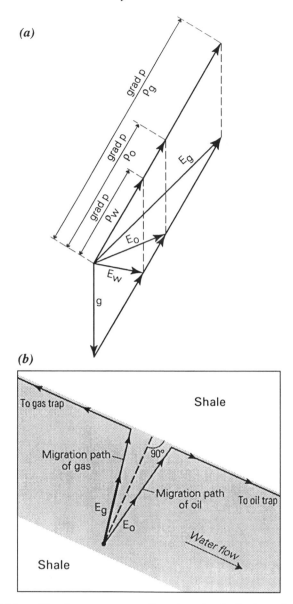

Figure 6.5 (a) Impelling forces on water, oil, and gas in a hydrodynamic environment. (b) Divergent migration of oil and gas in a hydrodynamic environment. From Hubbert (1953), reprinted by permission of the American Association of Petroleum Geologists.

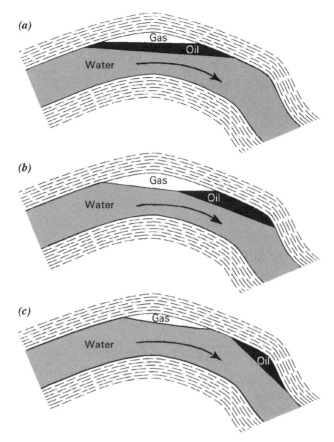

Figure 6.6 Oil and gas accumulations in a hydrodynamic environment. In (a), gas is entirely underlain by oil; in (b), gas is partly underlain by oil; and in (c), the gas and oil traps are entirely separated. The progression from (a) to (c) is due to increasing "hydrodynamic tilt" and can be interpreted either in terms of an increasing horizontal pressure gradient (increasingly vigorous groundwater flow) or in terms of varying hydrocarbon densities (i.e., the difference between ρ_w and ρ_o decreases from (a) to (c)). From Hubbert (1953), reprinted by permission of the American Association of Petroleum Geologists.

and they are presumably controlled by this regional-scale contact. On a more local scale, a commonly observed association between oil and gas fields and positive temperature anomalies (e.g., McGee and others, 1989) suggests that hydrocarbons often accumulate in areas where groundwater is flowing upward (see Chapter 3.4.1 for a more detailed discussion of the thermal effects of vertical groundwater flow).

Upward trending secondary migration often leads to the microbial degradation of oil, rather than its entrapment and preservation in a more useful form.

When oil is exposed to meteoric water at temperatures of less than about 70°C, it *biodegrades* to an asphaltic material that is unable to migrate. Thus in the North Sea, for example, where the top of the oil window occurs at about 3.2-km depth, much of the oil that has migrated to less than about 2.2-km depth has biodegraded (Blanc and Connan, 1994). The microbial processes involved are presumably similar to those that help to immobilize and (eventually) consume oil spills in the shallow subsurface.

6.4 Governing equations for immiscible multiphase flow

We will not derive the three governing equations for groundwater–oil–gas flow from first principles but will simply describe them by analogy to the variable-density groundwater flow equation previously derived in Chapter 1.5 (Eq. 1.32). For a rigorous derivation of water–oil–gas equations we refer the interested reader to petroleum engineering texts such as Aziz and Settari (1979).

As shown in Chapter 1.5,

$$\frac{\partial (n\rho_f)}{\partial t} = \nabla \cdot \left[\frac{\overline{k}\rho_f}{\mu_f} (\nabla P + \rho_f g \nabla z) \right] \tag{6.5}$$

is a fairly general form of the groundwater flow equation that accounts for the effects of variable fluid properties by calculating fluxes in terms of forces acting on the fluid ($\nabla P + \rho_f g \nabla z$) rather than hydraulic head, by allowing hydraulic conductivity to vary with fluid density and viscosity, and by posing changes in mass storage (the left-hand side) in terms of changes in the fundamental controlling parameters, porosity and fluid density, rather than specific storage and hydraulic head. This form of the groundwater flow equation is derived from the basic continuity equation (1.15) by assuming fully saturated conditions and inserting a fairly general form of Darcy's law. The vector operator ∇ indicates $\mathbf{i}\partial/\partial x + \mathbf{j}\partial/\partial y + \mathbf{k}\partial/\partial z$.

Equation 6.5 can be adapted to describe groundwater flow in the multiphase water–oil–gas system by introducing a *saturation* variable ($0 \leq S \leq 1$) to indicate that the pore space is not entirely occupied by water and a *relative permeability* variable (Chapters 3.1.3 and 3.1.7; $0 \leq k_r \leq 1$) to describe how water movement is impeded by the interfering presence of oil and gas, and by adding a subscript to the pressure variable to indicate that, because of capillary effects, the fluid-pressure gradients in water, oil, and gas will be different. Thus, in the absence of sources or sinks, the equation

$$\frac{\partial (nS_w\rho_w)}{\partial t} = \nabla \cdot \left[\frac{\overline{k}k_{rw}\rho_w}{\mu_w} (\nabla P_w + \rho_w g \nabla z) \right] \tag{6.6}$$

adequately describes variable-density groundwater flow in water–oil–gas systems. The equation describing the flow of oil is identical in form,

$$\frac{\partial(n S_o \rho_o)}{\partial t} = \nabla \cdot \left[\frac{\bar{k} k_{ro} \rho_o}{\mu_o} (\nabla P_o + \rho_o g \nabla z) \right]. \tag{6.7}$$

To describe the movement of the gas phase, one must often take into account gas carried in solution by oil as well as its movement as a separate phase. Thus

$$\frac{\partial [n(S_g \rho_g + S_o \rho_o C)]}{\partial t} = \nabla \cdot \left[\frac{\bar{k} k_{rg} \rho_g}{\mu_g} (\nabla P_g + \rho_g g \nabla z) \right. \tag{6.8}$$
$$\left. + \frac{\bar{k} k_{ro} \rho_o C}{\mu_o} (\nabla P_o + \rho_o g \nabla z) \right],$$

where C denotes the mass of dissolved gas per unit mass of oil (e.g., kg/kg). Sometimes the gas carried in solution by water must also be considered; this would require adding another term to the right-hand side of Eq. (6.8).

The concentration of gas in oil is usually based on independent empirical data, as are the relative permeabilities, which are generally described as nonhysteretic functions of the volumetric saturation(s). Thus Eqs. (6.6)–(6.8) comprise three equations in six unknowns (P_w, P_o, P_g, S_w, S_o, and S_g), and three constitutive relations are needed to complete the description of the system.

Capillary-pressure functions are also usually based on empirical data (e.g., Aziz and Settari, 1979), so that

$$P_o = P_w + P_{cow}, \qquad \text{where } P_{cow} = f(S_w, S_g) \tag{6.9a}$$

and

$$P_g = P_w + P_{cog}, \qquad \text{where } P_{cog} = f(S_w, S_g). \tag{6.9b}$$

In addition, the pore space must be entirely filled with water, oil, or gas, so that

$$S_w + S_o + S_g = 1. \tag{6.10}$$

Equations (6.5)–(6.10) now comprise a system of six equations in six unknowns and, given adequate empirical data, can be the basis for a reasonably complete description of groundwater–oil–gas flow in the subsurface.

6.5 Case studies

We will conclude this chapter by describing two simulation-based case studies of groundwater flow and petroleum migration that illustrate the state of the art as of this writing. Both studies employed numerical models that describe fluid flow and heat transport in a porous medium that is continually deformed by the deposition or erosion and compaction of sediments. Thus both are among the increasingly common but still-rare examples of flow-and-transport simulation in a dynamic geologic framework. Other examples of coupled deformation–flow–transport modeling were cited in Chapter 5.1.5 in the context of rapidly deforming *accretionary prisms*, and another excellent example in the hydro-carbon context is presented in Burrus and others (1992).

The Uinta basin study (McPherson, 1996) described here employed an ex-tended version of the Lawrence Berkeley Laboratories' TOUGH2 simulation model (Pruess, 1991) to solve equations describing groundwater flow (e.g., Eq. 6.5), oil migration (e.g., Eq. 6.6), and heat transport (e.g., Eq. 3.18) in three spatial dimensions. The less rigorous Los Angeles basin study (Hayba and Bethke, 1995) employed the proprietary model BASIN2 (Bethke and oth-ers, 1993) to solve equations of groundwater flow and heat transport in two spatial dimensions and then used the resulting temperature and groundwater flow fields to estimate patterns and rates of oil migration.

6.5.1 The Uinta basin

The Uinta basin (Figure 6.7) is located in northeastern Utah, at the northern edge of the Colorado plateau. The structural basin developed in latest Cretaceous and early Tertiary time (about 75 to 50 Ma). The basin is deepest to the north, adjacent to the Uinta mountains, where the thickness of Tertiary sediments approaches 6 kilometers. As described in Chapter 4.3.3, under present-day conditions, groundwater flow appears to depress temperature gradients near the elevated margins of the Uinta basin and enhance gradients near its axis (Chapman and others, 1984; Willett and Chapman, 1987). Oil is generated in the basin deep, mainly in the Green River Formation, an Eocene lake deposit that serves as both source rock and reservoir in the Uinta basin.

Sometime after the basin achieved maximum depth at about 30 Ma, the flanks were uplifted and eroded. The simulation study by McPherson (1996) showed that the timing of this uplift, which is not well known, has a major effect on the petroleum system. He considered two end-member models: "early uplift," with all of the uplift occurring before about 25 Ma, and "late uplift," with all of the uplift occurring after 10 Ma. Figure 6.8 shows simulated groundwater flow vectors and oil saturations for the early- and late-uplift models between 56 and 5 Ma. Both the early- and late-uplift models incorporated parameter sets that are consistent with empirically determined values from the Uinta basin (e.g., values of porosity as a function of effective stress, permeability as a function of porosity, thermal conductivity, total organic carbon content, and oil density and

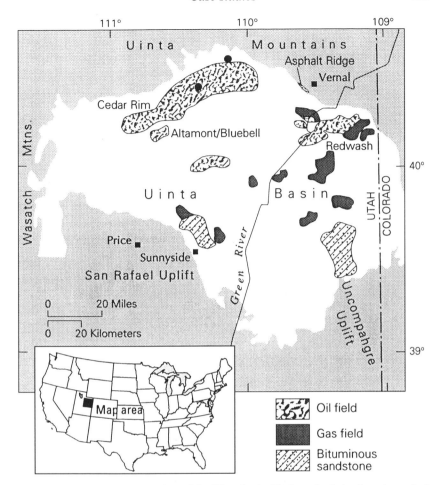

Figure 6.7 Map showing location of the Uinta basin, Utah, and relative locations of oil fields, gas fields, and oil sands. The simulation results shown in Figure 6.8 apply to the north–south cross section between the Uinta Mountains and the Sunnyside oil sands. After McPherson (1996).

viscosity; see McPherson (1996) for details). Because of a lack of empirical data, the relative-permeability function for lacustrine (source-rock) facies and the capillary–pressure relations for both alluvial and lacustrine facies are based on the parametric model presented by Parker and others (1987).

Although the early-uplift model (the left column of Figure 6.8) may be the more geologically plausible, the late-uplift model is more successful in representing the migration of hydrocarbons from the basin deep southward toward

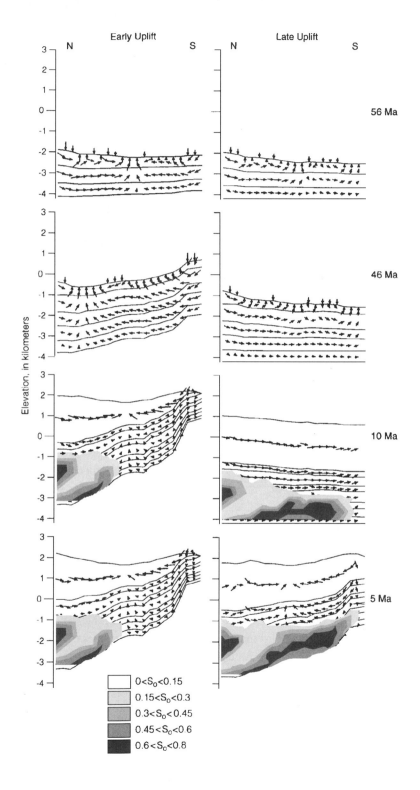

deposits such as the Sunnyside oil sands (Figure 6.7). The simulations revealed several unanticipated feedback effects. For example, if the basin flanks are uplifted and eroded too soon, the resulting reduction in temperature increases oil viscosity sufficiently to inhibit migration of oil to the south and east. Perhaps the most surprising feedback, however, is between surface-temperature history and permeability at depth. The oil generation rate is temperature dependent (e.g., Eq. 6.1), and the temperature at any point in the system depends on the surface-temperature history as well as geothermal factors. Furthermore, in the Uinta basin case, oil generation appears to cause elevated fluid pressures sufficient to induce *hydraulic fracturing* (Chapter 4.1.3). In McPherson's (1996) simulations, the early-uplift model, in conjunction with a particular surface-temperature history, causes large early-time overpressures. These induce hydraulic fracturing, which enhances intrinsic permeability, resulting in present-day overpressure values much lower than observed levels. An otherwise identical simulation with constant surface temperature results in lower permeabilities and higher values of overpressure.

The well-documented occurrence of overpressures is one of the more interesting aspects of the Uinta basin system and in fact had attracted considerable attention in the decade preceding McPherson's (1996) study. Fluid pressures in the Green River Formation locally exceed two thirds of the lithostatic level (e.g., >65 MPa at 4-km depth), and hydraulic head values are as much as 3 km above the maximum topography. Sweeney and others (1987) modeled oil generation profiles that are similar in form to the observed overpressure profiles and concluded that oil generation is a viable mechanism for the overpressure. Bredehoeft and others (1994) concurred, based on results of their three-dimensional hydrodynamic model. Bredehoeft and others' (1994) model considered water only, and McPherson's (1996) more complete model allowed him to identify controlling parameters for the overpressure. He found the necessary conditions to be: intrinsic permeability $\leq 5 \times 10^{-18}$ m^2, source-rock total organic carbon content ≥ 0.5 mass %, and oil viscosity $\geq 5 \times 10^{-4}$ kg/(m-s). Each of these values appears to be consistent with empirical data from the Uinta basin, and in fact the oil viscosity is likely substantially higher. McPherson (1996) agreed that oil generation is the dominant overpressuring mechanism but found that, because the presence of oil reduces the relative permeability to water, the effects of sediment compaction might also be significant during some periods of basin history.

Figure 6.8 Simulated groundwater flow vectors and oil saturations at selected times along a north–south cross section through the Uinta basin for "early uplift" (left) and "late uplift" (right) models of basin evolution. See Figure 6.7 for location of the cross section. After McPherson (1996).

6.5.2 The Los Angeles basin

The Los Angeles basin is a small (~50 by 75 km), actively subsiding rift basin that contains >9 km of sedimentary rocks deposited over the last 11 Ma. It is likely the world's most productive petroleum basin in terms of the hydrocarbons produced per unit volume of rock (Biddle, 1991).

The long-term rate of sedimentation and subsidence in the Los Angeles basin (nearly 1 mm/yr) is sufficient to cause significant advective perturbation of the thermal field, even in the absence of any groundwater movement (Chapter 4.3.1). Hayba and Bethke (1995) simulated two-dimensional groundwater flow and heat transport in the basin from 11 Ma ago to the present. Until about 0.5 Ma ago the basin was below sea level, so that there was no topographic drive for groundwater flow, and sediment compaction was the main driving force. The rates of compaction-driven flow would likely have been too small to have a dramatic effect on thermal gradients, but the generally upward and outward pattern of groundwater flow would have tended to counteract the thermal effects of subsidence to some extent. Under current (<0.5 Ma) conditions, there is a vigorous topographically driven groundwater flow system that may shift the position of particular isotherms vertically by nearly a kilometer, relative to their expected position in the absence of groundwater flow.

Having simulated the paleohydrology, Hayba and Bethke (1995) proceeded to consider the secondary migration of oil to the West Coyote oil field from probable source beds located about 13 km to the west-southwest. The rather slow rate of groundwater movement through the postulated *carrier beds*, combined with the low solubility of oil in water, seemed to preclude miscible transport over the time available. The one-dimensional rate of movement of immiscible oil, relative to the rate of groundwater flow in the same carrier bed, was then estimated in terms of the ratios of permeability, porosity, viscosity, and driving forces for oil and water:

$$\frac{v_o}{v_w} \sim \left(\frac{k'}{k}\right)\left(\frac{n}{n'}\right)\left(\frac{\mu_w}{\mu_o}\right)\left[1 + \frac{(\rho_w - \rho_o)g\frac{dz}{dL}}{\frac{d(P + \rho_w g z)}{dL}}\right], \tag{6.11}$$

where v is the *average linear velocity* or seepage velocity and the prime denotes conditions in the part of the carrier bed that transports oil. Equation 6.11 was derived from one-dimensional, multiphase versions of Darcy's law (see also Chapter 3.1.3) that can be written as

$$q_w = n v_w = -\frac{k_{rw} k}{\mu_w} \frac{\partial(P_w + \rho_w g z)}{\partial L} \tag{6.12}$$

for water and

$$q_o = nv_o = -\frac{k_{ro}k}{\mu_o}\frac{\partial(P_o + \rho_o gz)}{\partial L} \tag{6.13}$$

for oil, under the assumptions that (1) capillary forces act to segregate the oil into the most porous and permeable parts of the carrier bed, but can otherwise be ignored on the grounds of homogeneity, and (2) the various parts of the carrier bed are either completely saturated with oil or completely saturated with water, so that the relative permeabilities can be dropped.

The carrier beds through which oil likely migrates toward the West Coyote field dip rather steeply, so that the buoyant force acting on the oil might be 10 times as great as the hydrodynamic force acting on both phases. Thus reasonable assumptions about the four ratios on the right-hand side of Eq. (6.11) led to the conclusion that oil might have moved about 20 times faster than groundwater. Furthermore, the oil would need to saturate only a small volume of the carrier bed in order to transport the amount of oil in the West Coyote field over the time available.

Problems

6.1 A near-surface water table aquifer (15°C) and a deep confined aquifer (120°C) are separated by a 3-km-thick low-permeability lacustrine sequence. The hydraulic head in the water table aquifer is greater than that in the deep aquifer. Calculate how much of the lacustrine sequence is within the *oil window* (80 to 130°C), given vertical groundwater flow rates (specific discharge or Darcy velocity) of (**a**) 0.0 m/yr, (**b**) 0.001 m/yr, and (**c**) 0.1 m/yr. Assume steady-state conditions and that the thermal conductivity of the medium is 2.0 W/(m-K). (See Chapter 3.4.1.)

6.2 Hubbert (1953) stated specifically that $dz/dx = (\rho_w/(\rho_w - \rho_o))(dh/dx)$, where dz/dx is the slope of the oil–water interface and dh/dx is the slope of the potentiometric surface. Assume that ρ_w and ρ_o are 1,000 and 950 kg/m^3, respectively, that dh/dx measured perpendicular to the axis of an anticline is 0.01, and that the limbs of the anticline have a maximum dip of 3°. Will the anticline act as an oil trap?

6.3 Extend Eq. (6.8) to account for the gas dissolved in water as well as the gas dissolved in oil.

6.4 Derive Eq. (6.11) from Eqs. (6.12) and (6.13). State the assumptions made at each step of the derivation.

7

Geothermal processes

In this chapter we emphasize groundwater–rock systems in which groundwater flow and heat transport might be regarded as the primary coupled processes, that is, systems to which the heat transport theory outlined in Chapter 3 is generally applicable. We begin by discussing the Earth's heat engine and conductive heat losses, and we then address a variety of heat transfer problems to which fluid flow is critically important, including magma cooling and associated hydrothermal circulation; transport near the critical point of water; some more general multiphase processes; and the occurrence of hot springs and geysers. We will also briefly consider geothermal resources, volcanogenic ore deposits, and subsea hydrothermal systems. Elsewhere we have addressed advective heat transfer by groundwater in a generalized geologic context (Chapter 4.3) and as it relates to ore genesis in sedimentary basins (Chapter 5) and hydrocarbon maturation (Chapter 6.1).

7.1 Crustal heat flow

The mean conductive heat flow measured very near the Earth's surface is approximately 70 mW/m^2 (e.g., Chapman and Pollack, 1975). Correcting for the effects of hydrothermal circulation in the oceanic crust (Chapter 7.9.1) brings the mean global heat flux to 87 mW/m^2 (Pollack and others, 1993). Integrated over the surface of the globe, this amounts to a heat loss of more than 4×10^{13} W. The sources of this heat are not completely resolved, but the radioactive decay of isotopes of uranium, thorium, and potassium is certainly the most significant. Cooling of an originally hot Earth and the gravitational energy released by its density segregation may or may not be important sources, depending on the presumed mechanism of planetary accretion and the rate and timing of core formation (Verhoogen, 1980). It is of historical interest to recall that, prior to the discovery of radioactivity, many scientists believed that all of the current heat loss from the Earth was due to its continued cooling from an originally molten state. This erroneous assumption led Lord Kelvin to calculate the Earth's age as being only 20 to 30 million years, an estimate that was widely accepted in the latter half of the 19th century.

7.1.1 Measurement

Humans have recognized for millenia from observations of volcanoes and hot springs and in deep mines that temperature increases with depth in the Earth; however, the first actual measurements of the associated heat flow were not made until the late 1930s. By *Fourier's Law*, heat flow is the product of the temperature gradient and a coefficient known as *thermal conductivity*; that is, in three dimensions,

$$q_h = -K_m \nabla T, \tag{7.1}$$

where q_h is a vector. Thus q_h values can be obtained through measurement of K_m and ∇T. For practical purposes of measurement, the one-dimensional form of Fourier's Law shown previously as Eq. (3.10) is generally applicable, because under normal conditions, most variation in temperature occurs in the vertical dimension. Values of dT/dz are measured by lowering temperature probes down near-vertical drillholes, and K_m is usually estimated in the laboratory on the basis of measurements on drillcore, cuttings, and/or outcrop samples. Because K_m is a property of the bulk medium, measurements of intact, saturated drillcore are preferable; otherwise, K_m must be reconstructed on the basis of rock conductivity K_r and some independent estimate of porosity, using models such as Eq. (3.11). Note that Eq. (7.1) is written assuming that K_m is a scalar; in fact, some rocks are significantly anisotropic with respect to thermal conductivity (see, e.g., Chapter 4.3.1), so that the common practice of measuring K_m on randomly oriented cuttings can frequently lead to erroneous results.

Very accurate measurements of dT/dz and K_m are possible but do not necessarily translate to comparably accurate estimates of crustal heat flow at depth. For practical reasons, most heat flow estimates are obtained from drillholes <300 m deep. At such shallow depths, temperature profiles are often perturbed by such factors as terrain effects, which focus heat flow in valleys, increasing dT/dz near valley bottoms and decreasing dT/dz near ridge tops (e.g., Blackwell and others, 1980); long-term fluctuations in the land-surface temperature due to climatic changes and/or changes in land cover (e.g., Lachenbruch and Marshall, 1986); and groundwater movement. Each of these factors may be significant, even where their effects are subtle and difficult to detect.

Table 7.1 *Thermal conductivity data from the Cascade Range and*
adjacent areas in the northwestern United States.[a]

Lithology	Number of sites	Thermal conductivity mean (std. deviation) in W/(m-K)
<7 Ma volcanic rocks	17	1.54 (0.33)
>7 Ma lava flows (basalts and andesites)	14	1.65 (0.13)
>7 Ma tuffs and lahars	8	1.41 (0.17)
>7 Ma rocks, undifferentiated	43	1.49 (0.21)
<2 Ma sedimentary rocks	5	0.88 (0.19)
>2 Ma sedimentary rocks	16	1.31 (0.11)
granitic rocks	10	2.72 (0.24)

[a]From Ingebritsen and others (1994).

One factor that facilitates shallow heat flow estimates is that, for a given lithology, K_m at shallow depths may be fairly constant. For example, consider Table 7.1, which summarizes all reported thermal conductivity measurements from the Cascade Range of the northwestern United States. Within the Cascades the standard deviation of K_m for a given lithology amounts to only 10 to 20% of the mean value. In contrast, *terrain corrections* to the temperature gradient in this area of rugged topography are commonly as large as 30%. Thus, reasonable estimates of q_h might be based on a terrain-corrected dT/dz measurement and knowledge of the lithologic units encountered in the drillhole.

7.1.2 Lateral and vertical variations

On a global scale, there is a reasonably good correlation between the age of the Earth's crust and crustal heat flow. This relation is much clearer in the oceanic crust than in the continental crust; heat flow in much of the oceanic crust is approximated quite well by a model of simple cooling with time and distance from the mid-ocean ridge axes (e.g., Sclater and others, 1980). Average oceanic heat flow values range from about 50 mW/m^2 in the oldest ($\sim$200 Ma) oceanic crust to over 300 mW/m^2 in juvenile crust near the mid-ocean ridges (Figure 7.1a). The age–heat flow relation is much less defined on the continents, probably because the evolution of continental crust is much more variable than the simple, finite ridge-to-trench pathway followed by most oceanic crust. Nevertheless, on a global scale, there is a distinguishable relationship between the timing of the most recent tectonic activity and heat flow (Sclater and others, 1980). Mean continental heat flow ranges from about 40 mW/m^2 on the

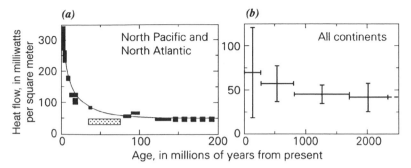

Figure 7.1 Relation between mean conductive heat flow and age for (a) well-sedimented areas in the North Pacific and North Atlantic and (b) all continents. In the continental case, "age" is defined roughly as the age of the most recent tectonic activity. After Sclater and others (1980).

stable cratons to about 70 mW/m² in Tertiary tectonic provinces (Figure 7.1b). Very generalized global heat flow maps created solely on the basis of assumed relationships between age and heat flow do not differ greatly from those that incorporate actual heat flow measurements (Chapman and Pollack, 1975).

To estimate heat flow and temperatures below the depth range explored by drillholes, one needs to estimate the variation in K_m and the abundance of radioactive isotopes with depth. Over large temperature ranges, K_m must be considered a temperature-dependent parameter. Sass and others (1992) showed that the temperature dependence of rock thermal conductivity to at least 250°C is well described by the relation

$$K_r(T) = \frac{K_r(0)}{1.007 + T(0.0036 - 0.0072/K_r(0))}, \qquad (7.2)$$

where $K_r(0) = K_r(25)[1.007 + 25(0.0037 - 0.0074/K_r(25))]$, $K_r(25)$ is the thermal conductivity at room temperature, T is the in situ temperature, and thermal conductivity is in W/(m-K). This relation was based on the data of Birch and Clark (1940) and verified by application to an independent set of data. It implies an inverse temperature dependence for most common rocks. The strength of the temperature dependence varies with the room-temperature conductivity; for a rather high room-temperature conductivity, $K_r(25) = 3.8$, the conductivity at 250°C, $K_r(250)$, is predicted to be ~2.65. For a moderate room-temperature conductivity of 2.4 a relatively modest decrease to $K_r(250) \sim 2.1$ is predicted.

Thermal conductivity describes a process known as *phonon conduction*, that is, the transfer of kinetic, rotational, and vibrational energy through a lattice. Thus the inverse temperature dependence can be qualitatively understood in terms of more erratic movement of phonons at a hotter, more excited state. The temperature dependence of K_r is complicated by the fact that at a temperature of about 600°C, *radiation* becomes an important heat transfer process. This process can be incorporated in standard models by defining a *radiative thermal conductivity* component with a direct temperature dependence (Clauser, 1988). "Radiative" thermal conductivity describes a process akin to blackbody radiation and is conceptualized in terms of separate material and photon phases. Because the "normal" conductivity depends inversely on temperature and "radiative" conductivity depends directly on temperature, the two effects may tend to cancel out when they are both represented by K_r. In experimental practice, it may be difficult to distinguish between "normal" and "radiative" conductivity at high temperatures.

Under steady-state, conduction-only conditions, the distribution of radioactive isotopes with depth dictates the vertical variation in heat flow, that is,

$$\frac{dq_h}{dz} = -A(z), \tag{7.3}$$

where A is a *heat production* parameter that depends on the concentration of radioactive isotopes. Within some physiographic provinces, heat flow data have been found to fit the linear relation

$$q_h = q_h^* + bA_o, \tag{7.4}$$

where q_h^* is termed the *reduced heat flow*, A_o is the heat production measured near the surface, and b is a constant (Lachenbruch, 1970). The linear relation defined by Eq. (7.4) is most likely to exist in areas where A_o can be regarded as representative of much of the crustal section, that is, where $A(z) \sim A_o$. This is most likely to be the case in igneous and metamorphic provinces where the crust is largely granitic. Where the linear heat flow–heat production relation is strong, q_h^* can be regarded as a reasonable estimate of heat flow at the base of the crust, that is, from the mantle.

7.1.3 Perturbations due to groundwater flow

As noted above, typical continental heat flow values lie in the range of 40 to 70 mW/m². This translates to temperature gradients of 20 to 35°C/km, given a typical thermal conductivity of 2 W/(m-K). However, in practice, heat flow observations cover a much larger range of <0 to $>1,000$ mW/m², and temperature gradients range from $<0°C/km$ to $>500°C/km$. Most of the larger departures from the typical continental heat flow range are due to the effects of groundwater flow. In areas of plentiful groundwater recharge, temperature profiles may be nearly isothermal or even negative to substantial depths (e.g., Figure 4.14, profiles 6, 8, 9, 11, and 13–15); the negative profiles are due to lateral flow of water that was recharged at higher elevations/lower temperatures. Such profiles translate to near-surface heat flow values ≤ 0 mW/m^2. Maximum temperatures in groundwater discharge areas are limited only by the boiling point–depth curve (Figure 3.4a), and in the immediate vicinity of hot springs may translate to heat flow values of $>1,000$ mW/m^2. In most of the remainder of this chapter we will be concerned with geologic systems in which the temperature regime is often profoundly affected by groundwater flow.

7.2 Magmatic–hydrothermal systems

Magmatic intrusions are the source of thermal energy to most of the Earth's high-temperature ($\geq 150°$C) hydrothermal systems, although a few high-temperature systems occur in areas of little or no volcanic activity and appear to be caused by deep circulation of meteoric water in areas of above-average conductive heat flow (e.g., Beowawe, Nevada: White, 1992).

7.2.1 Magmatic heat sources

The SiO$_2$ content of igneous rocks defines a continuum of rock types ranging from *mafic* (basalt/gabbro $<57\%$ SiO$_2$) to *silicic* (rhyolite/granite $>70\%$ SiO$_2$). Most high-temperature continental hydrothermal systems are related to inter-mediate-to-silicic magmas, which are less dense, more viscous, and have a lower melting point than mafic magmas (the rhyolitic solidus is approximately $800°$C, depending on the confining pressure and water content, versus approximately $1,150°$C for basalt). The denser mafic magmas are believed to originate in the mantle or lowermost crust, and the few high-temperature hydrothermal systems in mafic terranes either occur in areas with extraordinarily high intrusion rates (e.g., Iceland and Hawaii) or are related to layered gabbroic intrusions (e.g., the fossil system at Skaergaard, east Greenland). The less-dense silicic melts tend to accumulate in the upper crust (≤ 10 km depth); these may originate when mafic magmas in transit through the crust melt, or partly melt, the less-mafic host rock. Hildreth's (1981) influential models of lithospheric magmatism depict pods of silicic melt as being both shallower and more voluminous than their mafic parents (Figure 7.2). In general, pre-Quaternary (>2 Ma) magmas with volumes of less than about 1,000 km^3 will have cooled to ambient temperatures (Smith and Shaw, 1979), so that geothermally useful accumulations of heat in the upper crust are usually associated with Quaternary, silicic magmatism. This makes accurate mapping of rock age and composition a key element of geothermal exploration.

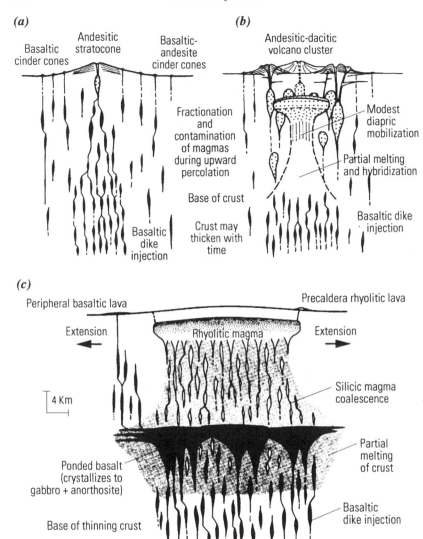

Figure 7.2 Styles of lithospheric magmatism: (a) basaltic to andesitic, (b) andesitic to dacitic, and (c) rhyolitic. Magmas are classified on the basis of SiO_2 content: basalt = <57% SiO_2, andesite = 57–62% SiO_2, dacite = 62–70% SiO_2, and rhyolite = >70% SiO_2. Note that in each case, magmatism is driven by basaltic dike injection near the base of the crust. After Hildreth (1981).

The amount of heat made available by a particular magma body depends upon its *latent heat of crystallization* and the degree of cooling. A 1 km^3 volume of silicic magma with a latent heat of crystallization of 270 kJ/kg (Harris and others, 1970), a density of 2,500 kg/m^3, and a heat capacity of 1 kJ/(kg-K) releases about 2×10^{18} J by cooling from an emplacement temperature of 800°C to an ambient temperature of 300°C, which might be regarded as a typical crustal temperature at 10-km depth. About 1/3 of this heat comes from crystallization and 2/3 from cooling. Steady intrusion and cooling of such magma at a rate of 1 km^3/m.y. translates to a heat flow of about 0.06 MW.

Conductive heat flow theory indicates that, if latent heat is neglected, maximum host-rock temperatures near an instantaneously emplaced magma body will be only one half the emplacement temperature (Carslaw and Jaeger, 1959; Lachenbruch and others, 1976). However, the mineral assemblages in contact with aureoles frequently indicate host-rock temperatures that approach the inferred emplacement temperature (e.g., Manning and others, 1993). Such departures from the instantaneous-source prediction are due to some combination of latent-heat effects, convection within the magma body, and/or convective resupply of magma from greater depths. In the context of numerical-simulation models the latter two effects are sometimes mimicked by assigning an artificially large latent heat of crystallization to the magma body (e.g., Norton and Taylor, 1979; Manning and others, 1993).

7.2.2 Heat transfer from magma to groundwater

Heat transfer between magma bodies and their associated hydrothermal systems is less well understood than heat transfer within the magma (e.g., Marsh, 1989) or hydrothermal systems (Chapter 3) themselves. A comparison between magmatic heat contents and hydrothermal heat discharge rates leads to a useful appreciation of the magnitude and rates of magma–groundwater heat transfer.

The heat discharged by individual *hot springs* ranges from <1 to nearly 100 MW in the Cascade Range (Ingebritsen and others, 1992) and up to 50 MW along nearby parts of the offshore Juan de Fuca Ridge (Bemis and others, 1993). Heat discharge from the greater Yellowstone, Wyoming, hydrothermal system is about 5,000 MW. The Yellowstone heat discharge rate is estimated on the basis of a hydrothermal fluid discharge rate of approximately 3,000 liters/s and geochemical indications that the hydrothermal fluids attained temperatures of at least 340°C (Fournier, 1989). The areally extensive Yellowstone system(s) may be related to a very large rhyolitic magma body such as that depicted in Figure 7.2c, but a comparable heat discharge rate of 4,250 MW at Grimsvotn Volcano, Iceland (Agustdottir and Brantley, 1994), is clearly related to more localized magmatism. If such large heat discharge rates are representative over geologic time, they imply that very large volumes of magma are being emplaced and cooled. We calculated in the previous section that intrusion of silicic magma at a rate of 1 km^3/m.y. would equate to about 0.06 MW, so that a steady heat discharge of 100 MW would correspond to intrusion at a rate of 1,700 km^3/m.y. Such volumes of magma are roughly equivalent to the largest known silicic eruptive units (Hildreth, 1981). Thus localized heat discharge rates $\geq$ 100 MW are very likely to be transient over geologic time scales of $> 10^5$–10^6 years or more, except perhaps in mid-ocean ridge settings such as Iceland (see Chapter 7.9).

Transfer of as much as 100 MW of heat from magma to groundwater by *conduction* implies that the heat transfer takes place over large surface areas and/or short distances. Let us consider the Lassen system in the Cascade Range of northern California as an example. There, the measured heat discharge from hydrothermal features totals > 100 MW (Sorey and Colvard, 1994). This heat is believed to be obtained from a silicic magma body emplaced at a depth of at least several kilometers. This body is too small to be resolved by seismic surveys, so that the heat transfer area is restricted to be no more than a few km^2. If we take the heat transfer area to be <5 km^2, the average conductive heat flux over that area must be >20 W/m^2. If we then assume a reasonable thermal conductivity of 2 W/(m-K) and a temperature difference of 500°C between the magma body (800°C) and circulating groundwater (300°C), then, by Eq. (3.10), the conductive length must be <50 m. This length might represent the thickness of a *conductive boundary layer* between the magma and the hydrothermal system, and this boundary layer would be expected to retreat inward as the magma body progressively crystallizes, cools, and cracks (cf. Lister, 1974).

7.2.3 Fluid circulation near magma bodies

The gravity (topographic) forces that drive most groundwater flow are substantially damped below a few kilometers depth in the crust, but magmatism introduces powerful new driving forces, and oxygen isotope studies of fossil hydrothermal systems document fairly vigorous circulation of meteoric waters within and adjacent to intrusions emplaced at midcrustal depths (e.g., Taylor, 1971; Norton and Taylor, 1979). Fluid circulation near magma bodies is driven by lateral variations in fluid density, by in situ changes in fluid density and porosity, and by the release of magmatic and metamorphic volatiles, as well as topography. Patterns of flow are controlled by these driving forces and by the distribution of permeability. In such environments the permeability field itself will be a complex, transient function of the initial host-rock permeability, strains caused by intrusion, and porosity changes resulting from fluctuations in pressure, temperature, and mineralization.

The first significant quantitative studies of fluid circulation near magma bodies were those of Norton and Knight (1977) and Cathles (1977). These pioneering studies, out of computational necessity, neglected every driving force for fluid flow except for lateral variations in fluid density, ignored two-phase phenomena, and assumed that fluid flow was quasi-steady over time. Nevertheless, they arrived at several important conclusions that are consistent with the results of later, more complete models, and they have strongly influenced our thinking with respect to magmatic–hydrothermal systems. Norton and Knight (1977) showed that, within the host rock, advective heat transport would be significant for permeabilities $\geq 10^{-17}\,\mathrm{m}^2$, but that higher host-rock permeabilities would not significantly shorten magma cooling times unless the permeability of the magma itself was also increased. They also demonstrated the feasibility of the large apparent *water–rock ratios* (Chapter 10.5.1) indicated by oxygen isotope data such as those of Taylor (1971).

By ignoring flow caused by in situ fluid density and porosity changes, Norton and Knight (1977) and Cathles (1977) forced a *convective* pattern of fluid flow, with an upflow of less-dense, heated fluid above the intrusion balanced by lateral inflow of denser fluid at depth (Chapters 3.5.3, 3.6). Their formulation also assumed that volumetric (rather than mass) flux is conserved at steady state, that is,

$$\nabla \cdot \boldsymbol{q} = 0; \tag{7.5}$$

as noted in Chapter 3.3, this is known as the *Boussinesq approximation* and allows the flow equation to be solved through a volume-based *stream-function* approach.

For relatively high values of host-rock permeability and/or near-steady-state conditions, the Norton and Knight–type formulation is reasonably appropriate. However, for relatively low values of host-rock permeability, in situ changes in fluid density are likely to be the dominant driving force over the first 10^3 to 10^5 years of magma cooling, and during this period, fluid flow will tend to radiate outward from the zones of maximum thermal expansion – that is, outward from the intrusion in all directions. Sammel and others (1988) applied a more complete model to the magma cooling problem, and at a time of 10^4 years they showed radially outward fluid flow in a low-permeability case ($k_x = 10^{-17}\,\mathrm{m}^2$, $k_z = 10^{-19}\,\mathrm{m}^2$) versus a convective pattern for an otherwise identical high-permeability case ($k_x = 10^{-15}\,\mathrm{m}^2$, $k_z = 5 \times 10^{-17}\,\mathrm{m}^2$). Hanson (1992) showed that for fluid density changes and production rates typical of metamorphic environments, fluid expulsion may temporarily dominate the convective flux for permeabilities as large as 10^{-16} m^2. Because the supply of pore fluid and magmatic/metamorphic volatiles is finite, at some point in the evolution of the system the radially outward pattern of flow will be superceded by a Norton and Knight–type convective pattern involving externally derived fluids.

Figure 7.3 shows a suite of simulation results from the U.S. Geological Survey's HYDROTHERM model (Hayba and Ingebritsen, 1994), which solves the fairly complete flow and transport Eqs. (3.15) and (3.16). We assumed a 900°C pluton (2.5-km high by 1-km half width) intruded into a homogeneous, isotropic medium at a depth of 2 km. The lateral and lower boundaries were impermeable and insulated; the top boundary was held constant at 20°C and 1 bar. The initial temperature gradient in the host rock was 20°C/km. At a host-rock permeability of 10^{-14} m^2 (Figure 7.3a), heat is advected from the top and sides of the magma body so rapidly that maximum temperatures in the hydrothermal plume above the magma scarcely exceed 200°C at a depth of 1 km (Figure 7.3e). At a host-rock permeability of 10^{-15} m^2 (Figure 7.3b), temperatures approach 300°C at 1-km depth, and two-phase flow develops in the hydrothermal plume and persists for several thousand years. At host-rock permeabilities $\leq 10^{-16}$ m^2 (Figure 7.3c), heat transport is mainly by conduction, and the maximum temperature at 1-km depth is only 170°C. At host-rock permeabilities $\leq 10^{-17}$ m^2, the initial pattern of flow is "expulsive" rather than convective. For host-rock permeability of 10^{-17} m^2 (Figure 7.3d), thermal pressurization causes fluid pressures to be significantly above hydrostatic for 10^4 years, and at 10^{-18} m^2, fluid pressures near the magma body would reach the lithostatic level. (See Hayba and Ingebritsen (1997) for additional details of such simulations and Chapter 4.1 for discussion of anomalous fluid pressures and hydraulic fracturing.)

7.2.4 Permeabilities in near-magma environments

The near-magma environment is characterized by high strain rates and large gradients in temperature and fluid composition, so that one would expect substantial variations in permeability over time. If host-rock permeabilities are sufficiently low to allow significant thermal pressurization, permeability may be episodic: When pore pressures exceed the least principal stress plus any tensile strength of the rock, they are relieved by *hydraulic fracturing*; when the resulting fractures seal, pressures can build up again, eventually leading to another fracturing episode. Because the supply of pore fluid and magmatic and metamorphic volatiles is limited, this pressurization/expulsion stage will have a finite

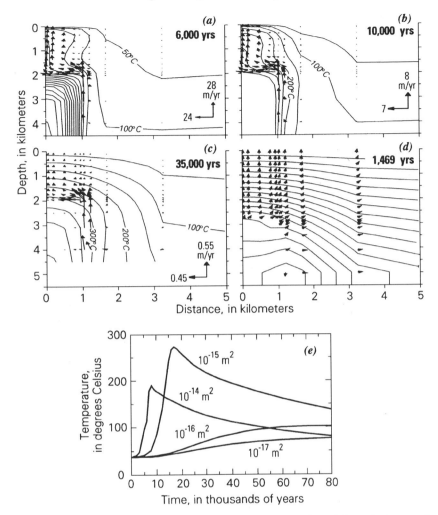

Figure 7.3 Fluid circulation near a magmatic intrusion at selected times for host-rock permeabilities of (a) 10^{-14} m^2, (b) 10^{-15} m^2, (c) 10^{-16} m^2, and (d) 10^{-17} m^2; (e) shows temperature histories at 1-km depth in the hydrothermal upflow plumes of (a) to (d). The temperature fields in (a) and (b) are advection-dominated; those in (c) and (d) are conduction-dominated. For (d) we show a simulation time that illustrates the "expulsive" pattern of fluid flow during the early stages of hydrothermal circulation.

duration. When the near-magma fluid supply is eventually depleted, convection cells dominated by meteoric fluids will develop, even in low-permeability host rocks. Many ore deposits in igneous environments seem to record hydraulic-fracturing episodes (e.g., Titley, 1990). However, oxygen isotope data commonly indicate interaction between large volumes of meteoric water and nearly molten rock (e.g., Taylor, 1990), and in such instances, host-rock permeabilities must have been large enough that the pressurization/expulsion stage was brief or nonexistent.

The rocks that host magmatic intrusions are likely to be much less permeable than compositionally equivalent rocks exposed in outcrop. At Kilauea volcano, Hawaii, for example, unaltered basalt flows exposed at the surface have permeabilities of $\geq 10^{-10}$ m^2, whereas basaltic rocks in the hydrothermal systems within the volcano's rift zones have permeabilities of $\leq 10^{-15}$ m^2 (Ingebritsen and Scholl, 1993). The relatively low permeability in the hydrothermal systems was determined by direct hydraulic measurements and also inferred from the fact that significant convective circulation seems to be absent at depth, despite high temperature gradients. The $> 10^5$ permeability difference between the near-surface rocks and those in the hydrothermal system is attributed to pervasive intrusion by less-permeable basaltic dikes and/or hydrothermal alteration. Both are probably important and are related by the role of recently emplaced dikes as heat sources that help drive alteration. The average permeability of the hydro-thermal systems at Kilauea ($\leq 10^{-15}$ m^2) is very similar to that inferred for the basaltic rocks that hosted the fossil hydrothermal system at the Skaergaard intrusion, East Greenland ($\sim 10^{-16}$ m^2: Manning and others, 1993). At Skaergaard, careful mapping had revealed several generations of intrusion-related porosity within a few hundred meters of the intrusion (Manning and Bird, 1991). However, numerical modeling revealed that paleotemperature estimates were best matched with a homogeneous permeability of 10^{-16} m^2, and the near-magma porosity/permeability enhancements were inferred to have been so minor or short-lived that they did not significantly affect heat transport (Manning and others, 1993).

7.3 Fluid flow and heat transport near the critical point

Near the *critical point* of water (Chapter 3.1.1), maxima in the thermal expansivity and heat capacity of the fluid nearly coincide with a minima in kinematic viscosity, thereby maximizing buoyancy forces and heat transport capacity while minimizing viscous-drag forces. Norton and Knight (1977) suggested that the near-critical extrema in fluid properties may control the overall style of hydrothermal fluid circulation, while noting that small differential pressure and temperature values would be required to adequately simulate the process. Somewhat later, in laboratory experiments involving a heated wire centered in a cylindrical vessel filled with water-saturated sand, Dunn and Hardee (1981) observed near-critical heat transfer rates as much as 70 times greater than the conduction-only rates for the same system (i.e., the Nusselt number $Nu \sim 70$;

see Chapter 3.5.1). They introduced the term *superconvection* in reference to near-critical effects.

The near-critical extrema in fluid properties create computational problems that have inhibited quantitative modeling, although Cox and Pruess (1990) attempted to reproduce the Dunn and Hardee results numerically. For models that use pressure and temperature as dependent variables, the near-critical extrema pose particularly difficult problems. In $P-T$ coordinates the critical point is at the vertex of the vaporization curve (Figure 3.1a) and represents a singularity in equations of state. For example, $c(P, T)$ diverges to ∞, and the partial derivatives of $\rho_f(P, T)$ diverge to $\pm\infty$ (Johnson and Norton, 1991). On the other hand, models such as the U.S. Geological Survey's HYDROTHERM (Hayba and Ingebritsen, 1994) avoid computational problems at the critical point by formulating the governing equations (Eqs. 3.15 and 3.16) in terms of pressure and enthalpy. In $P-H$ coordinates, two-phase conditions are represented as a region, rather than a single curve (Figure 3.1b), and the density, viscosity, and temperature of liquid water and steam merge smoothly to finite values at the critical point (Figure 3.1c; $P = 22.055$ MPa, $H = 2086.0$ kJ/kg, $\rho = 322$ kg/m^3, $\mu = 3.94 \times 10^{-5}$ Pa-s, and $T = 373.98°C$).

7.3.1 One-dimensional pressure–enthalpy paths

One-dimensional numerical simulations with the HYDROTHERM model indicate that permeability has a pivotal effect on near-critical pressure–enthalpy trajectories. We can explore near-critical effects in one dimension by simulating flow along a 1-km-long horizontal column, assigning various permeabilities and fixed $P-H$ values at the ends of the column. In this system, one-dimensional horizontal flow is driven by the fixed pressure drop between the high $P-H$ and low $P-H$ ends of the column. Medium thermal conductivity (K_m) is held constant at 2 W/(m-K). (Although there is some near-critical variation in the thermal conductivity of water, the resulting effect on K_m will be small for low-porosity media, because, as noted in Chapter 3 (Eq. 3.11), $K_m \sim K_r^{(1-n)} K_f^n$.) Arbitrary initial values of P and H are assigned to the interior of the column, and Eqs. (3.15) and (3.16) are solved iteratively until mass and energy fluxes reach a steady state, that is, until mass and energy entering the high $P-H$ end of the column equate with mass and energy exiting at the low $P-H$ end.

Figures 7.4a and 7.4b show results from two sets of one-dimensional experiments with different $P-H$ endpoints. Within each set, differences in permeability cause the flow path to take different trajectories that are distinguishable in both $P-H$ (Figure 7.4a) and $P-T$ (Figure 7.4b) coordinates. In set 1, the

One-dimensional simulations

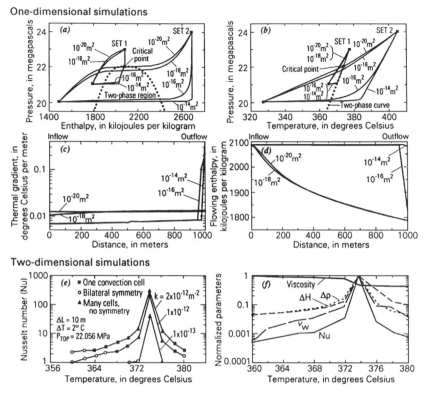

Figure 7.4 Results of one-dimensional simulations of pressure-driven flow near the critical point of water showing (a) pressure–enthalpy and (b) pressure–temperature trajectories for two sets of experiments and (c) temperature gradient and (d) flowing enthalpy as a function of distance along column for experiment set 1. Results of two-dimensional simulations of buoyancy-driven flow near the critical point of water showing (e) *Nu* as a function of T_{ave} for selected values of slab permeability k and (f) normalized values of ΔH, $\Delta \rho$, q_w, *Nu*, and μ as a function of T_{ave} for $k = 10^{-12}$ m^2.

high-permeability experiments nearly intersect the critical point, and in set 2 the low-permeability runs do so.

At low permeabilities ($\leq 10^{-18}$ m^2), heat transport by conduction dominates, and the *P*–*H* and *P*–*T* trajectories (Figures 7.4a and b) define a nearly constant temperature gradient (Figure 7.4c). At higher permeabilities ($\geq 10^{-16}$ m^2), advection dominates, and the cooling trajectories reflect a nearly constant *flowing enthalpy* (Figure 7.4d), as defined by

$$\frac{(q_w\rho_w H_w + q_s\rho_s H_s)}{(q_w\rho_w + q_s\rho_s)}. \tag{7.6}$$

Norton and Knight's (1977) early quantitative analyses of flow near cooling intrusions predicted a transition from conduction- to advection-dominated transport over essentially the same permeability range. The steep temperature gradients near the outflow boundary in the high-permeability cases (Figure 7.4c) are similar to those predicted by the analytical solution to an analogous problem (Chapter 3.4.1, Eq. 3.19).

One-dimensional experiments in which constant *P–H* boundaries were held much closer to the critical point also show little evidence of enhanced transport near the critical point (Ingebritsen and Hayba, 1994). These various one-dimensional experiments with flow driven by a fixed pressure drop can be regarded as roughly analogous to nonconvecting hydrothermal systems driven mainly by gravity (topography). In such systems it would appear that near-critical phenomena are much less important than variations in rock properties.

7.3.2 *Two-dimensional convection*

The one-dimensional pressure-driven experiments obviously precluded *convection*. We can use a two-dimensional ($\Delta L \times \Delta L$) vertical slab to explore buoyancy-driven convection near the critical point of water. The upper and lower boundaries of the slab are impermeable and isothermal (at T_{top} and T_{bot}), and the lateral boundaries are impermeable and insulated. Thermal conductivity is held constant at 2.0 W/(m-K), and the permeability of the slab is varied. The initial pressure at the top of the slab, P_{top}, is not held fixed, but the initial fluid density distribution is carefully prescribed so that P_{top} does not vary greatly during the course of a simulation. Because the boundaries of the slab are effectively impermeable, the initial mean fluid density within the slab ($\rho_w(P, H)$) has to be very close to the final "steady-state" value. Otherwise, a modest change in the mean value of $\rho_w(P, H)$ relative to the initial conditions would result in a significant pressure change throughout the slab. An iterative approach can be used to identify satisfactory initial conditions.

Results of the simulations are posed in terms of the Nusselt number, *Nu* (Chapter 3.5.1), versus the average temperature of the slab (T_{ave}, or $[T_{top} + T_{bot}]/2$). In this case the Nusselt number is determined as $q_{tot}/(K_m\Delta T/\Delta L)$, where q_{tot} is the horizontally averaged upward heat flux (conductive + advective) and ΔT is ($T_{bot} - T_{top}$).

The two-dimensional experiments demonstrate *superconvection* driven by large near-critical density differences and result in heat transfer enhancement by factors of $>10^2$ (Figure 7.4e). Maximum *Nu* increases with permeability, and the complexity of convective flow increases with increasing *Nu*: There is a systematic variation from unicellular convection at *Nu* < 2.5, to bilaterally symmetric cells at *Nu* ~ 2.5–5, to numerous, smaller cells at *Nu* > 5. *Nu* ~ 5 corresponds to a Rayleigh number (Chapter 3.5.3) of about 500, close to the theoretical value for transition to unsteady flow. Similar geometries have shown transitions to chaotic flow beginning at $Ra \sim 400$ (Kimura and others, 1986).

The near-critical heat transfer enhancements result from dramatic increases in the gradients in fluid enthalpy ($H_{bot} - H_{top}$, or ΔH) and density ($\rho_{max} - \rho_{min}$, or $\Delta \rho$) in the slab that occur as T_{ave} approaches the critical temperature (Figure 7.4f). The maximum volumetric fluid flow rate (q_w) in the slab is highly correlated with $\Delta \rho$, and Nu is highly correlated with the product of the mass flux and the enthalpy gradient ($q_w \rho_w \Delta H$). Nu is also proportional to the Ra value calculated using fluid properties at P_{top} and T_{ave}. The general correlation $Nu = 0.218 Ra^{0.5}$ (Combarnous and Bories, 1975) fits these results fairly well, but a closer correlation is obtained by an empirical fit to the HYDROTHERM results, which gives $Nu = 0.131 Ra^{0.521}$.

Other simulations (Ingebritsen and Hayba, 1994) indicate that sub-critical two-phase processes afford equally viable heat-transfer mechanisms. This makes sense on an intuitive level, because ΔH (in this case $H_s - H_w$) and $\Delta \rho$ ($\rho_{max} - \rho_{min}$) are both larger under two-phase conditions than they can be at or above the critical point itself.

The permeabilities required for superconvection in a 10-m $\times$ 10-m slab with $\Delta T = 2°C$ ($\sim 10^{-13}$ m^2; Figure 7.4e) are higher than those believed typical of near-magma environments (Chapter 7.2.4). Simulations with dimensions and permeabilities more representative of a magmatic–hydrothermal system ($\Delta L = 1$ km, $\Delta T = 100°C$, $P_{top} = 22.056$ MPa, $k = 10^{-16}$–10^{-15} m^2) led to $Nu \leq 6$. At temperatures near the critical point, the transition from brittle to ductile rheology (Fournier, 1991) and precipitation of silica (Fournier, 1983) would both act to eliminate permeability. Thus superconvection most likely occurs where strain rates are sufficient to maintain high permeability despite competing processes. This might be the case, for example, within active high-angle fault zones in volcanic environments.

7.4 Multiphase processes

Two-phase (steam–liquid water) flow is fairly common in shallow, high-temperature hydrothermal systems. Because the *critical pressure* for pure water is 22.06 MPa (Chapter 3.1.1), we would generally not expect two-phase conditions much deeper than 2.5 km (at 2.5-km depth, the hydrostatic pressure $\rho_w g h \sim 900$ kg/m^3 $\times$ 9.81 m/s^2 $\times$ 2,500 m $= 22.07$ MPa $= 220.7$ bars). However, in saline systems, phase separation can occur at pressures above the pure-water critical pressure, producing a relatively dilute vapor phase and a concentrated brine (e.g., Bischoff and Pitzer, 1989). The occurrence of heat pipes with *vaporstatic* pressure gradients can also extend the depth limit of two-phase conditions. At The Geysers, California, for example, pressures at a depth of 3 km are as low as 4.0 MPa. (See Chapter 3.7 for a discussion of heat pipes.)

At any depth, the presence of two phases will allow large and small-scale phase separation, with associated geochemical effects, and permits the occurrence of *vapor-dominated zones*. Two-phase mixtures also buffer pressure transmission and limit temperatures to the *boiling point–depth curve*. In this section we will discuss each of these points in turn.

7.4.1 Phase separation

Phase separation is a result of the density difference between steam and liquid water, which can cause the net forces acting on the two fluids to differ in direction as well as magnitude. In general, some degree of phase separation will occur if permeable zones exist that allow both vertical and horizontal movement of fluids and provide discharge outlets at different elevations. If we momentarily ignore the fact that hydrothermal circulation is not a *potential flow* (Chapter 3.1.3), then by assuming ρ_f is a function of pressure only we can use Hubbert's (1953) concept of *impelling force* to illustrate this point graphically. Neglecting capillary effects, we have

$$\phi_w = \int_{P_o}^{P} \frac{dP}{\rho_w(P)} + gz, \qquad \phi_s = \int_{P_o}^{P} \frac{dP}{\rho_s(P)} + gz,$$

$$E_w = -\text{grad } \phi_w \qquad\qquad E_s = -\text{grad } \phi_s \qquad (7.7)$$

$$= \mathbf{g} - \left(\frac{1}{\rho_{wo}}\right)\text{grad } P, \qquad = \mathbf{g} - \left(\frac{1}{\rho_{so}}\right)\text{grad } P,$$

where ϕ is potential per unit mass, E is the impelling force, $\mathbf{g}$ is gravitational acceleration, and ρ_{wo} and ρ_{so} are liquid and steam densities at a point of reference. Thus the impelling force is a vector quantity that defines the direction in which an element of fluid will tend to migrate.

Figure 7.5a depicts a high-temperature hydrothermal system in which phase separation takes place, and Figure 7.5b shows the impelling forces acting on elements of steam and liquid water within the zone of two-phase upflow. In Figure 7.5b the vectors defining $(1/\rho_{wo})$grad P and $\mathbf{g}$ sum to define E_w and the vectors defining $(1/\rho_{so})$grad P and $\mathbf{g}$ sum to define E_s. The topographic relief causes a lateral component to the pressure gradient which, along with the difference between ρ_{wo} and ρ_{so}, causes the impelling forces E_w and E_s to diverge. Liquid water tends to move laterally from point (S) in Figure 7.5a, and steam tends to rise.

Geothermal processes

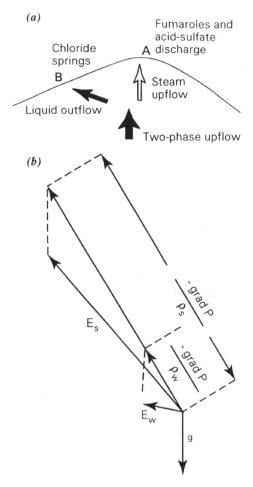

Figure 7.5 (a) Schematic diagram of a high-temperature hydrothermal system in which *phase separation* takes place; (b) impelling forces acting on elements of steam and liquid water in the area of phase separation (S). After Ingebritsen and Sorey (1985).

In mountainous terrain (e.g., at Lassen, California), this type of phase separation can occur on a scale such that liquid water discharges 10 km or more away from the main upflow zone. The phase separation process takes place on a smaller scale at high-temperature systems in gentler terrain, such as Broadlands and Wairakei in New Zealand (Grant, 1979; Allis, 1981).

Hot springs fed by the steam phase are low in chloride, gassy, and generally acidic. In contrast, hot springs fed by the residual liquid phase are relatively high in chloride, gas depleted, and have a near-neutral pH. The difference in chemistry between the steam-fed *acid-sulfate springs* and the liquid-fed *high-chloride springs* is attributed to the relative volatility of common constituents of thermal waters (e.g., White and others, 1971). Chloride and most other major ions have low volatility in low-pressure steam, whereas CO_2, H_2S, and other volatile constituents fractionate strongly into the vapor phase.

7.4.2 Vapor-dominated zones

Some two-phase hydrothermal systems include *vapor-dominated zones* within which steam is the pressure-controlling phase. Vapor-dominated zones may be extensive areally (to tens of square kilometers) and vertically (to more than 3-km depth) as at The Geysers, California, or they may be very localized, confined to a few fractures or fracture zones. Ingebritsen and Sorey (1988) identified three idealized types of vapor-dominated zones. Their model 1 (Figure 7.6a) has no significant liquid throughflow and involves an extensive vapor-dominated zone that is generally underpressured with respect to local hydrostatic pressure. In contrast, the vapor-dominated zones in models 2 and 3 (Figures 7.6b and c) are both "parasitic" to vigorous underlying flows of boiling liquid that feed hot springs at lower elevations. The vapor-dominated zone in model 2 is a smaller version of the underpressured vapor-dominated zone in model 1, whereas the localized vapor-dominated zones in model 3 are at pressures above local hydrostatic. Although each model has unique characteristics, they share some common features, including phase separation at pressures significantly greater than atmospheric, a zone in which steam is by far the more mobile phase, and *fumaroles* and *acid-sulphate* springs that result from the phase separation occurring at depth.

A classic paper by White and others (1971) distinguished hydrothermal systems with a vapor-dominated component from the more common *liquid-dominated systems*. They clarified the two-phase nature of vapor-dominated zones, which had previously been mistaken for single-phase steam reservoirs, and recognized the analogy between the natural steam–liquid counterflow (Figures 7.6a and b) and the industrial machines known as heat pipes (Chapter 3.7), which allow large net heat flux with little or no net mass flux.

Basic requirements for underpressured vapor-dominated zones such as that in model 1 (Figure 7.6a) include an intense, local heat source, restricted mass recharge, and an appropriate permeability structure. To exist, the underpressured vapor-dominated zone must be surrounded by low-permeability barriers that shield it from the normally pressured systems that overlie and surround it. Relatively high permeabilities, on the order of 10^{-15} m^2 or more, are required within the vapor-dominated zone itself, in order for the heat pipe process to be effective (Straus and Schubert, 1981). The permeability contrast at the boundaries of the vapor-dominated zone (Figure 7.6a) might be related to deposition of silica, calcite, or gypsum, as discussed by White and others (1971); to

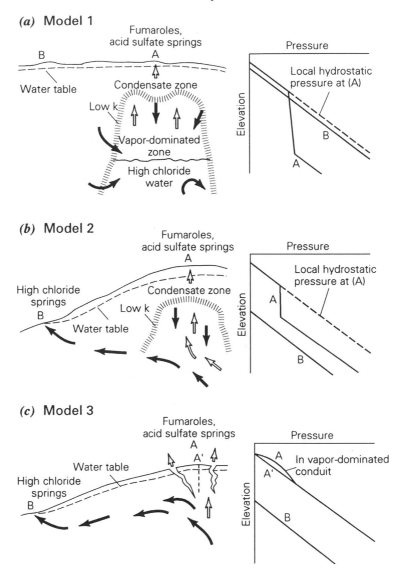

Figure 7.6 Conceptual models of three hydrothermal systems that include *vapor-dominated zones*. In (a) the vapor-dominated zone is large and there is limited liquid throughflow. In (b) and (c), liquid-dominated lateral flow links acid-sulfate springs at higher elevations with relatively high-chloride springs at lower elevations. From Ingebritsen and Sorey (1988).

argillization, to geologic structure and lithologic contrasts, or to a combination of these factors. At The Geysers, argillic alteration apparently helps to seal the top of the vapor-dominated zone (Hebein, 1985), which may be bounded laterally, at least in part, by mineralization along major fault zones (D. E. White, U.S. Geological Survey, written communication, 1986). At Larderello, Italy, vapor-dominated zones occur in carbonate rocks that are laterally and vertically isolated by low-permeability shales and sandstones (Cappetti and others, 1985). At systems such as The Geysers, two-phase conditions may persist to depths significantly greater than the 2.5-km limit suggested previously; because of the large thickness of the vaporstatic zone, subcritical fluid pressures (<22.06 MPa or 220.6 bars) may exist at unusually large depths.

Systems that involve large-scale phase separation and are similar to models 2 and 3 (Figures 7.6b and c) are relatively common in volcanic arcs. Gravity-driven flow is an essential component of these systems, whereas in model, 1 fluid circulation is controlled largely by density differences. Models 2 and 3 differ in terms of the nature and extent of vapor-dominated conditions, but that distinction may not be recognizable without drilling, as their surficial expression is very similar: Both include *fumaroles* and acid-sulfate springs at high elevations and relatively high-chloride springs at lower elevations.

In model 2 (Figure 7.6b), like model 1, phase separation occurs at pressures well below local hydrostatic, the pressure gradient within the vapor-dominated zone is near-vaporstatic, and a low-permeability aureole surrounds the vapor-dominated zone. Although pressures at depth below the acid-sulfate springs in model 2 are generally less than local hydrostatic, they are everywhere in excess of pressures at similar elevations beneath the high-chloride springs (Figure 7.6b), and that pressure difference drives the liquid-dominated lateral flow. The maximum thickness of the vapor-dominated zone is roughly constrained by the elevation difference between the acid-sulfate and high-chloride springs.

In model 3 (Figure 7.6c), phase separation takes place at pressures close to local hydrostatic and there is no requirement of a low-permeability aureole. The overall pressure gradient in the vapor-dominated conduits must be near hydrostatic, at pressures that are somewhat greater than those in the surrounding liquid-saturated medium.

Ingebritsen and Sorey (1988) found several conditions to be necessary for the evolution of model 2–like systems, including a period of liquid-dominated upflow and heating in the region that would eventually host the vapor-dominated zone, followed by some change in hydrogeologic conditions that caused liquid to drain from that region. In contrast, model 3–like systems required no changes in rock properties and boundary conditions, given an appropriate permeability structure and circumstances that would allow a large rate of steam upflow.

Although some natural systems can be roughly correlated with models 1–3 (Figure 7.6), most are significantly more complex, and many would be better represented as a combination of these models. For example, vapor-dominated conditions like those in model 3 are probably found locally in models 1 and 2, within vapor-dominated conduits that pass through their low-permeability aureoles.

7.4.3 Pressure transmission

In steam–liquid water systems, most changes in fluid volume are accommodated by boiling or condensation, and the *effective compressibility* of a two-phase mixture is about 30 times larger than that of pure steam at the same temperature and 10^4 times larger than that of liquid water at the same temperature. Grant and Sorey (1979) derived an empirical expression for effective compressibility of a steam–liquid water mixture that is accurate for pressures between 4 and 120 bars:

$$\beta_f = \left[\frac{(\rho_m c_m)}{n} \right] [1.92 \times 10^{-6} P^{-1.66}], \qquad (7.8)$$

where $(\rho_m c_m)$ is the volumetric heat capacity of the wetted rock given by $[(1 - n)\rho_r c_r + n S \rho_w c_w]$, c_w is defined as specific heat along the *saturation curve* (Figure 3.1a), and S is volumetric liquid saturation. The value of c is approximated by the *isobaric specific heat* in the case of both liquid water and rock and is assumed to be negligible in the case of steam. The values of the empirical constants apply for β_f in inverse bars, ρ in kg/m^3, c in J/(kg-K), and P in bars. At 250°C, and for values of $n = 0.10$, $\rho_r = 2{,}000$ kg/m^3, and $c_r = 1{,}000$ J/(kg-K), Eq. (7.8) gives $\beta_f = 0.9$/bar. Under the same conditions the compressibilities of pure steam and liquid water are only 0.03/bar and 1.3×10^{-4}/bar, respectively.

Fluid compressibility is one of the parameters that controls pressure transmission through a porous medium. For example, in a homogeneous medium the distance L over which "significant" pressure changes can propagate in time t is

$$L = (t D)^{1/2} \quad \text{for radial flow}$$

and

$$L = 2(t D)^{1/2} \quad \text{for linear flow}, \qquad (7.9)$$

where $D = k/[n\mu_f(\beta_f + \beta_r)]$ is the *hydraulic diffusivity*. These relationships define the time t at which the pressure change at L will be 1/10 of the pressure change at the pressure source or sink ($L = 0$). They can be derived from

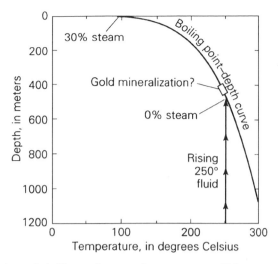

Figure 7.7 Hydrostatic boiling-point curve for pure water at Yellowstone National Park, Wyoming (surface boiling temperature 92.9°C), and hypothetical trajectory of a packet of rising fluid. Boiling-point curve from D. E. White (U.S. Geological Survey, written communication, 1973).

the appropriate line-source solutions (Carslaw and Jaeger, 1959). The potential for 10^4-fold variation in β_f between fully and partly saturated states clearly makes it a potentially controlling parameter. For example, any analysis of fluid-pressure response to magmatic intrusion (e.g., Elsworth and Voight, 1992) or geothermal-reservoir development (e.g., Ingebritsen and Sorey, 1985) would be critically dependent upon assumed values of β_f.

7.4.4 Boiling point–depth curves

Where liquid water is present, maximum temperatures at any depth are limited by the *boiling point–depth curve* (Figure 7.7). This curve is really a version of the pressure–temperature phase diagram for water (Figure 3.1a), with hydrostatic depth ($\rho_w g d$) being a surrogate for pressure. In areas of high-temperature hydrothermal discharge a packet of fluid may rise nearly isothermally until it intercepts the boiling-point curve at some depth. Above that depth, continuous boiling will cause it to follow the boiling-point curve to the surface.

The mass fraction of steam generated by adiabatic decompression and cooling along the boiling-point curve is given by

$$x = \frac{[H_w(T_1) - H_w(T_2)]}{[H_s(T_2) - H_w(T_2)]}. \tag{7.10}$$

The heat released by cooling from T_1 to T_2, $[H_w(T_1) - H_w(T_2)]$, supplies latent heat of vaporization, $[H_s - H_w]$. Values of $H(T)$ can be obtained from steam-table references such as Haar and others (1984). For typical hydrothermal discharge values of $T_1 = 250°C$ and $T_2 = 100°C$, the mass fraction $x = 0.30$. Because the *latent heat of vaporization* of water is very large, liquid rising from depth cannot be converted entirely to steam through an *adiabatic* process. In this case the rising fluid can generate only 30% steam by decompressing and cooling to the surface boiling point. Even a near-critical initial liquid temperature of 350°C will only generate about 55% steam. (See Figure 3.1c and Problem 3.1.)

The presence of salts and noncondensible gases in solution will change the shape of the boiling-point curve, and pressures in hydrothermal discharge areas may be significantly in excess of hydrostatic. Nevertheless, boiling-point phenomena make temperature–depth relations in discharge areas somewhat predictable. This fact is sometimes useful in mineral exploration, because current theories of epithermal ore deposits suggest that significant gold deposition may occur at the depth of initial boiling. Gold can be transported in bisulfide complexes, and H_2S tends to leave the system along with the first-generated steam. Thus one exploration strategy is to assume that gold might be found at the depth of initial boiling below "fossil" hot-spring silica deposits. That depth can be estimated on the basis of the inferred temperature and composition of paleofluids and the appropriate boiling-point curve.

A seminal and still frequently referenced paper by Cathles (1977) contained a misleading statement about boiling phenomena (p. 806–807): " ...liquid and vapor do not coexist over an appreciable volume [because] streamline P–T curves cross the [boiling-point] curve at high angles...". In fact, regardless of the angle of incidence, once the boiling curve is encountered the P–T trajectory cannot cross it until the entire latent-heat barrier is overcome. Thus, in nature, extensive two-phase zones are found in both vapor-dominated zones and hydrothermal discharge areas.

7.5 Hot springs

The existence of *hot springs* requires heterogeneous permeability. In a homogeneous medium with sufficiently low permeability, groundwater temperature at any depth will be close to that predicted from the regional temperature gradient, and springs will discharge at near-ambient temperatures. In a homogeneous medium with sufficiently high permeability, near-surface heat flow will be depressed in recharge areas and elevated in discharge areas, but spring discharge temperatures will generally not be elevated to hot spring–like values.

Forster and Smith (1989) found that they could simulate hot springs within topographically driven flow systems by invoking steeply dipping, relatively high-permeability conduits beneath topographic lows (Figures 7.8a and b). Their highest spring discharge temperatures were achieved by assigning a permeability of about 10^{-16} m^2 to the rock matrix surrounding a high-permeability conduit (Figure 7.8c). For lower "matrix" permeabilities, insufficient fluid was delivered to the high-permeability conduit. For higher matrix permeabilities, temperatures in the entire flow system were significantly depressed.

There is a strong parallel between Forster and Smith's (1989) result and the observation that many natural hot springs are found near the surface expressions of normal faults, which presumably act to facilitate deep, local circulation of meteoric water and/or interrupt the lateral flow of heated groundwater at depth. Normal faults may offset permeable horizons at depth, and they also tend to be associated with topographic lows that are the natural discharge areas for topographically driven groundwater flow systems.

In nonmagmatic areas, one can estimate a minimum depth of circulation for hot-spring waters on the basis of their discharge temperature and the regional temperature gradient. For example, waters from a 90°C spring in an area with a mean air temperature of 10°C and a geothermal gradient of 40°C/km must have circulated to a minimum depth of about 2 kilometers. Hot-spring discharge temperatures obviously cannot exceed the local boiling point, but in many cases, *chemical geothermometers* such as silica concentration (see, e.g., Fournier, 1981) indicate water–rock equilibration at much higher temperatures. In such cases the geothermometer temperatures can be used to estimate the depth of circulation.

Many thermal waters lack atomic bomb–related tritium (H^3) (Chapter 4.2.1), and therefore they must have risen through zones of young, relatively tritium-rich groundwater without mixing. The prevalence of unmixed hot-spring waters might be explained in terms of a slight excess pressure in the thermal upflow, relative to local hydrostatic pressure. Under such conditions, thermal water could flow out of an upflow conduit, but shallow, nonthermal water could not flow in to mix with the thermal upflow.

7.6 Geysers

Geysers are periodically discharging hot springs or fountains driven by steam or (less commonly) noncondensible gas. They are rare; perhaps 400–900 exist worldwide, of which approximately 200–500 occur in the great geyser basins of Yellowstone National Park (Rinehart, 1980; Bryan, 1995). For comparison, the number of steadily flowing hot springs worldwide is probably on the order of 10^5. Geyser eruptions typically pass through several stages (White,

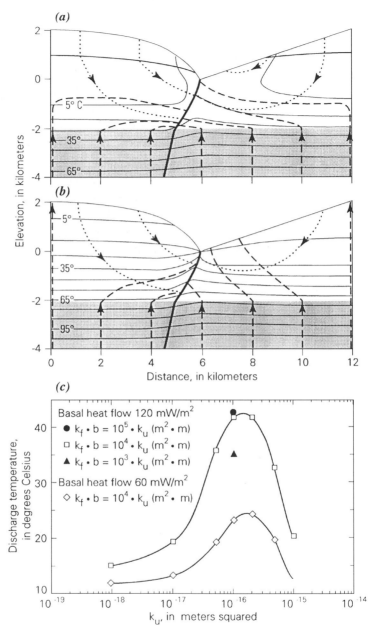

Figure 7.8 Topographically driven flow and heat transport showing the influence of a steeply dipping high-permeability conduit; (a) and (b) show temperatures (solid lines), heat-flow lines (dashed lines and arrows), and fluid flowlines (dotted lines and arrows). The upper, unshaded region has a permeability k_u of 10^{-15} m^2 in (a) and 10^{-16} m^2 in (b). In each case the lower, shaded region is essentially impermeable and the high-permeability conduit has a transmissivity $k_f b$ of $10^4 \times k_u$. (c) Temperature of fluid discharged from the high-permeability conduit for a range of values of k_u and $k_f b$ and two different values of basal heat flow. After Forster and Smith (1989).

1967): (1) initial overflow of liquid water at temperatures less than or equal to the local boiling point; (2) fountaining, liquid-dominated discharge; and (3) steam-dominated discharge of progressively decreasing intensity. Discharge then ceases for an interval.

Most geysers occur in areas where the water table is near the land surface, subsurface temperatures are at or near the boiling point to significant depths, and vertical pressure gradients are somewhat in excess of hydrostatic. Such conditions are most commonly found in the discharge areas of major hydrothermal systems. Maximum measured temperatures in drilled geyser areas generally exceed 170°C (White, 1967), and chemical geothermometry suggests that the water disharged from Old Faithful, Yellowstone, has boiled adiabatically from a temperature of about 205°C (Fournier, 1969). At Yellowstone (surface boiling temperature ∼93°C), this temperature corresponds to a hydrostatic depth of about 190 m (Figure 7.7). However, research drilling in the thermal areas of Yellowstone has revealed vertical pressure gradients consistently in excess of 110% of boiling-point hydrostatic and ranging as high as 147% of hydrostatic (White and others, 1975).

An interesting characteristic of geysers is their apparent sensitivity to small strains in the Earth. Fairly compelling data document geyser responses to co-seismic strains on the order of 1 microstrain (e.g., Silver and Vallette-Silver, 1992), barometric strains of about 0.1 microstrain (White, 1967), and perhaps diurnal tidal strains of about 0.01 microstrain (Rinehart, 1972). More speculative correlations have been suggested between geyser activity and preseismic strains of perhaps 10 nanostrain (Silver and Vallette-Silver, 1992) and long-period tidal strains of about 1 nanostrain (Rinehart, 1972; see White and Marler, 1972 for a rebuttal). The strain sensitivity is generally documented in terms of eruption frequency, one of the easiest geyser characteristics to measure. Ingebritsen and Rojstaczer (1993) showed that some of the responsiveness to strains ≥0.1 microstrain might be explained in terms of changes in the hydraulic properties of the geyser conduit and the surrounding rock matrix, or in terms of surface loading-induced water-level differences between the geyser conduit and the (presumably) less-compliant matrix.

Geyser behavior can be mimicked fairly effectively using the governing equations for groundwater flow and heat transport written as Eqs. (3.15) and (3.16), despite the implicit assumptions of Darcian flow and local rock–water thermal equilibrium (Chapter 3.1.7). Here we will briefly describe the results of some HYDROTHERM simulations involving a 200-m-deep, high-permeability conduit embedded in a less-permeable two-dimensional matrix (Figure 7.9a). The upper boundary was maintained at a pressure of 1 bar and an enthalpy corresponding to a temperature of 100°C. The lateral boundaries were maintained at hydrostatic pressures and boiling-point enthalpies. The conduit is hydraulically

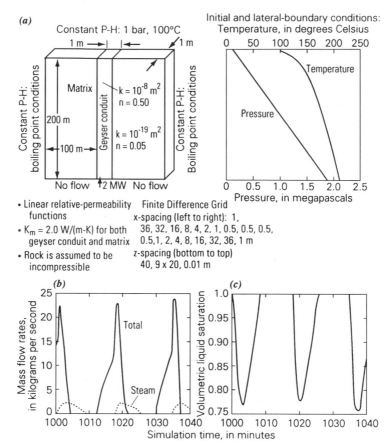

Figure 7.9 (a) Two-dimensional model used in numerical simulations of *geysering*. Note that although there is free exchange of heat between geyser conduit and matrix, the permeability of the matrix is so small that there is negligible exchange of mass. The linear relative-permeability functions (Figure 3.2) are $k_{rw} = (S_w - 0.3)/0.7$ and $k_{rs} = 1 - k_{rw}$, where k_{rw} and k_{rs} are relative permeabilities to liquid water and steam, respectively, and S_w is volumetric liquid saturation. (b) Mass flow rates and (c) volumetric liquid saturation at 10-m depth for geyser cycles obtained using the parameters in (a).

isolated from the surrounding matrix and is supplied with a constant flux of heat from below (see the parameter values in Figure 7.9a). These boundary conditions are most appropriate for geysers that discharge through near-constant-level pools and are fueled by steam rising from depth. Imposing the pure-heat condition at the lower boundary makes the upper boundary the only source of mass, so that if the heat flux is sufficient to generate a mobile steam phase (that is, greater than about 1 W/m²), there are only two types of solutions that can conserve mass and energy. One is a heat-pipe solution involving counterflow of liquid and steam, with constant flow rates such that the latent heat difference between the phases is just sufficient to accommodate the basal heat flux (Chapter 3.7). The other possibility is a periodic discharge, with the direction of the mass flux at the upper boundary reversing at (semi-)regular intervals.

For particular combinations of conduit permeability and basal heat input the behavior of this system is quite geyserlike. In Figure 7.9, results are posed in terms of total mass and steam flow rates (Figure 7.9b) and volumetric liquid saturation (Figure 7.9c). These quantities are evaluated at 10-m depth in the fracture zone, the depth to the center of the uppermost active finite-difference block (Figure 7.9a). The discharge is periodic, with a fairly steady period of 17–18 minutes developing after a few cycles (Figure 7.9b). At the beginning of each eruption there is a period of liquid-only discharge; at mid-eruption, there is two-phase flow; and, finally, there is a period of steam-only flow that persists as long as there is a steam phase at 10-m depth (Figure 7.9c). Thus the eruption stages mimic those observed at natural geysers. The simulated mid-eruption flow is only about 3% steam by mass but is about 98% steam by volume, because there is an 800-fold density difference between the two phases at ~2 bars (10-m hydrostatic depth).

7.7 Geothermal resources

Given the global-average temperature gradient of about 25°C/km (Chapters 4.3.1, 7.1.3), one can calculate that the heat stored in the upper few kilometers of the Earth's crust would be sufficient to supply the world's consumption of energy indefinitely. But in general, such calculations have little practical relevance because, given current technology and economic conditions, successful exploitation of geothermal energy requires that it be concentrated well above "background" levels. Essentially, all exploitation of geothermal resources to date has occurred in the discharge areas of moderate- to high-temperature hydrothermal systems, where energy is naturally concentrated.

Although many space-heating applications involve moderate fluid temperatures, conventional electrical-power generation requires high temperatures ($\geq$150°C). In order to generate electrical power, the steam fraction of a two-phase mixture produced from drillholes is separated and used to operate a low-pressure turbine. For environmental reasons and to maintain reservoir pressures, both the separated liquid and steam condensate are subsequently injected in other drillholes. *Binary* systems use *heat exchangers* and a second fluid with a lower boiling point to lower the practical temperature limit for electrical-power generation to about 100°C.

In terms of emissions, geothermal electric power is a much more benign technology than fossil-fuel power. Emissions of CO_2 and S compounds per unit of power produced are as much as an order of magnitude lower at geothermal plants than at oil- or coal-fueled plants (Armannsson and Kristmannsdottir, 1992). The most problematic environmental impact of geothermal development is often the effect on naturally occurring thermal features, which have their own intrinsic value. The pressure declines associated with fluid production tend to dry up the hot springs associated with liquid-dominated systems, although such effects can sometimes be mitigated by fluid-injection schemes. The same pressure declines can cause large, transient increases in steam discharge from vapor-dominated zones, due to increases in steam saturation and mobility (Allis, 1981; Ingebritsen and Sorey, 1985).

The largest existing complex of geothermal power plants exploits the large vapor-dominated system at The Geysers in northern California. Most wells tapping vapor-dominated zones (Chapter 7.4.2) produce superheated steam only, which makes power generation and fluid disposal unusually easy. At The Geysers, installed electrical capacity as of 1990 was nearly 2,000 MW_e but, because of overproduction and the resulting declines in reservoir pressure, actual power output was about 400 MW_e less. Total installed geothermal capacity in the United States as of 1994 was 2,979 MW_e (e.g., Fridleifsson and Freeston, 1994), sufficient to supply about three million people (slightly over 1% of the population) at the approximate U.S. demand rate of 1 MW_e per 1,000 people. Although the United States remains the world's largest geothermal producer, several countries with lower per-capita energy demands obtain much larger proportions of their total electricity from geothermal plants. As of 1994, installed geothermal electric power totaled 6,300 MW_e in 21 countries, with an annual growth rate of about 4%, and direct use of geothermal energy totaled 11,400 MW_t in 40 countries, with an annual growth rate of about 10% (Fridleifsson and Freeston, 1994). (Here we are using the subscripts "e" and "t" to distinguish megawatts of electrical power from megawatts of thermal energy. Both units are J/s $\times 10^6$, but MW_t do not correlate directly with MW_e, as substantial energy is lost in the conversion even under the most favorable circumstances. Elsewhere in the text, MW denotes MW_t.)

7.8 Ore deposits

Although commercial extraction of heat from active hydrothermal systems has been growing steadily over the past few decades, extraction of minerals from fossil hydrothermal systems continues to have a much larger economic significance and provides a major practical impetus for research on hydrothermal systems. Many ore deposits are localized in vein networks that once hosted hydrothermal-fluid circulation. The most important of these *hydrothermal ore deposits* involve silver and gold and the sulfides of copper, tin, lead, zinc, and mercury.

In most particular cases the sources of the fluids and solutes that formed a hydrothermal ore deposit are incompletely known. Oxygen- and hydrogen-isotope data indicate that meteoric waters were present in most magmatic–hydrothermal environments. Oxygen isotope exchange between meteoric waters and igneous rocks results in low-^{18}O rocks, and such rocks appear to be rather ubiquitous in areas of hydrothermal mineralization (see Chapter 10.5.1 and, e.g., the overviews by Taylor, 1979, and Criss and others, 1991). There is also abundant chemical and isotopic evidence for contribution of magmatic fluids and solutes to ore-forming hydrothermal systems (e.g., Hedenquist, 1992). It seems likely that, in low-permeability host rocks, a magmatic component might dominate an early "expulsive" stage of fluid flow (Chapter 7.2.3) and that the magmatic signature might be overprinted by that of meteoric fluids as the heat source wanes. Lead isotope studies have proved to be particularly helpful in determining the source of metals for some deposits, since the isotopic composition of lead released by magma is distinguishable from that of lead scavenged from various host rocks (e.g., Doe and Zartman, 1979).

Regardless of the source of fluids and solutes, many types of economic ore deposits seem to require relatively saline fluids; *epithermal* deposits are a notable exception, with typical fluid salinities less than 2 weight percent. The solubility of most ore minerals is quite low but can be greatly enhanced by the formation of complex ions in relatively chloride- or sulfur-rich fluids. Ore deposits occur where metals in solution are induced to precipitate by (1) temperature changes, (2) pressure changes, (3) reactions between the hydrothermal fluids and adjacent wall rock, or (4) the mixing of solutions with different compositions (e.g., Skinner, 1979). Many *copper-porphry* deposits occur in near-magma environments, where reactive magmatic fluids encounter large gradients in both temperature and chemical composition (*pyritic shell* in Figure 7.10). Many *epithermal gold* deposits seem to be localized by boiling, which has the effect of reducing the sulfur concentration in the residual fluid; this can be regarded as an instance of precipitation due to pressure change (*polymetallic veins* in Figure 7.10; see also Chapter 7.4.4). There are few well-documented examples of precipitation due to mixing, but mixing would seem most likely to occur along the laterally flowing limbs of gravity- or density-driven hydrothermal systems (e.g., at Creede, Colorado: Plumlee, 1989; Hayba, 1993).

Chapter 5 was devoted to the important subset of ore deposits formed in sedimentary basins with no associated igneous rocks. The particular emphasis

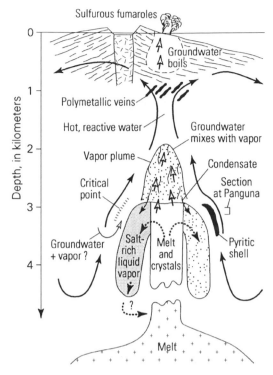

Figure 7.10 Model of ore genesis developed by Eastoe (1982) on the basis of the Panguna deposits, Papua New Guinea. From Sawkins (1990).

there was Mississippi Valley–type (MVT) stratabound lead–zinc deposits and sediment-hosted uranium deposits.

7.9 Subsea hydrothermal systems

In the past century, perhaps only the plate tectonic revolution has had a greater impact on our understanding of Earth processes than the discovery and exploration of *subsea hydrothermal systems*. In fact, these systems exist as a consequence of plate tectonics, discharging mainly on or near the *mid-ocean ridge* spreading centers and near the toes of the *accretionary prisms* of sediment that overlie subduction zones. Hydrothermal activity near the mid-ocean ridges is critically important to both the Earth's thermal budget and to global geochemical cycles. Furthermore, the discovery of associated ecosystems based on chemosynthetic bacteria has greatly affected thinking in the biologic sciences (see, e.g., the review by Lutz and Kennish, 1993). Although the scope of this book is generally limited to continental processes, we will consider subsea hydrothermal systems briefly here, focusing on thermal aspects of the mid-ocean ridge (MOR) systems, their impact on ocean chemistry, and the quantitative description of MOR circulation. Much of this discussion owes to a recent review article by Lowell and others (1995). Accretionary prisms are briefly described in Chapter 4.1.5.

In the mid-1960s, hydrothermal brines and metalliferous sediments discovered in the Red Sea provided the first direct evidence of hydrothermal activity associated with an oceanic ridge. Subsequently, experiments reacting seawater and basalt at $200°C$ showed dramatic release of heavy metals and H_2S, consistent with the presence of massive sulfide deposits on the seafloor (Bischoff and Dickson, 1975). Low-temperature hydrothermal fluids were eventually sampled at the Galapagos spreading center in 1977, and their chemistry was used to predict the existence of a sulfurous, high-temperature ($\geq 350°C$) hydrothermal end member whose composition agreed with the earlier experimental studies (Edmond and others, 1979). In 1979, such high-temperature fluids were actually found discharging from vents dubbed *black smokers* at latitude $21°N$ on the East Pacific Rise, just south of the Gulf of California. Although 90% of the mid-ocean ridge system has still not been systematically explored, at least 25 other high-temperature vent fields have since been identified.

Chemical energy, rather than solar (photosynthetic) energy drives rich MOR hydrothermal ecosystems; reduced inorganic compounds (H_2S, S^0, $S_2O_3^{2-}$, NH_4^+, Fe^{2+}, NO_2^-, Mn^{2+}) from hydrothermal vents provide energy for chemosynthetic bacteria that in turn support populations of clams, mussels, and tubeworms and, indirectly, populations of predators and scavengers. Faunal biomass estimates for the MOR ecosystems exceed even those for productive estaurine ecosystems, and the multitude of new species discovered there have required designation of 90 new genera, 20 new families, and a single new phylum, Vesimentaria, for the giant tube worm (Lutz and Kennish, 1993). Many of these species are able to tolerate levels of dissolved metals and gases previously considered to be toxic.

7.9.1 Importance to the Earth's thermal budget

As noted in Chapter 7.1.2, age–heat flow relations such as those depicted in Figure 7.1a ($q_h \propto t^{-1/2}$, where t is the age of the crust) approximate heat flow in much of the oceanic crust quite well. However, they substantially overpredict heat flow in young, poorly sedimented oceanic crust. Pronounced heat flow deficits relative to the $q_h \propto t^{-1/2}$ model occur in poorly sedimented crust for crustal ages up to 65 Ma. A number of workers recognized that the discrepancy between theory and observation (as well as between observations in the well- and poorly sedimented areas) could be explained by advective heat loss via hydrothermal circulation in the younger, poorly sedimented areas. Examination of this hypothesis with various data sets and crustal-cooling models has consistently led to the result that hydrothermal circulation near the MORs accounts for 20–25% of the Earth's total heat loss (e.g., Williams and Von Herzen, 1974; Sclater and others, 1980; Stein and Stein, 1994). About 30% of the hydrothermal heat loss may occur in <1 Ma crust (Stein and Stein, 1994) or, given spreading rates of a few centimeters per year, within a few tens of kilometers of the MOR axis itself.

The hydrothermal fluids responsible for the total seafloor advective heat loss of $\sim 10^{13}$ W are predominantly seawater, with its original composition modified to a variable extent by water–rock interaction and, in the near-MOR environment, by *phase separation* (Chapter 7.4.1) and by addition of magmatic fluids. The volume of fluid flow required to support this heat loss can be calculated as $10^{13}/[c_w(T_2 - T_1)]$, where c_w is the heat capacity of the fluid and T_2 and T_1 are the hydrothermal outflow and inflow temperatures, respectively (see Wolery and Sleep, 1976). The appropriate value of T_1 is the deep ocean temperature of 3–4°C, and reasonable mean values for T_2 likely lie in the range of 20–200°C. (Lower-temperature, *diffuse* discharge accounts for more hydrothermal heat flux than the 350–400°C black-smoker discharge: Lowell and others, 1995.) Thus hydrothermal discharge rates are likely in the range of 1×10^7 to 2×10^8 liters/s, sufficient to cycle the entire volume of the oceans ($\sim 1.4 \times 10^{21}$ liters) in 0.2 to 4 Ma.

7.9.2 Influence on ocean chemistry

Such geologically rapid turnover rates imply that hydrothermal circulation likely has a profound influence on ocean chemistry. For comparison, the mean residence time of the dominant solute species sodium and chloride, calculated as the mass present in the ocean divided by the mass delivered by streams, is approximately 100 Ma (e.g., Drever, 1982). Historically, it was believed that the solute content of the oceans results from simple accumulation: Streams bring salt to the sea, where pure water evaporates, so that the present salt content represents the contribution of streams over the entire history of the Earth. Late 19th-century estimates of the age of the Earth made on this basis were fairly compatible with Lord Kelvin's estimates based on the cooling of an originally hot Earth (see Chapter 7.1 for Kelvin's approach). Since the recognition of the plate tectonic cycle and the more accurate establishment of the Earth's age through radiometric dating, ocean chemistry has generally been assumed to be in a (roughly) steady state, with the addition of solutes by streams balanced by subtraction through sedimentation, diagenesis, and subduction. However, the oceanic concentrations of certain elements still required large, incompletely identified sources or sinks to achieve a mass balance. Many of these solute-balance difficulties were eliminated by seawater–basalt interaction experiments (e.g., Bischoff and Dickson, 1975) and the nearly contemporaneous observations of the composition of hydrothermal fluids. In particular, it now seems nearly certain that much of the magnesium and sulfur delivered by streams is removed by interaction with moderately hot basalt. Furthermore, hydrothermal sources provide much of the calcium required to precipitate the excess bicarbonate delivered by streams. Without hydrothermal sources and sinks of solutes, the oceans might actually be dominantly sodium bicarbonate with a pH near 10, rather than dominantly sodium chloride with a pH of about 8 (MacKenzie and Garrels, 1966).

7.9.3 Quantitative description

The understanding of subsea hydrothermal systems is poorly developed relative to continental systems. This is partly due to the paucity of data and partly due to the lack of suitable numerical models for chemically reactive, high-temperature, multiphase, rapidly deforming systems.

For continental hydrothermal systems we often have detailed two-dimensional information from geological mapping, some significant knowledge of the third dimension from drillholes, relatively accurate measurements of near-surface mass and heat fluxes, and (less frequently) some knowledge of the time variation in the chemistry of and heat discharge from hydrothermal vents. Because of the difficulty of accessing and making measurements at seafloor systems, our knowledge of their space and time dimensions is spotty, and most of the available flux measurements have large error bars. Fundamental hydrogeologic parameters such as permeability (Chapter 1.2) have rarely been measured in a MOR environment and are usually estimated only by inference.

As noted in the preface, any complete quantitative description of hydrothermal circulation at the MOR must invoke a relatively complex, multiphase form of Darcy's law (Chapter 3.1.3), because gradients in salinity between vents indicate that there is often active *phase separation* (Chapter 7.4.1) between a relatively dense, saline brine and a less-saline vapor. In pure-water systems, phase separation can occur only below the critical point for pure water (Chapter 7.3; $T = 374°C$, $P = 22.06$ MPa), and the relatively low critical pressure for pure water would preclude boiling in most subsea environments. However, in saline systems the critical point is elevated (e.g., to $T \sim 400°C$, $P \sim 30$ MPa for seawater) and phase separation can occur above the critical

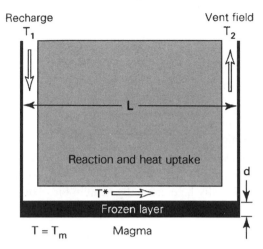

Figure 7.11 Single-pass model of a *MOR-crest hydrothermal system* above a crystallizing magma chamber. Heat is transferred from the magma chamber to circulating hydrothermal fluid at temperature T^* through a *conductive boundary layer* of thickness d. L is the lateral distance of flow across the top of the magma chamber and T_1 and T_2 are the hydrothermal recharge and discharge temperatures, respectively. From Lowell and Germanovich (1994).

point as well as below it, with *supercritical boiling* generating a highly saline brine phase and a low-salinity vapor phase (Bischoff and Rosenbauer, 1984; Bischoff and Pitzer, 1989). Large gradients in salinity and temperature (approximately 2–400°C) further dictate that any complete model of subsea hydrothermal systems must include both solute and heat transport. We must expect that the flow systems are highly transient; the exceptionally high rates of heat discharge from individual vents can only be explained as the result of rapid crystallization and cooling of large volumes of magma (Chapter 7.2.1), so that the intensity and spatial distribution of heat sources must vary with time. Precipitation and dissolution of minerals must also cause continuous variations in porosity and permeability, because the extreme variations in fluid composition and temperature make for an highly reactive chemical environment. Moreover, as permeability, flow rates, and temperatures wax and wane, near-vent rates of thermomechanical deformation are likely large enough to substantially affect permeability (Germanovich and Lowell, 1992). On a longer time scale, plate movement away from the MOR itself (another mode of deformation) will also influence the overall pattern of fluid circulation.

Despite these complexities, very simple quantitative models can sometimes result in useful estimates of the behavior of MOR systems. For example, Lowell and Germanovich (1994) used *single-pass* or *U-tube models* (Figure 7.11) to elucidate some aspects of the behavior of MOR-crest hydrothermal systems, including the magmatic heat supply (see Problem 7.6). However, as of this writing, existing numerical-simulation models are inadequate to fully describe the physiochemical behavior of MOR hydrothermal systems. Reactive-transport models like those described in Chapter 2.5 have not been developed to deal with high-temperature multiphase flow. High-temperature multiphase models such as the HYDROTHERM model (Chapter 3.1) do not include reactive transport or even salinity. Models developed to deal with deforming media (e.g., Chapter 4.1.5) generally do not include reactive transport or high-temperature multiphase flow.

Problems

7.1 Use a one-dimensional version of Fourier's law to construct a geotherm (draw a temperature–depth profile) for a 15-km-thick crust, assuming that the mean surface temperature is $10°C$, that the near-surface conductive heat flow q_{ho} is 80 mW/m^2, that the thermal conductivity K_m is described by Eq. (7.2) with $K_r(25) = 2.0$ W/(m-K), and that thermal conductivity and heat flow vary with depth according to Eqs. (7.2) and (7.3). (These values of crustal thickness, surface temperature, near-surface heat flow, and thermal conductivity are reasonably appropriate for the Basin and Range Province of the western United States.) Use a value of 2×10^{-6} W/m^3 for the near-surface *heat production A_o*, and assume that this value is representative of the entire crust (that is, $A(z) \sim A_o$). What is the predicted temperature at the base of the crust? Does this value seem reasonable? What is the reduced heat flow q_h^* at the base of the crust?

7.2 In the north-central Oregon part of the Cascade Range volcanic arc, heat discharge by hot springs amounts to about 1 MW per kilometer of arc length. Assuming that this value represents a long-term average and that the heat is entirely supplied by cooling magma bodies, calculate the magmatic intrusion rate required to support the heat discharge. Use particular values for magma density (2,500 kg/m^3), heat capacity (1.25 kJ/kg-K), and *latent heat of crystallization* (270 kJ/kg), and experiment with a reasonable range of initial and final magma temperatures.

7.3 Assume a homogeneous, isotropic, two-dimensional system, a horizontal pressure gradient of 1 bar/100 m, and a vertical pressure gradient defined by ρ_{wo}. Evaluate ρ_{wo} and ρ_{so} at $250°C$ and then use Eq. (7.7) to calculate the vertical angle between hypothetical steam- and liquid-flux vectors. By how much does the steam vector depart from the vertical? By how much does the liquid vector depart from the vertical?

7.4 A dike is intruded into a 250°C geothermal aquifer. Use Eqs. (7.8) and
(7.9) to estimate the timing of the resulting pressure pulse at a well 200-m
distant, assuming **(a)** a liquid-saturated ($S = 1.0$) aquifer and **(b)** a two-
phase ($S = 0.90$) aquifer, and assigning aquifer properties of $k = 10^{-13} \, \text{m}^2$,
$n = 0.10$, $\rho_r = 2{,}000 \, \text{kg/m}^3$, $c_r = 1{,}000 \, \text{J/(kg-K)}$, and $\beta_r = 1 \times 10^{-5}/\text{bar}$.
Assume in each case that pressure–temperature conditions in the aquifer
fall on the vaporization (saturation) curve.

7.5 Use Figure 7.7 to estimate the depth at which a 300°C upflow of liquid
water will first boil. Use Eq. (7.10) to determine the mass fraction of steam
that will be generated by *adiabatic decompression* of this fluid to surface
conditions.

7.6 Use the single-pass circulation model of Figure 7.11 to calculate some
characteristics of a *MOR hydrothermal vent system* that steadily discharges
350°C fluid at a rate of 10 kg/s. Assume that the width of an underlying
magma chamber perpendicular to the MOR axis is 1 km, that the vent
field is 1-km long parallel to the MOR axis, that $K_m = 2 \, \text{W/(m-K)}$, $T_1 =
4°C$, $T^* \sim T_2 = 350°C$, and $T_m = 1{,}200°C$, and that the crystallizing and
cooling magma supplies heat at a rate of $3 \times 10^{18} \, \text{J/km}^3$ (see Chapter
7.2.1). Calculate **(a)** the steady hydrothermal heat discharge in MW, **(b)**
the thickness d of the conductive boundary layer (see Chapter 7.2.2), and
(c) the rate of downward migration of the boundary layer (cf. Lister, 1974)
required to continually access high-temperature, newly crystallized rock.

8

Earthquakes

In this chapter we explore the relationship between hydrogeology and seismic activity. We begin by addressing the role of high pore-fluid pressure in fault movement, first defining *effective stress* and *Coulomb's law of failure* and then describing examples of *induced seismicity*, which clearly demonstrate the principle of effective stress. We then proceed to consider various possible fluid-pressure distributions at seismogenic depths, earthquake-related hydrologic phenomena, and the influence of seismicity on crustal permeability. The discussions of anomalous fluid pressures and hydraulic fracturing in Chapters 4.1.2 and 4.1.3 are also relevant here, as is the discussion of the *stress–heat flow paradox* of the San Andreas fault in Chapter 4.3.5.

8.1 Effective stress

The concept of effective stress was first proposed by the Czech-American civil engineer Karl Terzaghi (Terzaghi, 1925), who is credited with founding the discipline of soil mechanics. Effective stress can be understood in terms of the force balance on a horizontal plane in a one-dimensional fluid-saturated geologic medium (Figure 8.1). The downward force or *total stress* acting on the plane, σ_T, is due to the weight of the overlying medium (both solids and fluids). Part of this weight is borne by the interconnected solid matrix and is termed the *effective stress*, σ_e. The remainder is borne by the interconnected fluid phase and is equivalent to the fluid pressure, P_f, so that the force balance on the horizontal plane is

$$\sigma_T = \sigma_e + P_f; \qquad (8.1)$$

recall that both stress and pressure are defined as a force per unit area. Under conditions in which the total stress remains approximately constant, any

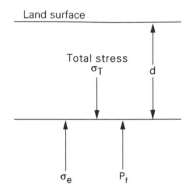

Figure 8.1 Total stress, effective stress, and fluid pressure on a horizontal plane located at
depth d below the land surface/water table in a one-dimensional, fluid-saturated geologic
medium.

variation in fluid pressure is balanced by a change in effective stress, so that

$$dP_f = -d\sigma_e. \tag{8.2}$$

Ignoring for the moment any tensile strength of the rock, we see that the approximate upper limit for fluid pressure is the total stress σ_T, or

$$P_f = \sigma_T; \tag{8.3}$$

if fluid pressure reaches this level, the fluid bears the entire weight of the overlying medium, the effective stress on the horizontal plane goes to zero, and
horizontally oriented partings will develop as the fluid "floats" the overlying
medium. In this one-dimensional context (Figure 8.1), then, the overburden
weight or *lithostatic load*,

$$P_f = \sigma_T = [\rho_r(1 - n) + \rho_f n]gd, \tag{8.4}$$

defines the criterion for *tensile failure* of the rock mass, where ρ_r and ρ_f
are average density values over the depth range d and n is the porosity. At
shallow to midcrustal levels the lithostatic pressure gradient is typically
$\sim$25 MPa/km depth (250 bars/km depth), whereas the "normal" hydrostatic
gradient is $\sim$10 MPa/km depth (100 bars/km depth).

In the real, multidimensional world, fluid-induced failure actually tends to
occur at fluid pressures that are substantially less than the total overburden
pressure (see Chapter 4.1.3). The prospects for failure in the upper crust are
much enhanced if fluid pressures reach even 1.25 times hydrostatic, or approximately 0.5 times lithostatic (Rojstaczer and Bredehoeft, 1988). Furthermore,
hydraulically induced fractures are often oriented (sub)vertically, rather than

horizontally. The preferred fracture orientation can be understood in terms of the normal state of stress in the Earth, which can be described in terms of three mutually orthogonal *principal stresses*, σ_1, σ_2, and σ_3. In general, one of the principal stresses will be nearly vertical, at least in regions with gentle topography. The other two principal stresses must thus be nearly horizontal. In a normal-faulting environment the *greatest principal stress*, σ_1, is vertical, and hydraulic fracturing will occur in (sub)vertical planes that are orthogonal to the *least principal stress*, σ_3 (Figure 4.2). In the (more common) reverse- or thrust-faulting environment, σ_1 is horizontal and (sub)horizontal fracturing will indeed tend to occur at fluid pressures close to the total overburden pressure (e.g., Hubbert and Willis, 1957). To avoid possible confusion in terminology, we should note that in rock mechanics the term *hydrostatic* is often used to denote the condition $\sigma_1 = \sigma_2 = \sigma_3$. Moreover, because differential pressures are needed to cause deformation, the "hydrostatic" component of stress, $(\sigma_1 = \sigma_2 = \sigma_3)/3$, is sometimes subtracted from the principal stresses to define the *deviatoric stresses*. Here, we will continue to use *hydrostatic* only to denote fluid-pressure conditions such that $P_f \sim \rho_w g d$.

The concept of effective stress is relatively intuitive in one dimension (Figure 8.1) but can readily be extended to three dimensions, because the fluid pressure at a particular depth or location acts equally on any arbitrarily oriented plane. The effective stresses in each of the principal-stress directions can thus be defined as $\sigma_1 - P_f$, $\sigma_2 - P_f$, and $\sigma_3 - P_f$. In two or three dimensions, stress is actually a tensor, like hydraulic conductivity or permeability (Chapter 1.2.2, Eq. 1.10), but we will assume that our coordinate axes are aligned with the principal stress directions (e.g., σ_{11}) so that the off-diagonal terms of the tensor (e.g., σ_{12}) become zero.

Some controversies have surrounded this rather simple concept of effective stress. By the 1970s it was generally accepted that an effective normal stress component would have the form $\sigma_i - A P_f$, where A is some coefficient, but there was vigorous debate regarding the correct value of A. The resolution of the conflict was to recognize two separate interpretations of the concept of effective stress. The first is that the effective stress governs the values of stress and pore pressure at which laboratory samples fail under compression in the *triaxial tests* discussed in the following section. For this purpose, and in general for the purposes of this book, empirical data show that A is indistinguishable from 1. The other interpretation is that the effective stresses are linear combinations of stress and pore pressure that can be described in terms of an effective stress tensor without explicitly including pore pressure in the equations. For this purpose, expressions for A are easily obtained and are not exactly equal to 1 (e.g., Nur and Byerlee, 1971). A comprehensive reference on the concept of effective stress is provided by Berryman (1992).

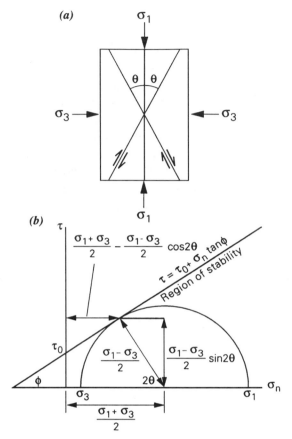

Figure 8.2 (a) Relation between principal stresses and fractures in a sample brought to failure in a *triaxial test* and (b) representation of the failure conditions on a *Mohr diagram*.

8.2 Coulomb's law of failure

Hydraulic fracturing results from a pulling apart or *tensile failure* of the rock mass, but natural earthquakes are caused by sliding or *shear failure* along a plane. Many data on shear failure under compressive stress are available from the results of *triaxial tests*. (In the Earth below about 100- to 200-m depth all stresses are compressive.) In triaxial tests a piston apparatus is used to place a cylindrical volume of rock under a directed compressive stress that defines σ_1, while the value of σ_3 is controlled by the fluid pressure in a surrounding vessel. For any given value of σ_3, the value of σ_1 can be increased until the sample fails along a plane located at some angle θ to the axis of compression (Figure 8.2a). For many geologic media $\theta \sim 30°$.

For a uniform confining pressure σ_3 it is possible to plot the stress conditions at failure on a two-dimensional *Mohr diagram*, a graphic representation of the state of stress introduced by the German engineer Otto Mohr in 1882. The *Mohr circle* defines the normal (σ_n) and shear (τ) stresses on the failure plane itself, provided that σ_1, σ_3, and θ are known, as they are in the case of failure in triaxial tests (Figure 8.2b). A Mohr circle with radius $(\sigma_1 - \sigma_3)/2$ is centered on the abscissa at $(\sigma_1 + \sigma_3)/2$. A line is drawn from the center of the circle at an angle of 2θ relative to the abscissa. The intercept between this radial line and the circle defines σ_n and τ on the failure plane, that is,

$$\sigma_n = \frac{\sigma_1 + \sigma_3}{2} - \left[\frac{\sigma_1 - \sigma_3}{2}\right] \cos 2\theta \tag{8.5}$$

and

$$\tau = \left[\frac{\sigma_1 - \sigma_3}{2}\right] \sin 2\theta. \tag{8.6}$$

Under compression the values of σ_1 and σ_3 are both taken as positive, as shown on Figure 8.2b. Under pure tension, σ_1 is positive and σ_3 is zero. Under pure shear, σ_1 is positive and σ_3 is negative. Thus the position of the Mohr circle along the abcissa indicates the type of stress state. The size of the circle indicates the magnitude of the difference between σ_1 and σ_3 and is a measure of the *differential stress*.

For numerous experiments, conditions at failure are well described by *Coulomb's law of failure*,

$$\tau = \tau_0 + \sigma_n \tan \phi, \tag{8.7}$$

where ϕ is called the *angle of internal friction* and is often about 30°. Thus $\tan \phi$, which is also known as the *coefficient of friction*, typically has a value of about 0.6. The French military engineer and physicist Charles Augustin de Coulomb predicted this relationship in the 1770s. At that time it was already known that

$$\tau = \sigma_n \tan \phi \tag{8.8}$$

described the behavior of noncohesive materials such as sand; Coulomb's insight was to add the constant τ_0 to represent the intact shear resistance of a cohesive material. The Coulomb failure line described by Eq. (8.7) bounds a *region of stability* on Figure 8.2b. Mohr circles that touch the failure line describe conditions required to induce failure. Although Coulomb's law of failure describes many experimental results well, it remains a fundamentally empirical

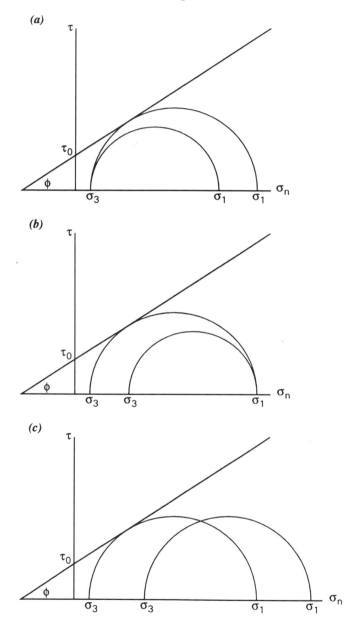

Figure 8.3 Mohr diagrams showing failure conditions induced by (a) increasing the greatest principal stress σ_1, (b) decreasing the least principal stress σ_3, and (c) increasing the internal pore-fluid pressure P_f. In each case the σ_1 and σ_3 values plotted on the abcissa are actually the effective stresses, that is, $\sigma_1 - P_f$ and $\sigma_3 - P_f$.

relation, because it is not yet possible to predict macroscopic failure behavior from molecular-scale behavior or from first principles (e.g., Scholz, 1990; Marder and Fineberg, 1996).

An apparent discrepancy between the failure behavior of *drained* and *undrained* fluid-saturated samples troubled early students of rock mechanics. In "drained tests" on saturated material, applied stresses are increased slowly and water is allowed to escape from the sample, and the results were similar in most respects to those obtained for dry samples. "Undrained tests," in which the contained water is retained by an impermeable jacket, gave apparent ϕ values of $\sim 0°$ (i.e., the failure criterion of Eq. (8.7) was defined by a horizontal line). By the 1950s it was recognized that the apparent discrepancy between drained and undrained behavior disappeared if, in the case of saturated samples, *effective stresses* were plotted on the abcissa of the Mohr diagram (i.e., $\sigma_1 - P_f$, $\sigma_3 - P_f$). (See Hubbert and Rubey (1959) and, in particular, their summary of earlier work.)

To summarize, then, in triaxial tests any sample can eventually be made to fail by increasing σ_1 with σ_3 held constant (Figure 8.3a) and/or by decreasing σ_3 with σ_1 constant (Figure 8.3b). Fluid-saturated samples can also be brought to failure by increasing the internal fluid pressure P_f. Varying σ_1 or σ_3 can induce failure by increasing the *differential stress* on the sample, an effect expressed on the Mohr diagram as an increase in the size of the Mohr circle. Increasing fluid pressure does not affect the shear stress but induces failure by decreasing the effective strength of the sample. This is expressed by shifting the circle toward the origin without changing its size (Figure 8.3c).

8.3 Induced seismicity

The principle of effective stress and results of laboratory failure experiments with fluid-saturated samples combine to suggest that high pore-fluid pressures should reduce the amount of work required for tectonic deformation. Many natural mechanisms have been suggested for maintaining elevated pore-fluid pressures (e.g., Hanshaw and Zen, 1965; Neuzil, 1995), but pore-fluid pressures can also be affected by human intervention, most directly by the injection or withdrawal of fluids through wells. Furthermore, injection through wells represents a fluid source that is usually both larger and better constrained than the natural *geologic forcing* rates discussed as possible causes of high fluid pressure in Chapter 4.1.

In this section we describe examples of seismicity directly induced by injection of fluid at depth. The *Rocky Mountain arsenal* example was essentially an uncontrolled experiment. Injection of liquid waste at a depth of 3.6 km just northeast of Denver, Colorado, unexpectedly generated earthquakes as large as Richter **M** 5.5 that were felt in the Denver and Boulder metropolitan areas. Shortly after a relationship between the timing of waste injection and the earthquakes was disclosed to the general public by geologist David Evans in November, 1965, the waste-injection program was discontinued. In contrast, the *Rangely, Colorado*, example was a controlled experiment. Earthquakes were intentionally induced by fluid injection at >1.7 km depth under conditions such that the the actual fluid pressures and stress state were known and the earthquake hypocenters could be precisely located.

From studies of injection-induced seismicity only, where subsurface fluid pressure changes are relatively large, one might conclude that the Mohr–Coulomb failure criteria and effective stress law are adequate models of earthquake occurrence. But many earthquakes are also induced by the filling of surface reservoirs, or by interannual or shorter-term fluctuations in reservoir levels that translate to pressure changes as small as 0.1 to 1.0 bars (e.g., Roeloffs, 1988a). For example, most shallow (<10 km) earthquakes in the Aswan, Eygpt, area are clearly related to short-term changes in the water level behind the Aswan High Dam (Awad and Mizoue, 1995). Much of the surface reservoir–induced seismicity cannot readily be explained in terms of Mohr–Coulomb failure due to reduced effective stress.

8.3.1 The Rocky Mountain arsenal

In 1962–1966 the U.S. Army injected approximately 6.25×10^8 liters of waste fluid from munitions production into fractured Precambrian gneiss beneath the Denver basin. The unanticipated result was a swarm of over 1,500 recorded seismic events in 1962–1967 (Figure 8.4). The three largest earthquakes (Richter **M** ≥ 5) actually took place in 1967, after waste injection had ceased. The 1962–1967 earthquake *epicenters* were consistently located in a rectangular, northwest-trending, 10-km by 3-km area roughly centered on the injection well, and the earthquake *hypocenters* were located approximately between the injection depth of 3.6 km and 7 km depth. (An earthquake epicenter is the point on the Earth's surface directly above the *focus* of an earthquake; focus and *hypocenter* both refer to the point in the subsurface where the earthquake rupture begins. See, for example, Bolt (1993) for basic definitions of these and other earthquake-related terms.)

The persistence of the earthquake swarm after injection had ceased was attributed to the continued outward migration of the pressure front (Healy and others, 1968). That is, fluid pressures at some distance from the well continued

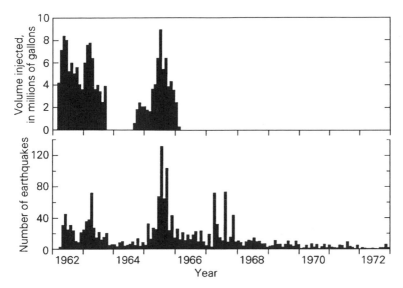

Figure 8.4 Time series of (a) waste-fluid injection and (b) seismicity at the Rocky Mountain arsenal, 1962–1972. From Hsieh and Bredehoeft (1981).

to increase significantly for some time after injection ceased, despite pressure declines in the more immediate vicinity of the well.

Hsieh and Bredehoeft (1981) later developed quantitative models of the fluid pressure history in the injection zone, using the distribution of earthquake hypocenters to define the dimensions of the "reservoir" and the long-term (9-year) pressure decline in the injection well to define its hydraulic properties. They modeled the system behavior using analytical solutions to Eq. (1.36) under appropriate *boundary conditions*, and they concluded that the earthquakes were likely triggered by a rather small pressure buildup of 3.2 MPa (32 bars). That is, earthquakes were apparently confined to areas where the pressure buildup relative to preinjection conditions exceeded 32 bars. Migration of earthquake epicenters away from the well could be related to the outward migration of this critical pressure increase.

Prior to waste injection, fluid pressures in the injection zone were probably well below the hydrostatic value computed relative to the land surface. The initial pressure in the injection zone is not known, but the best estimate of ~270 bars at 3.6-km depth translates to a water level ~920 m below the land surface, and even after the "critical" pressure increase of 32 bars the medium was apparently failing under subhydrostatic fluid-pressure conditions (a pressure gradient of ~83 bars/km relative to the land surface, versus the "normal" hydrostatic gradient of ~100 bars/km). One implication of the Hsieh and

Bredehoeft (1981) analysis, then, is that the gneissic rocks beneath the Rocky Mountain arsenal were already very close to failure prior to waste injection, so the earthquakes there might eventually have occurred spontaneously. The existence of *critically stressed faults* (faults near incipient failure) has since been documented in a wide variety of tectonic settings (e.g., Zoback and Healy, 1984). Thus, enhanced pore pressure at depth might be expected to generate earthquakes in many areas. (The importance of critically stressed faults to fracture permeability is discussed in Chapter 8.6.2.)

8.3.2 Rangely, Colorado

The Denver-area earthquakes induced by waste injection at the Rocky Mountain Arsenal gradually ceased after the injection was discontinued in 1966 (Figure 8.4). The Denver earthquakes were consistent with the hypothesis of Mohr–Coulomb failure under conditions of reduced effective stress, but the hypothesis could not be established conclusively there owing to incomplete knowledge of fluid pressures and the state of stress at depth and imprecise earthquake locations. The proximity to Denver and the large size of some of the induced earthquakes made any further activity at the Rocky Mountain arsenal impossible. Thus a group of scientists from the U.S. Geological Survey who had studied the Denver earthquakes sought an area where fluid pressure–induced seismicity could be monitored under more controlled conditions. They shortly identified the Rangely oil field in northwestern Colorado as a promising candidate area (Raleigh and others, 1976).

At Rangely, water had been injected at high pressure for secondary recovery of oil since 1957. The injection took place in a Pennsylvanian/Permian sandstone at >1.7 km depth, and by 1967, fluid pressures near the injection wells were as large as 290 bars, about 1.7 times the predevelopment, hydrostatic pressure. Seismometers about 80 km to the northwest at Vernal, Utah had detected seismicity in the vicinity of Rangely since first being installed in 1962.

In 1969 the U.S. Geological Survey installed a local seismic network at Rangely. The results of hydraulic fracturing (Chapter 4.1.3) allowed the least principle stress σ_3 to be measured and σ_1 to be calculated ($\sigma_3 \sim 314$ bars, $\sigma_1 \sim 552$ bars). The field operator, Chevron Oil Company, supported the study by sharing the results of their own periodic, field-wide pressure surveys and by following a particular injection/withdrawal schedule in part of the field.

The precise earthquake-focus locations from the local seismic network showed that induced seismicity at Rangely was occurring on the only known fault in the oil field, a modest east-northeast-trending feature with a total vertical displacement of only 10 to 15 meters. One group of earthquake hypocenters was located within the injection horizon at 2.0- to 2.5-km depth and another group was occurring near 3.5-km depth in the underlying crystalline basement. Intentional cycling of fluid pressures at depth confirmed a "critical" fluid pressure of ~257 bars for failure, a value consistent with effective stress/Mohr–Coulomb theory (Eq. 8.7), given the established values of σ_3 and σ_1 and an experimentally determined ϕ value of ~39° and assuming that τ_0 is negligible.

Ongoing (1976–) induced seismicity at the Renqiu oil field in northern China (Zhao and others, 1995) appears to be similar in many respects to that documented at Rangely. At Renqiu, water injection at 3- to 4-km depth and less than 433 bars pressure is causing swarms of earthquakes ($M < 4.5$) near the water-injection wells at 4- to 5-km depth. As at Rangely, a clear cause-and-effect relationship has been demonstrated by stopping and restarting water injection. In the Chinese case, most of the earthquakes are concentrated in parts of the oil field with relatively low permeability and porosity, where injection is more apt to cause significant fluid-pressure increases.

An interesting feature of the Rangely results was that the fluid-pressure distribution actually prevented large ($M > 3.1$) induced earthquakes. Whereas parts of the east-northeast-trending fault close to the injection wells were weakened by fluid-pressure increases, nearby sections of the fault were actually strengthened by extraction of fluid and the resulting fluid-pressure decreases. This limited the length of the fault zone liable to failure and thus the size of potential earthquakes. This particular aspect of the Rangely experiment led the investigators to propose that the behavior of great earthquake faults such as the *San Andreas fault* in California might conceivably be controlled by deep, strategically placed injection/production wells (Raleigh and others, 1976). They estimated that an average San Andreas slip rate of 2 to 3 centimeters per year could be accommodated by **M** 4.5 earthquakes at 6-month intervals. Earthquake magnitude is generally related to the rupture length, and **M** 4.5 quakes would require a rupture length of about 5 kilometers. Thus, appropriately sized quakes might be induced by very deep wells spaced about 2.5 km apart. The ends of each 5-km interval would be strengthened by fluid withdrawal while fluid was injected at the center of the interval to induce a controlled quake. Raleigh and others (1976) recognized that the technical feasibility of this proposal depended on the permeability structure of the fault zone, the existing state of stress and fluid pressures, and other material properties of the fault zone. However, factors other than expense and a lack of measurements at seismogenic depths have kept this idea from being implemented. It has since been recognized that small increases in fluid pressure may trigger larger-than-expected earthquakes, as in the case of surface reservoir–induced seismicity (Roeloffs, 1988a). In general, our incomplete understanding of the processes governing natural earthquakes precludes such grand experiments in earthquake control.

8.4 Fluid pressures at seismogenic depths

Because there is no direct information on the fluid-pressure regimes associated with natural (i.e., noninduced) seismicity, hypotheses about fluid pressures at seismogenic depths are necessarily inference- or model-based. In this section we describe several hypotheses with respect to the fluid pressures associated with fault movement, proceeding generally in chronological order. For a comprehensive overview of the role of fluids in faulting we refer the reader to the collection of papers edited by Hickman and others (1995).

Hubbert and Rubey (1959) introduced the concept of fault movement due to Mohr-Coulomb failure of a fluid-saturated geologic medium. They focused specifically on overthrust faulting and aimed at resolving the mechanical problem of moving very long fault blocks over nearly horizontal surfaces. Without consideration of pore-fluid pressures, such movement appears to be precluded by the limited strength of the overthrust block and/or the large frictional resistance to low-angle sliding.

Much more recently, a consensus has developed that the mechanical behavior of the San Andreas fault (and presumably other transform or strike-slip faults) is profoundly influenced by the presence and distribution of high pore-fluid pressures. This realization has developed through detailed investigation of the so-called stress–heat flow paradox of the San Andreas fault (Chapter 4.3.5). If the San Andreas had the relatively high frictional strength typical of the upper crust, its movement should generate significant frictional heat. In fact, there is no detectable heat flow anomaly associated with the fault. As of this writing, the preferred explanation is that the San Andreas is mechanically weak, and that this weakness is due to high fluid pressures at depth. A number of models for the fluid-pressure distribution in and around the fault zone have been proposed, and we choose three of these to illustrate the range of possibilities. Irwin and Barnes (1975) suggested that the San Andreas operates in an environment where ambient fluid pressures are maintained at elevated levels by generation of metamorphic fluids. Byerlee (1990) suggested that the high fluid pressures are local to the fault zone, rather than ambient, and result from the sealing of locally derived fluids within the fault zone. Rice (1992) suggested continual upwelling of overpressured fluids within the fault zone, with associated lateral leakage of such fluids away from the fault zone.

An important caveat to the discussion in this section is that, except in geothermal areas (Fournier, 1991), there is as yet little convincing evidence for greatly elevated pore pressures at midcrustal depth in crystalline rocks, whereas there is good evidence for hydrostatic pore pressures at depths as great as 12 kilometers (Zoback and Zoback, 1997).

8.4.1 Hubbert and Rubey

Seminal, comprehensive studies of *thrust faulting* by Hubbert and Rubey (1959; Rubey and Hubbert, 1959) first applied Terzaghi's (1925) concept of *effective stress* to the problem of faulting. In thrust faulting, part of the brittle crust is thrust over adjacent crust at a low dip angle, typically 10 to 30°. Thrust faults commonly cause older rocks to overlie younger rocks and are thus also called low-angle *reverse faults*. In the *thrust belts* associated with continental collisions (e.g., the Himalaya, Alps, Appalachia, etc.), the upper crust is thickened by displacements on a series of thrust faults. Individual thrust sheets can extend over hundreds of kilometers, as in the southern Appalachians of the United States. Even such large blocks seem to have maintained much of their mechanical integrity while moving.

Without consideration of fluid-pressure effects, there is no satisfactory mechanical explanation for the low-angle thrusting and/or sliding of large, thin, relatively intact sheets of rock. If one assumes that a "dry" rock sheet is moved by a horizontal force, its own strength can readily be shown to be insufficient to withstand the required force, even if a fairly low value is assumed for the coefficient of friction along the fault plane (Smoluchowski, 1909). That is, the sheet would be deformed internally before being moved. If one assumes that the movement is due to gravitational sliding, the strength of the sheet ceases to be a consideration, but a rather large slope (~30°) is required to initiate movement (Figure 8.5a). Field relations generally preclude such slopes.

As described in Chapter 8.1, the greatest principal stress σ_1 is horizontal in reverse- or thrust-faulting environments. The least principal stress σ_3 is vertical, and (sub)horizontal failures will occur when the fluid pressure approaches the total overburden pressure. Thus the concept of *effective stress* affords a viable mechanism for reducing the frictional resistance to overthrusting. Hubbert and Rubey (1959) showed that the effective stress concept can be derived from direct application of Archimedes' principle of buoyancy, and that the result for a porous solid (i.e., a geologic medium) is identical with that for a completely enclosed solid immersed in liquid. They proceeded to show that, given elevated fluid pressures, very large blocks can slide down arbitrarily small slopes (Figure 8.5b).

Hubbert and Rubey (1959; Rubey and Hubbert, 1959) further noted the common occurrence of elevated ($\gg$hydrostatic) fluid pressures in sedimentary

(a)

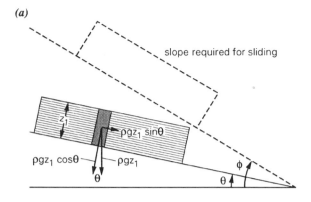

(b)

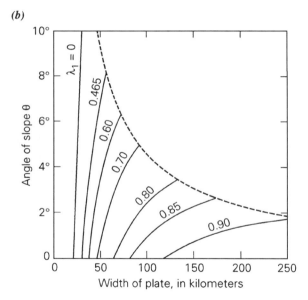

Figure 8.5 (a) Normal and shear stresses on the base of a block inclined at angle θ, and angle ϕ required for sliding. Hubbert and Rubey (1959) pointed out that tests of both rocks and unconsolidated materials consistently give angles of internal friction $\phi \sim 30°$ and that this is also the mean angle of sliding friction of rock on rock. Thus the angle of tilt of the surface must be raised to about 30° for the block to slide. From Hubbert and Rubey (1959). (b) Width of plate that can be pushed downslope for various values of slope angle θ and fluid pressure/overburden pressure ratio λ, assuming "intermediate" values for the crushing strength of the plate and the coefficient of friction. The dashed curve truncating the solid curves represents a total relief of 8 km. From Rubey and Hubbert (1959). Reprinted by permission of Geological Society of America.

basins. In Chapter 4.1 we discussed various causes of elevated fluid pressure. Here, it is useful to recall that, with certain exceptions (e.g., chemical osmosis), significantly elevated fluid pressures are transient phenomena related to ongoing geologic processes (e.g., compaction of sediments). If these processes cease, pressures must more-or-less gradually decay toward the hydrostatic level. The occurrence of overpressuring depends on the effective rate of fluid production by various geologic processes (*geologic forcing* in Chapter 4.1.1), the size of the affected domain, the intrinsic permeability of the medium (Chapter 1.2), the density and viscosity of the fluid and, in multiphase systems, on relative permeabilities (Chapter 3.1.3) as well.

8.4.2 Irwin and Barnes model for the San Andreas

At some plate boundaries the rigid crustal plates slide past one another along *transform faults*. Most such faults are associated with the mid-ocean ridge (MOR) system, are parallel to its spreading direction, and act to segment the MOR, making it discontinuous. Less commonly, transform faults offset continental crust; the best known example is the San Andreas fault of California, which accommodates differential movement between the Pacific and North American plates.

Over geologic time the San Andreas fault moves at an average rate of a few centimeters per year. Minimum right-lateral offset along the fault is about 600 km over about the last 15 Ma. Discrete segments of the fault accommodate the relative plate motion in diverse ways, ranging from continuous, nearly aseismic fault creep to infrequent, catastrophic earthquakes. Irwin and Barnes (1975) suggested that this diverse behavior might be related to along-strike variations in the geologic structure and associated fluid-pressure regimes. Berry (1973) had already documented the occurrence of high fluid pressures in the general vicinity of the fault and suggested that they facilitated creep, but he did not attempt to explain along-strike variations in fault behavior. (*Fault creep* is fault motion that is not obviously associated with earthquakes.)

In west-central California, fault creep and frequent, small-magnitude earthquakes occur along the segment of the San Andreas between Parkfield and San Juan Bautista (Figure 8.6a). This "creeping" segment of the fault is not believed to be subject to great earthquakes. Along the segments south of Parkfield and north of San Juan Bautista there is little or no fault creep and less low-level seismicity, and most of the plate motion is in fact accommodated by infrequent, large-magnitude earthquakes. These variations in fault behavior may possibly be related to the composition of the eastern wall of the fault. Between Parkfield and San Juan Bautista, the San Andreas juxtaposes the mainly granitic Salinian

(a)

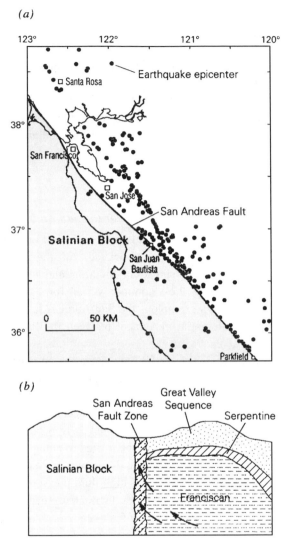

(b)

Figure 8.6 (a) Map of west-central California showing location of the San Andreas fault, the Salinian block, and earthquake epicenters during the second quarter of 1972. The general distribution of epicenters is similar to that which would be expected over the long term. After Irwin and Barnes (1975). (b) Schematic section across the San Andreas fault between Parkfield and San Juan Bautista showing postulated movement of metamorphic fluids from the Franciscan assemblage toward the fault zone. From Irwin and Barnes (1975).

block to the west (Figure 8.6) with metamorphic rocks of the Franciscan assemblage that are overlain by sandstones, shales, and basal serpentine of the Great Valley sequence (Figure 8.6b). The Franciscan rocks produce abundant, CO_2-rich metamorphic fluids (e.g., Barnes, 1970; Barnes and others, 1975). Irwin and Barnes (1975) suggested that the Great Valley sequence has relatively low vertical permeability and inhibits the escape of these fluids, so that fluid pressures can increase within the Franciscan and the fluids will tend to migrate toward the fault zone, which is assumed to have a higher vertical permeability (Figure 8.6b). South of Parkfield and north of San Juan Bautista, the Franciscan assemblage is not capped by the Great Valley sequence, so that the metamorphic fluids can escape freely to the surface without generating high fluid pressures.

Barnes and others (1984) later published a map showing a strong spatial correlation between the worldwide distribution of CO_2-rich springs and zones of seismicity. This motivated Bredehoeft and Ingebritsen (1990) to examine some aspects of the Irwin and Barnes (1975) hypothesis, albeit in a global context rather than the particular context of the San Andreas fault. The CO_2 flux from deep in the Earth's interior to the surface has been estimated by numerous investigators to be in the range of 10^{12} to 10^{13} moles per year. If the tectonically active area is taken to be 1% of the Earth's surface (1.3×10^6 km^2), and a global CO_2 flux of 3×10^{12} moles/yr is distributed over that area, the CO_2 flux per unit area amounts to $\sim$0.003 kg/(s-km^2). Under midcrustal pressure–temperature conditions, this rate of pure CO_2 upflow would be sufficient to generate near-lithostatic fluid pressures if crustal permeability is $\leq 10^{-20}$ m^2. There is likely to be some H_2O present as well, but the limited solubility of CO_2 at midcrustal pressure–temperature conditions implies that the volumetric flow rate q required to transport a given flux of CO_2 in solution or as a two-phase mixture would be somewhat greater than the flow rate required to transport the same CO_2 as a separate phase. Thus the limiting permeability value of 10^{-20} m^2 suggested for a CO_2-only system may be reasonable for hydrous systems as well. Bredehoeft and Ingebritsen (1990) concluded that available data cannot preclude the Irwin and Barnes (1975) hypothesis about CO_2 or CO_2-rich fluids as a source of elevated fluid pressures. Although in situ permeabilities of $\leq 10^{-20}$ m^2 have rarely been measured, they may be within the realm of expectation deep in the crust.

In their analysis, Bredehoeft and Ingebritsen (1990) solved a system of equations similar to Eqs. (3.17) and (3.18) using equations of state for pure CO_2. This relatively complicated approach was required in order to simulate hydraulic fracturing as a result of excess CO_2 pressure. However, they could have arrived at the same limiting value of permeability through a simple nondimensional analysis such as that described in Chapter 4.1.2. The density and viscosity of

CO_2 in the pressure–temperature range of their experiment are fairly constant and not greatly different from the values for H_2O ($\rho_{CO_2} \sim 500$ to $1{,}000$ kg/m^3; $\mu_{CO_2} \sim 0.5$ to 1.5×10^{-4} kg/(m-s)).

8.4.3 Byerlee and Rice models for the San Andreas

In the 1980s a comprehensive investigation of the *stress–heat flow paradox* of the San Andreas fault inspired other models for the fluid-pressure distribution in the vicinity of the fault. The stress–heat flow paradox was discussed in Chapter 4.3.5 and we will only reiterate some of the key points here. Briefly, displacement rates along the San Andreas are large enough that frictional heat generation should cause a local heat flow anomaly unless the fault has an unusually low frictional resistance of about 10 MPa, or 100 bars (Brune and others, 1969). However, relatively abundant, high-quality heat flow data show that no such anomaly exists (Lachenbruch and Sass, 1980). Results of deep drilling ($\sim$3.5-km depth) along the southern San Andreas at Cajon Pass appear to confirm that the fault moves at very low shear stresses and therefore must be extremely weak relative to the surrounding crust (Zoback and Healy, 1992). The current consensus is that this weakness is likely due to elevated pore-fluid pressures within the fault zone.

Studies in the 1980s further revealed that the greatest principal stress σ_1 in the nearby crust is approximately normal to the trace of the San Andreas (e.g., Zoback and others, 1987). Thus the right-lateral transform movement of the great fault is taking place in a generally compressional environment. In fact, folds and thrust faults striking roughly parallel to the San Andreas accommodate a component of convergent motion at the plate boundary and are largely responsible for building the coastal mountain ranges of southern and central California (see Figure 4.13a for typical topographic profiles across the Coast Range).

Under conditions of fault-normal compression there are mechanical problems with the Irwin and Barnes (1975) fluid-pressure model, which essentially proposed that the fault exists in an ambient environment of elevated fluid pressures. In their model, fluid pressures in the adjacent crust on at least one side of the fault may approach lithostatic because of metamorphic-fluid generation. Pressures at equivalent depths within the fault zone itself are actually somewhat lower, allowing lateral flow of metamorphic fluids toward the San Andreas (Figure 8.6b). However, failure under approximately fault-normal compression actually requires either a decrease in fluid pressures away from the fault or, given uniform fluid pressures, an extraordinarily low mechanical stength for the fault itself (tan $\phi \sim 0.1$, versus ≥ 0.6 in most rocks). (Recall that, for uniform values of P_f and tan ϕ, Mohr–Coulomb theory requires that the angle between σ_1 and the failure plane be $\leq 45°$; see Figure 8.2.) Decreasing fluid pressure toward the San Andreas would actually act to strengthen the fault zone itself relative to the adjacent crust.

Both Byerlee (1990, 1993) and Rice (1992) have proposed models in which the distribution of fluid pressures is essentially opposite to that invoked by Irwin and Barnes (1975). In their models, fluid pressures are elevated in the fault zone, sufficient to weaken the fault, and are nearly hydrostatic outside the fault zone. Thus the direction of lateral fluid flow at depth is away from the San Andreas, rather than toward the fault. Rice (1992) hypothesized that fluid pressures in the fault zone are maintained by a source of fluids near the brittle–ductile transition at the base of the seismogenic crust. The rate of production of these fluids and the permeability distribution are such that fluid pressures in the fault zone remain at $\gg$hydrostatic levels despite some loss of fluid to the surface and to the surrounding crust. Byerlee (1990, 1993) supposed that the original porosity of the fault zone is the only source of fluids and that the fault zone is perfectly sealed off from the surrounding crust, so that the finite supply of original pore fluid can be conserved indefinitely.

From a purely hydrogeologic standpoint, both the Irwin and Barnes (1975) and Rice (1992) models seem fairly plausible. The discharge of "deep" geothermal, metamorphic, and connate fluids is in fact often focused at fault zones, and we saw in Chapter 4.1 how distributed sources of fluid can elevate fluid pressures. The lack of a fluid source in the Byerlee (1990, 1993) model poses greater hydrogeologic problems. Over the $\sim$15 Ma lifespan of the San Andreas, loss of the original porosity of the fault zone represents a *geologic forcing* (Chapter 4.1.1) of only $\sim$2 × 10^{-16}/s, assuming that an initial porosity of 0.10 is completely eliminated. Thus for a fault zone $\sim$1-km thick a hydraulic conductivity K of $\leq 2 \times 10^{-13}$ m/s would be required to maintain significantly elevated fluid pressures (Chapter 4.1.2, Eq. 4.4). If the fluid is pure water and the temperature at seismogenic depths is $\sim$250°C, this translates to an intrinsic permeability k of $\leq 2 \times 10^{-21}$ m^2. As previously noted in the context of the Irwin and Barnes (1975) model, such values are very near the lower limit of what has been measured but do not seem inconceivable in the midcrust. Byerlee (1990) actually invoked a threshhold hydraulic gradient (i.e., non-Darcian flow; see Chapter 1.1.1) to explain retention of fluid in the fault zone, but it is not clear that this was strictly necessary to his model. Devolatilization in a regional-metamorphic setting, the process invoked by Irwin and Barnes (1975) as a cause of high fluid pressures in the Franciscan assemblage, can occur at rates of up to 3×10^{-15}/s (Table 4.1) and therefore permits a somewhat larger limiting value of permeability. Because they invoke focused upflow in the fault zone, both the Irwin and Barnes (1975) and Rice (1992) models imply that the vertical permeability in the fault zone is at least 10^2 to 10^3 times larger than the vertical permeability in the nearby crust. The Byerlee model, in contrast, implies that permeability in the fault zone itself is relatively low.

Regardless of the merits of the Byerlee (1990, 1993) and Rice (1992) models from a hydrogeologic perspective, their mechanical effects are identical and can apparently explain the gross behavior of the San Andreas system. Rice (1992) showed that if fluid pressure and stress state vary continuously between the fault zone (high fluid pressure) and adjacent crust (hydrostatic fluid pressure), and if both regions have identical strength properties, the fault zone can be critically stressed for strike-slip failure while the adjacent crust is critically stressed for thrust failure (see his Figures 3 and 4 and associated discussion).

The alternative of a mechanically weak fault ($\tan \phi \sim 0.1$) in an environment of uniform fluid pressures appears less tenable. Laboratory data permit $\tan \phi$ as low as ~ 0.2 for montmorillinite, a component of fault gouge (Morrow and others, 1992), as opposed to values of ≥ 0.6 for most rocks. Chrysotile serpentine is another unusually weak mineral that is considered to be a common fault-zone constituent in central and northern California. The coefficient of friction for chysotile is comparably low at standard temperature and pressure ($\tan \phi \sim 0.2$), but it has been shown to be substantially larger at elevated temperatures, exceeding 0.5 at 290°C (Moore and others, 1996). Thus, under hydrostatic conditions, even the lowest permissible values of $\tan \phi$ are unable to account for the behavior of the San Andreas as it is currently understood.

A great deal of work has been done on rock friction in the past ten years or so, much of it aimed at understanding the behavior of faults like the San Andreas. An important caveat to this discussion is that many researchers have come to believe that a single value of $\tan \phi$ is inadequate to describe the earthquake process. Instead, the friction coefficent may depend on the rate of slip and/or the amount of time that the slipping surfaces have been in contact. Laboratory measurements indicate that variations in $\tan \phi$ due to these two variables are small, but the measurements have not been made under in situ conditions or at seismic slip rates. A slip rate- and time-dependent frictional behavior could emerge from cyclic variations in porosity caused by porosity creation during earthquakes followed by compaction between earthquakes (Segall and Rice, 1995).

8.5 Earthquake-induced hydrologic phenomena

Earthquakes often cause large changes in spring and stream discharge, water levels in wells, and geyser behavior. Coseismic and postseismic hydrologic changes are abundant and well documented. Many preseismic or precursory hydrologic changes have also been reported, but in most cases the documentation has not been sufficient to convince the scientific community or to rule out explanations unrelated to earthquakes. The suggestion of precursory hydrologic changes is tantalizing, because the ultimate goal of most earthquake research is to reduce the hazard posed by major events, and some warning of impending activity would be very useful in that regard. We tend to agree with Roeloffs (1996) that "the documentation of a few of these reports [of precursory hydrologic changes] is now approaching levels that require them to be given scientific credibility." However, in this section we will focus on the better-documented coseismic and postseismic hydrologic changes, for which plausible physical mechanisms have been advanced. Regardless of whether the various reported precursory changes are in fact "real," their physical basis is poorly understood.

We refer those readers who are particularly interested in hydrologic precursors to the summary by Roeloffs (1988b). It is also worth noting the relatively well-documented hydrogeochemical anomalies associated with the more recent, disastrous **M** 7.2 earthquake at Kobe, Japan. The chloride and sulphate contents of wells about 20 km east of the Kobe epicenter increased steadily from August 1994 to just before the earthquake on 17 January, 1995 (Tsunogai and Wakita, 1995). The radon content in a well about 30 km northeast of the epicenter peaked dramatically about 9 days before the eruption (Igarashi and others, 1995). Gas ratios in mineral springs 220–230 km east of Kobe changed markedly in September 1994 and again on 17 January, 1995, coincident with the earthquake (Sugisaki and others, 1996). The hydrology of the Kobe area is well known and carefully monitored in part because the groundwater there is regarded as highly suitable for drinking and for brewing sake. In fact, the preseismic changes in chloride and sulfate content were reconstructed by analyzing dated, bottled mineral water distributed in the domestic market. It seems likely that the well-documented hydrologic precursors at Kobe, viewed in conjunction with similar precursory anomalies from the **M** 7.0 Izu–Oshima earthquake in 1978, will revitalize interest in hydrologic earthquake precursors (Silver and Wakita, 1996).

8.5.1 Streamflow and springs

Many of the quantitative data on streamflow response to seismicity are derived from the U.S. Geological Survey's extensive hydrologic monitoring network in the tectonically active western United States. There are many examples of postseismic increases in stream and spring discharge, some instances of springs going dry, and a very few examples (e.g., Waller, 1966) of decreased stream discharge. The hydrologic disturbances associated with fairly recent moderate-to-large earthquakes in California are particularly well documented (Sorey and Clark, 1981; Rojstaczer, 1994; Roeloffs and others, 1995; Quilty and others, 1995).

In some instances the streamflow data are actually sufficient to permit estimates of the total "excess" streamflow derived from a particular seismic event. The **M** 7.5 Hebgen Lake earthquake (17 August, 1959) apparently produced about 0.5 km^3 of water, the **M** 7.3 Borah Peak earthquake (28 October, 1983) about 0.3 km^3 of water, and the **M** 7.1 Loma Prieta earthquake (17 October, 1989) only about 0.01 km^3 of water (Muir-Wood and King, 1993; Rojstaczer and others, 1995). These excess flows were estimated by comparing the actual postearthquake stream hydrographs to synthetic hydrographs that approximate what the flow would have been in the absence of an earthquake.

As nearly as can be determined, the streamflow increases are often essentially coseismic. The U.S. Geological Survey gaging stations generally record streamflow every 15 minutes, and flow increases tend to be observed within 15 minutes at those stations that are not temporarily disabled by the ground motion. Streamflow may continue to increase for a few days and then gradually decay toward the prequake baseline condition over several months.

Selected examples of stream- and spring-discharge increases associated with California earthquakes are shown in Figure 8.7. After the 1989 Loma Prieta event, many streams within about 50 km of the epicenter showed 4- to 24-fold increases in discharge (e.g., Figure 8.7a). There was no clear relation between the magnitude of the increase and proximity to the epicenter. The Loma Prieta event occurred during a prolonged California drought, so that some water managers viewed the unexpected "earthquake water" as a boon. However, the excess streamflow decayed fairly rapidly, and at Big Trees (Figure 8.7a), for example, the total earthquake-induced streamflow proved to be only about 20% of the total annual flow (Rojstaczer and Wolf, 1994).

A series of earthquakes as large as **M** 6.3 on 25–28 May, 1980, caused temporary increases in the discharge (Figure 8.7b) and turbidity of hot springs in the Long Valley caldera of east-central California. These earthquakes had other obvious effects on the hydrothermal system, including emptying and refilling of boiling pools and temporary increases in fumarolic activity (Sorey and Clark, 1981).

Two mechanisms have been invoked to explain earthquake-related changes in spring and stream discharge: near-surface permeability enhancements (e.g.,

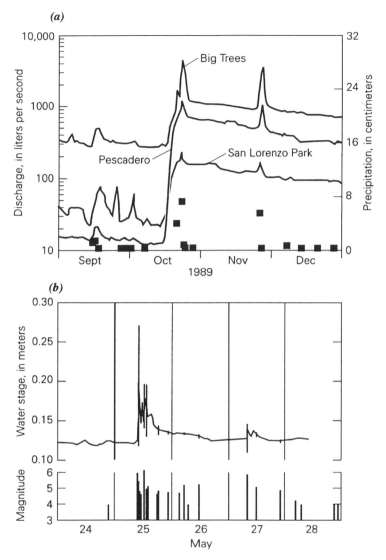

Figure 8.7 (a) Stream discharge in response to the 17 October, 1989 Loma Prieta earthquake at Pescadero (~50 km from epicenter), Big Trees (~20 km from epicenter), and San Lorenzo Park (~30 km from epicenter and ~20 km upstream from Big Trees). Streamflow values are daily means and precipitation values are daily totals. From Rojstaczer and Wolf (1994). (b) Water stage at the weir on Little Hot Creek in Long Valley, California, and record of **M** > 4 earthquakes recorded at Mammoth Lakes on 24–28 May, 1980. From Sorey and Clark (1981).

Rojstaczer and Wolf, 1992) and expulsion of water from depth due to elastic compression (e.g., Muir-Wood and King, 1993). We will discuss these alternatives in Chapter 8.6.

8.5.2 Well behavior

Coseismic and postseismic changes in water levels in wells are commonly observed. These changes can generally be attributed to one of three causes. (1) Large (as much as 20 m), near-field (perhaps <50 km from the epicenter) water level declines can sometimes be related to near-surface permeability enhancement due to ground motion (Rojstaczer and Wolf, 1994). (2) Measureable water level changes at distances of up to a few hundred kilometers from the epicenter can often be directly related to the magnitude of the crustal strain produced by slip on the fault (e.g., Roeloffs, 1988b). Finally, (3) most of the coseismic water level oscillations observed at larger distances are resonance phenomena caused by particular well-formation configurations that act to amplify a very small crustal strain signal (e.g., Cooper and others, 1965). In this section we will discuss the direct elastic strain response and the resonance behavior; near-surface permeability enhancement will be discussed in Chapter 8.6. (Strain is a measure of deformation: In one dimension $\varepsilon = \Delta L / L_0$, where L_0 is the initial length, and in three dimensions $\varepsilon = \Delta V / V_0$, where V_0 is the initial volume.)

We will take advantage of the analogy between the way that the water level in a well responds to tectonic strain and the response to the strain caused by *Earth tides* (Roeloffs, 1988b). If rock-grain compressibility is negligible, then the tidal response of *confined systems* (Chapter 1.3) is approximately

$$\Delta h = \frac{\Delta \varepsilon}{(\beta_w n \rho_w g)}, \qquad (8.9)$$

where h is the water level or *hydraulic head* in the well, $\Delta \varepsilon$ is the incremental volume strain associated with the Earth tide, and β_w is the compressibility of water (Bredehoeft, 1967). The largest Earth tides cause volumetric strains of $\sim 10^{-8}$, and the associated water-level response of confined systems is typically a few centimeters. Note the analogy between Eq. (8.9) and the *specific storage* previously defined by Eqs. (1.21), (1.22), and (1.28). *Unconfined systems* generally do not respond to Earth tides because the fluid content of such systems varies with water table elevation in addition to porosity and fluid density. One can interpret the lack of strain sensitivity of unconfined or water table systems as being due to their large *storage coefficients* (0.05 to 0.25, versus <0.005 for confined systems) (see Chapter 1.5.2).

The strain signal associated with the various Earth tides is well known (e.g., Melchior, 1978), so that the strain response of a well can be calibrated on the basis of its tidal behavior. Roeloffs (1996) suggested that the typical *volumetric strain efficiency* of a well in a confined system is 50 cm per μ strain, so that a strain signal as small as 2×10^{-9}, nearly 10 times smaller than the tidal signal, might be detected if water levels can be accurately determined to within ~ 1 millimeter. Some wells are instrumented to act as volume strain meters, particularly in California and Japan, and those wells may in fact detect strains of this magnitude. However, in most hydrologic-monitoring wells, fluctuations smaller than the daily tidal signal are likely to pass unnoticed.

Earthquakes create both transient *dynamic strain* associated with seismic waveforms and permanent *static strain* due to the dislocation associated with fault rupture. The coseismic strain field can be calculated from elastic half-space models if the size and orientation of the fault zone and the magnitude of the slip are known (e.g., Okada, 1992). The dynamic strains decrease with distance from the rupture according to $r^{-3/2}$ or r^{-2} and the static strain decreases according to r^{-3}, where r is the rupture length. Even for large events such as the 28 June, 1992 **M** 7.3 Landers, California, earthquake the static strain changes will be less than the daily tidal strain at distances of more than a few hundred kilometers (Hill and others, 1993). Thus large ($\gg 1$ cm) coseismic water level changes observed more than a few hundred kilometers from the rupture cannot be directly related to the coseismic volume strain predicted by elastic half-space models.

In fact, such large, distant water level changes occur when tiny *dynamic strain* signals are amplified by resonance at periods comparable to those of seismic Rayleigh waves (Cooper and others, 1965). Coseismic water level oscillations associated with the great (**M** 9.25) Alaskan earthquake of 27 March, 1964 attracted considerable scientific attention. Large water level fluctuations were observed at numerous wells in eastern Canada, the central and eastern United States, and Puerto Rico (for example, water level variations of 4.6 meters were measured in a U.S. Geological Survey well near Perry, Florida). Cooper and others (1965) noted that well–aquifer systems have the essential features of a seismograph:

a mass (the column of water in the well plus some part of the water in the aquifer), a restoring force (the difference between the pressure head in the aquifer and the displaced water level in the well), and a damping force (the friction that accompanies the flow of water through the well and the aquifer).

They proceeded to demonstrate that fluid-pressure fluctuations in highly permeable aquifers will be greatly amplified in wells of favorable dimensions. For large values of aquifer transmissivity ($T = LK$), the maximum amplification frequency depends almost entirely on the height of the water column in the well.

8.5.3 Geysering

Geysers are periodically discharging hot springs driven by steam and/or non-condensible gas. Their general characteristics and behavior were discussed previously in Chaper 7.6. They deserve further consideration here, because fairly compelling data indicate strong response to coseismic strains on the order of 1 μstrain. More speculative correlations have been suggested between geyser activity and much smaller preseismic strains (e.g., Silver and Vallete-Silver, 1992).

The eruption frequency of the "Old Faithful" geyser in Calistoga, California, changed in conjunction with each of the last three **M** 6–7 earthquakes in central California (Silver and Vallette-Silver, 1992). These earthquakes were centered ≥ 130 km from Calistoga, and associated *static strains* at Calistoga were on the order of 0.1–1 μstrain (10^{-7}–10^{-6}), with *dynamic strains* about one order of magnitude greater (Hill and others, 1993). Following the 1975 Oroville and 1989 Loma Prieta events, there was a marked decrease in eruption frequency, with a gradual recovery toward the preeruption frequency over a period of months. After the 1984 Morgan Hill event there was a change from a unimodal to a bimodal eruption pattern. Although there is more subtle evidence for relatively small preseismic changes in the frequency of the Calistoga geyser, these preseismic changes were well within its normal range of variation and therefore would not constitute useful precursory signals. Many of the geysers at Yellowstone responded to the **M** > 7 earthquakes at Hebgen Lake (1959, distance ~50 km) and Borah Peak (1983, distance ~230 km) (Marler and White, 1977; Hutchinson, 1985). After the Hebgen Lake event, the frequency of many semiregular geysers increased, long-dormant geysers erupted, and many hot springs erupted for the first time. Response to the more distant Borah Peak event was relatively slow, subtle, and variable.

Ingebritsen and Rojstazcer (1993, 1996) simulated geyser behavior using the governing equations for groundwater flow and heat transport written as Eqs. (3.15) and (3.16). Their simulation results provide a framework for considering the response of natural geysers to small, external strains. They suggest that eruption frequency should exhibit varying degrees of sensitivity to the permeability, porosity, and length of the geyser conduit; the permeability of the surrounding rock matrix; and recharge rates.

Coseismic and postseismic changes in geyser frequency might be explained in terms of permeability changes caused by strong ground motion. At locations distant from the earthquake source, these dynamic strains are significantly larger than the static strains. The simulation results suggested that increases in the permeability of the geyser conduit itself or in the permeability of the surrounding rock matrix should increase eruption frequency, and increases in effective conduit length should decrease eruption frequency. Because the dynamic shear strains caused by distant earthquakes are small (Hill and others, 1993), increases in effective conduit length might be related to reopening of a network of existing fractures, rather than creation of new fractures.

Ground motion and associated fracture creation or reopening were invoked to alter permeability because, in an initially high-permeability environment, the changes in fracture aperture associated with seismically induced static strains of ≤ 1 μstrain would not cause large changes in permeability. Significant elasticity-related changes in the permeability or porosity of the geyser conduit itself seem particularly unlikely. Because the simulation results suggest that the permeability of the surrounding matrix might be 10^3–10^4 times lower than that of the geyser conduit itself, it could conceivably be affected by an elastic response. For instance, if the matrix has a fracture permeability given approximately by $k = Nd^3/12 = 10^{-11}$ m^2 (Snow, 1968), with the fracture spacing $N = 0.02/m$ and the fracture aperture $d = 0.002$ m, and a strain of 1 μstrain is wholly accomodated by this fracture set, then the matrix permeability would change by about 7%. Such a change in intrinsic permeability would be an upper bound, since we are assuming that the fractures are favorably oriented relative to the imposed deformation and have essentially an infinite compliance. The possibility remains that static strains could affect relative permeabilities to liquid water and steam (Chapter 3.1.3) much more significantly by causing minor changes in saturation.

The simulation results presented by Ingebritsen and Rojstaczer (1993, 1996) provide a reasonable context for explaining the coseismic response of geysers and perhaps also their response to barometric strains. However, they seem unable to explain responsiveness to tidal strains or to the very small preseismic strains permitted by the standard elastic half-space models. The simulated geyser systems showed some signs of chaos but were not fully chaotic; they might be more responsive if they were fully chaotic, that is, if slight differences in the coupled variables influenced subsequent time-integrated system behavior (e.g., eruption frequency). It is not yet clear whether the frequency of natural geysers is chaotic. Alternatively, enhanced sensitivity might be caused by the phenomenon of *metastable water*. It has been observed in both field (R. O. Fournier, U.S. Geological Survey, written communication, 1994) and laboratory contexts (Steinberg and others, 1982) that liquid water in hot-spring and geyser conduits can be superheated above its saturation temperature, because some energy is required to nucleate a vapor phase. Under such metastable conditions, very minor changes in pressure might retard or accelerate boiling and thus have a systematic effect on geyser frequency.

8.6 Effect of earthquakes on crustal permeability

Earthquakes can create or significantly enhance fracture permeability, and the static state-of-stress can have a controlling influence on the permeability of existing fractures. We will describe the hydrologic changes associated with the 1989 Loma Prieta, California, event in terms of earthquake-induced permeability changes. We then conclude this chapter by summarizing a study that documented a relationship between in situ stress and fluid flow in fractured, faulted rock.

8.6.1 Analysis of the Loma Prieta case

As noted in Chapter 8.5.1, two mechanisms have been invoked to explain earthquake-related changes in spring and stream discharge: near-surface permeability enhancements and expulsion of water from depth due to elastic compression. A thorough review by Muir-Wood and King (1993) concluded that many earthquake-induced hydrologic changes can in fact be explained in terms of processes occurring at midcrustal depths. However, the various hydrologic changes observed after the **M** 7.1 Loma Prieta event on 17 October, 1989 are more readily explained as a consequence of near-surface permeability increases.

The hydrologic changes associated with the Loma Prieta earthquake were extremely well documented, due in large part to the presence of about twenty U.S. Geological Survey streamflow-gaging stations within about 50 kilometers of the epicenter. Many of the gaged streams showed earthquake-induced 4- to 24-fold increases in streamflow (e.g., Figure 8.7a). The ionic concentration of the stream water generally increased, but the ratios of various ionic constituents remained nearly constant. Meanwhile, the water table dropped by $\sim$4 m in highland areas of the affected basins within several weeks after the earthquake (Rojstaczer and Wolf, 1994; Rojstaczer and others, 1995).

The increased streamflow, constant ionic composition, and declining water table can all be explained in terms of a roughly order-of-magnitude increase in near-surface permeability. In fact, the increased streamflow is well fit by a simple Darcian model of groundwater flow induced by permeability enhancement beneath a hillside, given *hydraulic diffusivity* (Chapter 7.4.3, Eq. 7.9) values of 2.0 to 2.6×10^{-2} m^2/s (Rojstaczer and others, 1995). Such diffusivity values are typical of the sandstone and siltstone that dominate the local near-surface lithology (e.g., Roeloffs, 1996, Figure 14). In contrast, Rojstaczer and Wolf (1994) pointed out that

[i]f, for example, expulsion of overpressured fluids at depths of 5 km were responsible for the response of the streams (a rise to peak flow within several days of the earthquake), then the hydraulic diffusivity of the pathway would be on the order of [10^2 m^2/s]. This value is orders of magnitude above that which has been inferred or observed in the crust...

Such values of diffusivity are typically observed only in carbonate aquifers and unconsolidated sands and gravels and thus seem unlikely to apply to mid-crustal depths. The constant ionic composition of the stream water also seems incompatible with a midcrustal fluid source.

8.6.2 State-of-stress and the orientation of conductive fractures

In many geologic media, fractures have a controlling influence on permeability, and the distribution of fractures in drillholes can be mapped using devices such as acoustic televiewers. However, the permeability of the various fractures cannot be determined remotely, and many of the remotely sensed fractures prove to be nonconductive. The subset of conductive fractures can only be identified through more direct means such as hydraulic testing in packed-off intervals of the drillhole or by examination of detailed downhole-flowmeter and/or temperature logs.

Recently, Barton and others (1995) showed that the subset of conductive fractures can often be predicted quite well on the basis of fracture orientation relative to the principal stresses. They considered data from granites and granodiorites encountered by the 3.5-km-deep Cajon Pass scientific drillhole (Chapter 4.3.5), rhyolites and tuffs encountered by a >2-km-deep exploratory well at Long Valley, California, and tuffs penetrated by a >1.7-km-deep drill-hole at Yucca Mountain, Nevada. In each case the magnitude and orientation of the principal stresses (σ_1, σ_2, and σ_3) were known, the entire set of fractures was determined from image data, and the subset of conductive fractures was determined from perturbations seen in precision temperature logs. The stress data were normalized by the total overburden pressure to permit comparison over large depth ranges. The study results showed that, at each site, most of the conductive fractures are oriented such that they are critically stressed for failure. That is, they lie above the Coulomb failure curve for tan $\phi > 0.6$, as shown on the left column of Figure 8.8. In contrast, the nonconductive fractures lie below the failure curve, as shown in the right column of Figure 8.8. One important inference is that potentially active faults are the most important conduits for fluid flow in many fractured-rock environments.

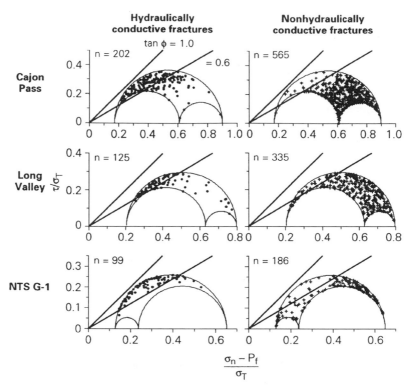

Figure 8.8 Normalized shear stress versus normalized effective normal stress for hydraulically conductive (left) and nonconductive (right) fractures at Cajon Pass, Long Valley, and Yucca Mountain. The stresses are normalized by the total overburden pressure σ_T. Coulomb failure curves are shown for $\tan \phi = 0.6$ and 1.0, with cohesion τ_0 assumed to be zero. In Figures 8.2 and 8.3 we used two-dimensional Mohr diagrams to represent the results of triaxial tests. Here, *three-dimensional Mohr diagrams* are required to represent the data from natural, three-dimensional systems. The outer Mohr circle is still defined by σ_1 and σ_3, and the two inner Mohr circles intersect at the value of σ_2. (See Jaeger and Cook (1979) for the construction of three-dimensional Mohr diagrams.) After Barton and others (1995).

Problems

8.1 At 10-km depth below the land surface the greatest principal stress σ_1 is vertical and has a magnitude of 2,500 bars. The least principal stress σ_3 is 1,500 bars. **(a)** Assuming that failure conditions are described by Eq. (8.7) with $\tau_0 = 200$ bars and $\phi = 30°$, how high must the fluid pressure be to allow failure? How does the fluid pressure at failure relate to **(b)** the "normal" hydrostatic condition of ~ 100 bars/km and **(c)** to the lithostatic condition of ~ 250 bars/km? **(d)** A 250-m-thick shale unit occurs at 10 km depth and is generating petroleum from kerogen (Chapter 6.1) at a rate of 10^{-14} m^3/(m^3-s) (Table 4.1). Hydraulic heads above and below the shale are approximately hydrostatic. How low would the shale permeability have to be in order for natural hydraulic fracturing to occur? Does this value seem reasonable? (Part (d) was devised by Professor Roy Haggerty for his students at Oregon State University.)

8.2 The injection of $\sim$6.25 $\times$ 10^8 liters of waste fluid over a four-year period induced a swarm of seismic events in an approximately 10 km $\times$ 3 km $\times$ 3.5 km volume of Precambrian gneiss below the Rocky Mountain arsenal (Chapter 8.3.1). **(a)** Assume that the waste fluid acted as a distributed source over this volume of $\sim$10^2 km^3, and use Eq. (4.4) to calculate the limiting hydraulic conductivity K required in order for injection at a rate of $\sim$1.6 $\times$ 10^8 liters/year to cause significant increases in fluid pressure. (Assume that the waste fluid had the properties of pure water at standard conditions.) **(b)** Is the value calculated using Eq. (4.4) consistent with the transmissivity value of $\sim$2 $\times$ 10^{-5} m^2/s (Hsieh and Bredehoeft, 1981) determined by hydraulic testing of the Precambrian gneiss? (Recall that $T = LK$: Chapter 1.5.3.) **(c)** If not, suggest possible reasons for the discrepancy.

8.3 Draw a Mohr–Coulomb diagram (e.g., Figure 8.2) to demonstrate that the oil-reservoir rocks at the Rangely, Colorado site should fail at a "critical" fluid pressure of $\sim$257 bars, given the independently determined values of σ_1 (552 bars), σ_3 (314 bars), and ϕ (39°), and assuming that τ_0 is negligible. (See Chapter 8.3.2.)

9

Evaporites

Evaporites are sediments deposited from natural waters that have been concentrated as a result of evaporation. The source waters can be either marine or continental in origin. Evaporite deposits are important sources of gypsum, halite, sylvite, and other economically important minerals. The genesis of these deposits is controlled partly by the movement of groundwater into and/or out of basins that are hydrologically closed with respect to surface-water outflow. During the evolution of an evaporite deposit, both the quantity (fluxes) and quality (chemistry) of local groundwater can control salt deposition.

Substantial thicknesses of evaporites are buried in some sedimentary basins. There, they act as important barriers to groundwater flow, but they are also susceptible to dissolution by groundwater. Buried evaporites can migrate slowly upward through overlying sediments in the form of salt domes. The interaction between groundwater and salt domes results in variable-density convection adjacent to the dome and diagenesis at the top of the dome. In this chapter we organize and discuss the role of groundwater in terms of these three different stages of evaporite evolution: formation, burial, and diapiric rise.

9.1 Evaporite formation

In this section we introduce the theoretical evaporite sequence that would be produced by closed-basin evaporation of seawater, and we compare this sequence with those actually observed in nature. We then proceed to show how the discrepancies can be explained by the role that groundwater plays in adding and removing solutes from evaporite basins. We also discuss the role of groundwater in continental evaporite deposits.

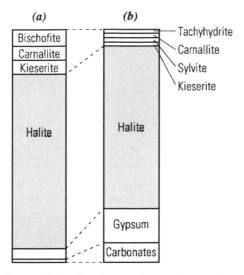

Figure 9.1 Approximate relative abundances of evaporite salts found in (a) the theoretical evaporation of seawater (Harvie and others, 1980) and (b) an average of some major marine evaporite deposits around the world. Modified from Borchert and Muir (1964).

9.1.1 The marine evaporite problem

It has long been known that by simply evaporating seawater one can obtain approximately the same salts that are observed in most major evaporite deposits (Usiglio, 1849). However, in spite of the fact that both the theoretical and observed evaporation sequence will produce calcium sulfate, halite, and the more soluble salts of calcium and magnesium chloride, some notable deviations occur (Figure 9.1). The differences between theory and observation include (1) the relative abundance of observed calcium sulfate, (2) the absence of magnesium sulfates in many deposits, (3) the presence of sylvite in some deposits, and (4) the presence of tachyhydrite in some deposits. Theory predicts the precipitation of the sulfate minerals, but in amounts that differ from typical deposits. Theory also predicts that sylvite and tachyhydrite should not precipitate, yet they are observed in many deposits. These deviations have been of interest to geochemists for many years and have inspired many theories (Hardie, 1984). Most recent theories involve groundwater either as a source or sink of solutes during the evaporation process.

Thermodynamics allows the prediction of mineral precipitation sequences during evaporation (Hardie and Eugster, 1970). An improved understanding of the thermodynamics of brines (Pitzer, 1973, 1975; Harvie and Weare, 1980; Harvie and others, 1982, 1984) has led to more detailed predictions of brine

Table 9.1 *Major element concentrations in average seawater.[a]*

Element	Concentration (mg/kg)
calcium	411
magnesium	1,290
sodium	10,760
potassium	399
chloride	19,350
bicarbonate	142
sulfate	2,710
bromide	67
strontium	8
boron	4.5
fluoride	1.3
silica	0.5–10
oxygen	0.1–6
pH	8.22

[a] After Drever (1982).

Table 9.2 *Some prominent minerals found in evaporite deposits.*

Name	Formula	Name	Formula
anhydrite	$CaSO_4$	hexahydrite	$MgSO_4 \cdot 6H_2O$
bischofite	$MgCl_2 \cdot 6H_2O$	kainite	$KMgClSO_4 \cdot 3H_2O$
bloedite	$Na_2Mg(SO_4)_2 \cdot 4H_2O$	kieserite	$MgSO_4 \cdot H_2O$
calcite	$CaCO_3$	mirabilite	$Na_2SO_4 \cdot 10H_2O$
carnallite	$KMgCl_3 \cdot 6H_2O$	polyhalite	$K_2MgCa_2(SO_4)_4$
dolomite	$CaMg(CO_3)_2$	sylvite	KCl
glauberite	$Na_2Ca(SO_4)_2$	tachyhydrite	$Mg_2CaCl_6 \cdot 12H_2O$
gypsum	$CaSO_4 \cdot 2H_2O$	thenardite	Na_2SO_4
halite	$NaCl$	trona	$Na_3H(CO_3)_2 \cdot 2H_2O$

evolution, including predictions for the specific case of seawater (Harvie and others, 1980). If one calculates the reaction path for the evaporation of standard seawater (Table 9.1), one obtains a fairly precise estimate of the sequence and amounts of minerals that should precipitate. These minerals (Table 9.2) are all salts composed of the major ions in seawater. Such a calculation indicates the following amounts of salts would precipitate after cumulative evaporation of 1 km of seawater: 44 mm of dolomite, 88 cm of gypsum, 12.6 m of halite, 99 cm of kieserite, 1.75 m of carnallite, and 2.23 m of bischofite (Sanford and Wood, 1991). In addition, 2 mm of glauberite, 11 mm of polyhalite, 4 mm of hexahydrite, and 2 mm of kainite would precipitate between halite and kieserite but would eventually redissolve if they remain in contact with the brine. To perform these calculations, assumptions must be made regarding whether such back-reaction of minerals will occur, whether gypsum or anhydrite will be the calcium-sulfate phase, and whether calcite or dolomite will be the carbonate phase. Results obtained under different assumptions vary slightly but do not change the overall reaction path.

As indicated previously, certain conspicuous differences exist between marine mineral assemblages predicted on the basis of standard seawater and those observed in real deposits. The first noticeable difference is that a major potash ore mineral, sylvite, is absent from predicted assemblages. The Paradox Basin (Hite, 1961) and Michigan Basin (Nurmi and Freedman, 1977) are classic marine evaporite deposits that include significant amounts of sylvite. Even a significant variation in the composition of seawater is not sufficient to change the predicted mineral suite and cause sylvite to appear (Holland, 1984), although sylvite may be secondary in some deposits (Wardlaw, 1968). Another soluble salt that is observed though not predicted is tachyhydrite. This very soluble salt is present in large quantities in certain evaporite deposits in Brazil (Wardlaw, 1972) and Thailand (Hite and Japakasetr, 1979). The mineral suites present in these deposits cannot be predicted by evaporating seawater, but they could be explained if significant diagenetic reactions or groundwater provided additional solutes. The common magnesium sulfate salts also often contradict theory. Many evaporite deposits are "$MgSO_4$-poor" in spite of theoretical predictions that a significant mass of magnesium sulfates should precipitate (Hardie, 1990). One way these mineral anomalies can be accounted for is by considering groundwater either as a source or sink for solutes, and the rest of this section will be devoted to this topic.

9.1.2 Groundwater inflow

The suite of minerals precipitated during the evaporation of a brine depends primarily on its chemical composition (i.e., the ionic ratios of the solutes), which in turn depends on the solutes supplied by the source water. If pure seawater is the only source of water and the basin is hydrologically closed, then the theoretical calculations for evaporation of seawater should be applicable. If other

significant sources of solutes are present, however, then the sequence of minerals that precipitates may be different. Given the wide range of potential source-
water compositions, this would seem to allow an easy solution to the discrepancies described above – for example, one might assume that terrestrial (e.g.,
river) water provides the solutes required for the observed mineral suite. Some
nonmarine evaporites are in fact clearly identifiable as such (Chapter 9.1.5), but
other evaporites clearly have a dominantly marine signature (Hardie, 1984).
Furthermore, given the relatively high concentrations of solutes in seawater
with respect to typical groundwater or river water, it would take an unusually
large proportion of the latter to impact the overall solute composition. This suggests that there may be rather special conditions under which addition or loss
of solutes can account for the observed mineralogy. Two such conditions have
been proposed recently to account for some of the mineralogical anomalies,
but in order to explain these hypotheses we must first introduce the topic of the
chemical divide (Hardie and Eugster, 1970).

The sequence in which minerals are precipitated during the evaporation of
any brine is a function of the solute ratios, rather than the absolute solute
concentrations. As precipitation of a specific salt begins, one of the elements
will be limiting in amount, such that the amount precipitated cannot exceed the
input of the element in shortest supply. This limiting element is then continually
consumed in the precipitation, whereas the other element or elements in the salt
continue to increase in concentration in an amount proportional to the excess
above that needed to precipitate the salt. When equilibrium is exceeded with
respect to any given mineral, different reaction paths will be taken, depending
upon which element is limiting. At each point of mineral equilibria, therefore,
there is a *chemical divide* at which the evolving brine must continue along one
of at least two divergent reaction paths.

The ratios of the solutes in the source water will control which elements are
limiting and, thus, which paths are taken. Each path leads to a different mineral
equilibrium further along the path and ultimately to a different overall suite
of minerals. (This solute evolution applies to the major ionic solutes subject
to mass- and charge-balance constraints, i.e., not to silica or redox species.)
The first divide will be most significant in brine evolution and usually occurs at
either calcite or gypsum precipitation. For seawater, the first divide occurs when
gypsum precipitates (Figure 9.2). Depending upon whether calcium or sulfate
is limiting, that elemental concentration will decrease as the brine evolves, and
other solute ratios (e.g., Ca/Mg) will then be affected. Thus if certain elements
are added or subtracted from standard seawater, the resulting evaporite mineral
suite might be more consistent with observation.

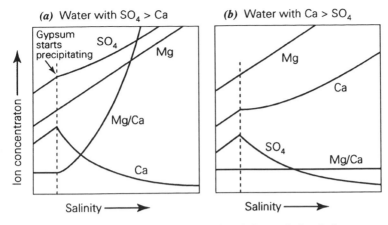

Figure 9.2 The chemical divide concept as it affects brine evolution during evaporation for an initial water with concentrations of (a) sulfate greater than calcium and (b) calcium greater than sulfate. From Hardie (1987).

9.1.3 CaCl₂ brines

One major discrepancy between theoretical and observed evaporite deposits is the abundance of magnesium sulfate salts. Ancient potash evaporites can be categorized in terms of a $MgSO_4$-rich group with mineral assemblages predicted by evaporating modern seawater and a $MgSO_4$-poor group with mineral assemblages that cannot be derived from evaporating modern seawater. The $MgSO_4$-poor deposits outnumber the $MgSO_4$-rich deposits by a ratio of greater than three to one (Hardie, 1990). The $MgSO_4$-rich deposits are usually characterized by the presence of the post-halite minerals polyhalite, kieserite, or kainite. Carnallite and bischofite are commonly present as the most soluble phases. The $MgSO_4$-poor deposits are usually characterized by the post-halite minerals sylvite, carnallite, and sometimes tachyhydrite. Famous examples of $MgSO_4$-rich deposits include the Caspian Sea (Zharkov, 1984), the Delaware Basin of New Mexico (Jones, 1972), the Zechstein deposits of Northwestern Europe (Harvie and others, 1980), and the Mediterranean Sea (Kuehn and Hsu, 1978). Famous examples of $MgSO_4$-poor deposits include the Michigan Basin (Nurmi and Freedman, 1977), the Rhine Graben (Meriaux and Gannat, 1980), the Qaidam Basin of China (Casas and others, 1992), and the Dead Sea (Zak, 1974).

One geologic setting that appears to be ideal for evaporite formation is the active rift or transtensional strike-slip basin (Hardie, 1990). These evaporite basins are active, hydrologically closed, and arid, regardless of latitude, due to flanking mountain ranges that create orographic deserts (Figure 9.3). They

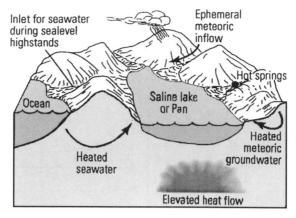

Figure 9.3 Conceptual model of a transtensional strike-slip basin with evaporite formation. Upwelling $CaCl_2$ waters represent a major fraction of the solute input. After Hardie (1990).

are usually associated with newly initiated branches of existing ocean basins, such that there is usually a local source of seawater available to fill the developing topographic depressions. At some point in time, failure of the rift and/or isolation from the ocean prevents the evolving evaporite deposit from being completely overrun and redissolved by seawater. Active rift systems are often associated with geothermal anomalies that drive the convection of hot brines. Hot springs in many rift basins worldwide have calcium chloride ($CaCl_2$) signatures similar to that of oceanic-rift brines (Whitehead and Feth, 1961; White and others, 1963). In each case the circulating brines were likely created by seawater interacting with basalts to produce a calcium-chloride dominated chemistry (Chapter 5.1.2; Hardie, 1983). Upwelling $CaCl_2$ brine is a potential source of solute that could significantly alter the reaction paths of evaporating brines.

The addition of a $CaCl_2$-dominated water to seawater during evaporation would cause changes in the reaction path that could explain some of the anomalies observed in the rock record. Many closed-basin hot springs have a dissolved solids content near if not greater than that of seawater (Hardie, 1990), so that a moderate volumetric flux of hot-spring water could change the overall chemistry and reaction path of the mixed brine. Adding excess calcium and chloride to evaporating seawater would cause two major changes. First, if the calcium concentration exceeded the sulfate concentration on a molar basis, then sulfate would become the limiting element during gypsum precipitation. As concentrations increased beyond gypsum equilibrium, then, according to the chemical divide rule, calcium concentrations would increase and sulfate concentrations would decrease. These trends are the opposite to those for normal seawater (compare Figures 9.2a and 9.2b). Second, as the brine progressed beyond halite saturation toward the more soluble salts, it would never become supersaturated with respect to the magnesium sulfates, because of the low sulfate concentrations. Thus the resulting mineral assemblage would be $MgSO_4$-poor. Moreover, if enough calcium were present, tachyhydrite (rather than bischofite) would eventually precipitate after carnallite. Some of the major discrepancies between the theoretical and observed evaporite assemblages can thus be explained by the addition of a $CaCl_2$ water.

9.1.4 Magnesium depletion

The fact that the original composition of seawater needs to be enriched in calcium with respect to sulfate in order to produce $MgSO_4$-poor evaporite deposits has long been recognized (Braitsch, 1971). This modification to seawater alone, however, cannot cause sylvite to occur as a primary precipitate – yet sylvite occurs in many $MgSO_4$-poor deposits.

When an evaporating brine exceeds halite precipitation, competition occurs among the chloride minerals sylvite, carnallite, and bischofite as to which will precipitate. For standard seawater evaporation, the abundance of magnesium in the brine will cause potassium to be taken out as carnallite ($MgKCl_3$-$6H_2O$) rather than as sylvite (KCl). If the brine is somehow depleted in magnesium, however, it is possible for sylvite equilibrium to be reached sooner than that of carnallite and for sylvite to precipitate in favor of carnallite. Such a reduction in magnesium will also cause tachyhydrite ($CaMg_2Cl_6$-$12H_2O$) to precipitate in favor of bischofite ($MgCl_2$-$6H_2O$), which could explain the presence of tachyhydrite in some deposits.

One process that may lead to both Ca enrichment and Mg depletion is dolomitization of carbonate rocks flanking the basin (Williams-Stroud, 1994). The conceptual model proposes that water levels in the basin will on average be slightly below sea level because of evaporation, such that the water level gradient will cause seawater to continually migrate through (and dolomitize) the carbonate rocks between the ocean and the evaporite basin (Figure 9.4). There is some doubt that seawater alone can dolomitize limestone, but the seawater in this model may have been additionally concentrated by evaporation after overwashes onto the exposed land surface. Alternatively, some small fraction of the Mg-rich brine in the basin may be convecting through the limestone, giving up some Mg in the process. The Paradox Basin of Utah and Colorado, which has a conspicuous absence of $MgSO_4$ minerals and a relative abundance of sylvite, is one locality where this process may have operated (Williams-Stroud, 1994).

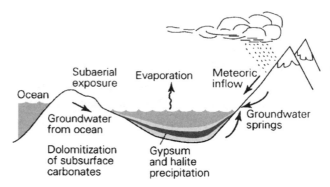

Figure 9.4 Conceptual model of an evaporite basin where Mg-depleted seawater that has contributed to dolomitization provides an important fraction of the solute input. After Williams-Stroud (1994).

A second process that may lead to both Ca enrichment and Mg depletion is temporal variation in fluxes through the mid-ocean ridges (Hardie, 1996). Throughout the Phanerozoic, $MgSO_4$-poor evaporite deposits can be correlated with high calcite/aragonite precipitate ratios, which are also thought to be caused by a high Mg/Ca ratio in seawater. Periods dominated by $MgSO_4$-poor deposits include the Cambrian through Mississippian and the Jurassic through the Paleogene. Models of seawater chemistry that account for river water, mid-ocean ridge water, calcite precipitation, and silica precipitation demonstrate that a 10% change in the mid-ocean ridge flux can lead to significant changes in the ratios of the major ions in seawater (Spencer and Hardie, 1990). The fact that mid-ocean ridges do produce $MgSO_4$-poor water suggests that temporal changes in associated flux may affect evaporite deposits. (See Chapter 7.9 for a brief discussion of hydrothermal circulation at the mid-ocean ridge.)

9.1.5 Continental evaporites

Up to this point, our discussion has focused on evaporites of marine origin. However, there are many ancient evaporite deposits with salt assemblages so different from that predicted from seawater evaporation that a nonmarine (continental) source of solutes is the only viable explanation (Hardie, 1984). Moreover, many evaporite deposits are forming today in continental environments in climates where evaporation exceeds precipitation (e.g., Last, 1989). Determining the origin of an ancient evaporite deposit (marine or nonmarine) is usually done by examining both the regional geologic setting and the mineral assemblage of the deposit. The potential evaporite mineral assemblage for any natural water or brine can easily be computed (Bodine and Jones, 1986). Examination of many natural waters indicates that, upon evaporation, the precipitation of certain salts can reveal the nature of their environments (Jones and Bodine, 1987). The minerals trona, glauberite, mirabilite, thenardite, polyhalite, and bloedite, for example (Table 9.2), rarely form from marine waters and are typical of nonmarine deposits.

There are many examples of continental evaporite deposits in which groundwater plays an important role in the salt development by contributing solutes through local springs. The Amadeus basin of central Australia contains a chain of playa–salt lakes more than 100 km long that is fed mainly by groundwater discharge (Jankowski and Jacobson, 1990). The Lake Magadi rift basin, Kenya, contains the largest sodium-bicarbonate deposit in Africa and is fed mostly by springs draining the surrounding volcanic terrane (Jones and others, 1977). The Great Salt Lake, Utah, contains a significant fraction of evaporated NaCl spring water (Spencer and others, 1985). Pilot Valley, Utah, is fed by peripheral springs that are discharging what is likely brine recirculated from beneath the main playa floor (Duffy and Al-Hassan, 1988). Saline lakes of the Southern High Plains of Texas have solutes whose origin is groundwater of the local Ogallala Formation (Wood and Jones, 1990). Groundwater outflow also significantly controls the lake salinities (Wood and Sanford, 1990). Some of these lakes have been a commercial source for sodium sulfate.

9.1.6 Groundwater outflow

Models developed to explain evaporite deposits usually describe a topographically closed basin with the exception of, in the case of marine evaporites, a narrow strait connecting the basin to the sea. Solute- and water-flux components of such a model include water influx and associated solute concentrations (Q_i and C_i) and water outflux and associated solute concentrations (Q_o and C_o) (Figure 9.5). Outflow continually removes brine either by density-driven flow through the strait in an undercurrent, in a marine model (King, 1947), or through groundwater seepage in a nonmarine model (Adams and Rhodes, 1960). Rainfall (Q_p and C_p) and surface runoff (Q_s and C_s) can add components of water and solute fluxes. Evaporation removes water at a rate Q_e that is larger than the precipitation rate Q_p. The concentration of the water in the basin, C_o, increases above the input concentration, C_i, but continuous redilution of the evolving brine by inflow limits the basin-water solute concentration.

To quantify the conceptual model of a basin with groundwater outflow, we will assume that the evolving brine stays well-mixed within the finite region occupied by the brine lake. The well-mixed brine also includes that brine present in the pore water in the top several meters of sediment. Thus in cases when

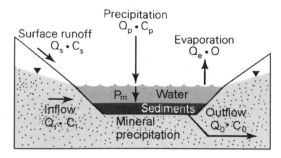

Figure 9.5 Conceptual model of a leaky basin showing major solute-flux components.

the surface water is shallow, and may dry up altogether, the volume of brine considered in the calculations does not go to zero but remains a finite volume of pore water. This assumption of a well-mixed brine reduces the model to a lumped-parameter equation that can be used to describe the evolution of the brine composition over time. In real systems, spatial gradients in solute concentrations can develop, but including them would create a significantly more complicated model, and such complexity is not necessary to understand the overall evolutionary path a brine will follow.

By combining the solute fluxes of the conceptual model (Figure 9.5), a first-order differential equation for the solute mass balance is obtained:

$$Q_pC_p + Q_sC_s - Q_iC_i - Q_oC_o - P_m = \frac{d(VC_O)}{dt}, \tag{9.1}$$

where P_m is the mineral precipitation rate [M/t], V is the volume occupied by the basin water, and t is time. Dimensionless parameters can be introduced to simplify the analysis, including the length of time relative to that required to evaporate one basin volume, $Q_et/V = t_o$, the ratio of water outflow to water inflow (leakage ratio), $Q_o/Q_i = Q_r$, and the ratio of the basin solute concentration to the input (and initial) concentration, $C_o/C_i = C_r$. For cases in which there is no mineral precipitation, surface runoff and atmospheric precipitation are negligible, and Q_e and Q_i are nearly identical (the leakage is relatively small), the solution to the solute mass-balance equation (9.1) can be written as

$$C_r = \frac{1}{Q_r} - \left(\frac{1}{Q_r} - 1\right)\exp[-Q_rt_o]. \tag{9.2}$$

From Eq. (9.2) it can be observed that at initial time $t_o = 0$ the concentration ratio equals one, but that at infinite time (steady state) the concentration ratio is equal to the inverse of the leakage ratio. Thus the increase in brine salinity relative to the inflow salinity is limited by the ratio of the fluid inflow to outflow.

At steady state, a small concentration in a large inflow will be balanced by a large concentration in a small outflow. The exponential term in Eq. (9.2) determines how quickly a brine will approach steady state, and calculations indicate that for leakage ratios of 0.10, 0.01, and 0.001, approximately 30, 300, and 3,000 basin volumes are required to reach a virtual steady state.

For a basin with a water depth of only a few meters and a net evaporation rate of 1 m/yr, the time required to evaporate one "basin volume" is a few years. The average residence time of the solutes will be the water residence time multiplied by C_r. If the salinity of a marine-evaporite brine were 10 times that of seawater (not uncommon), the solute residence time would be ten times the water residence time. Such a short water residence time relative to the salt residence time would support the assumption of a well-mixed basin. Consider also that for a basin that has only 1% leakage, steady state would be reached in about 1,000 years. Given the thousands to millions of years required for thick evaporite deposits to form, the basin must generally be near steady state with respect to water and solute fluxes. Changes in mineral precipitation over time would occur only as leakage conditions in the basin change. Moreover, as salts are deposited and compacted, the permeability of the basin floor is likely to decrease (see Chapter 9.2), causing a decrease in the leakage and the appearance of soluble salts near the final stages of the basin development.

In using Eq. (9.2) we assume that no mineral precipitation is occurring. In reality, the concentrations of many different solutes evolve simultaneously in a basin, causing a variety of different minerals to precipitate (Eugster and Jones, 1979). Multiple solutes and minerals generate a coupled set of mass-balance equations that is too complex for analytical solutions, and thus numerical methods must be used to analyze the problem. Such an analysis was performed by Sanford and Wood (1991) using seawater as an input to the geochemical code PHRQPITZ (Plummer and others, 1988), which contains the virial-coefficient approach of Pitzer (1973, 1975) for calculating the mineral thermodynamics and thus can be applied to brines. The leakage term in the analysis was handled by continually adding seawater salts to an evolving brine and mixing with standard seawater. Output from the model predicted the precipitation of soluble salts and the evolution of the brine to a condition of steady state with respect to leakage.

For a leakage ratio of 1 part in 1000, for example, a basin will act virtually closed hydrologically and brine evolution paths will resemble that of a completely closed basin (Figure 9.6a). The intermediate salts of glauberite, polyhalite, hexahydrite, and kainite are predicted to precipitate in small quantities along with the major salts of gypsum, halite, kieserite, carnallite, and bischofite. An invariant point is reached after about 150 basin volumes evaporate

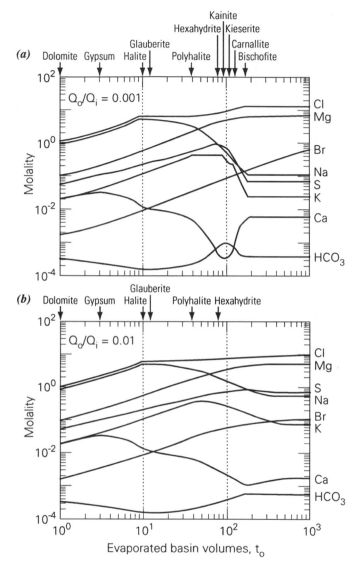

Figure 9.6 Brine evolution predicted by the evaporation of seawater with leakage ratios of (a) one part in one thousand and (b) one part in one hundred. After Sanford and Wood (1991).

and long before steady state is achieved with respect to leakage. This is the point at which the ratios of the solutes in the inflow water are being matched by the ratios of the solutes in the various minerals being precipitated. For a leakage ratio of 1 part in 100, however, steady state with respect to leakage (at about

300 basin volumes) is reached before the invariant point (Figure 9.6b). For this system, the brine never reaches supersaturation with respect to kieserite, carnallite, or bischofite because it is continually being diluted with the inflow water that replaces the leakage. At steady state for this case, polyhalite and hexahydrite will continue to precipitate. Thus large amounts of these intermediate minerals could accumulate in spite of their predicted scarcity in the closed-basin models.

Two examples can be offered where such intermediate minerals exist in significant quantities in nature but are not predicted in significant amounts in closed-basin models. The first is the presence of sylvite in many deposits. As discussed in Chapter 9.1.2, if the inflow water is seawater with an elevated Ca/Mg ratio, sylvite will be predicted to precipitate in the mineral assemblage. Sylvite will also, however, be predicted to redissolve in favor of carnallite as the brine salinity continues to rise. A small amount of leakage from the basin could cause a steady state salinity to be reached after sylvite supersaturation but before its reundersaturation, leading to significant thicknesses of sylvite being deposited.

The second example is the presence of mirabilite in the saline lakes of the southern High Plains of the United States. Some of the ephemeral saline lakes of West Texas and eastern New Mexico contain mirabilite ($Na_2SO_4 \cdot 10H_2O$) deposits that are economically viable. An analysis of the evaporation paths of Ogallala aquifer water in these lakes predicts the precipitation of mirabilite, but this later dissolves forming glauberite ($Na_2Ca(SO_4)_2$) and halite (NaCl) (Wood and Sanford, 1990). Leakages of about 1 part in 1000, however, yield steady-state conditions and predictions of meter-thick deposits of mirabilite without the precipitation of more soluble salts – exactly what is observed in some lakes. Hydrologic measurements below these lakes yield typical leakage rates of 1 mm/year (Sanford and Wood, 1995) – a value that is about 1/1000th of the evaporation rate in the area – and this supports the leakage hypothesis.

9.2 Bedded evaporites

Once evaporites are deposited in layers and start to undergo burial, their physical hydraulic properties begin to change dramatically. Crystals that sink from a brine will initially have a porosity somewhere between 40 and 50 percent. As burial proceeds, compaction will usually reduce this porosity to less than a few percent at 300-m depth (Sonnenfeld, 1984). The interstitial brine is forced upward opposite to the net downward movement of the sediments. Below 500-m depth, porosity is usually less than 1 percent, but the exact value is also a function of the mineralogy.

Permeability is a function of porosity, and it decreases upon burial at an even more accelerated rate than porosity. Uncompacted salt crystals can have the same permeability as an unconsolidated clastic deposit with the same equivalent hydraulic grain radius. Thus salt crystals can have permeabilities approaching 10^{-7} m^2, but after 500 m of burial the permeability may be reduced to the order of 10^{-21} m^2. This extreme change in permeability is due to the plasticity of salt under deformation and its tendency to recrystallize easily under conditions of increasing pressure and temperature. This behavior causes the salt to squeeze out the in situ brine and exclude virtually all remaining pore spaces.

The fact that beds of evaporite are virtually impenetrable to groundwater is demonstrated by the existence of numerous evaporite beds that are hundreds of millions of years in age. Being highly soluble, these deposits would have dissolved long ago had there been substantial groundwater movement. For this reason, salt deposits have been considered seriously as sites for long-term disposal of toxic and radioactive wastes. Much of the available information about salt permeability and dissolution has come from studies surrounding these potential disposal sites (Bredehoeft, 1988). An example is the Waste Isolation Pilot Project (WIPP) site of New Mexico. In situ hydraulic tests there reveal hydraulic conductivity values for undisturbed bedded halite to be about 1×10^{-14} m/s (Beauheim and others, 1993). At such low conductivities, evaporite beds also become exceptionally efficient confining beds. The effect of confining beds on regional groundwater flow systems was discussed in Chapter 4.1.

In spite of the fact that salt can form confining layers of very low permeability, there is widespread evidence for the subsurface dissolution of bedded salt. Most subsurface bedded salts are bounded by regions where groundwater flow has evidently dissolved away much or all of the deposit. Near the land surface, such dissolution can create evaporite-karst features that are not uncommon throughout the United States (White, 1988) and the world. Deeper in the subsurface, fractures or faults can concentrate fluid flow to create local dissolution features. Void spaces caused by salt dissolution can propagate upward through the overlying rocks until depressions eventually appear at the land surface (Davies, 1989).

There is evidence that the depth dependence of halite permeability may reverse once burial becomes deep enough. Experimental data show that the intercrystalline dihedral angles of halite vary with pressure and temperature (Lewis and Holness, 1996). For angles less than 60°, porosity becomes predominantly interconnected along crystal triple junctures, whereas at angles greater than 60°, porosity is virtually isolated. Halite dihedral angles become continually more acute at higher pressures and temperatures until, at equivalent burial depths of

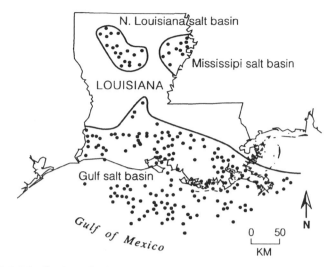

Figure 9.7 Distribution of salt domes in Louisiana and adjacent parts of the Gulf of Mexico. From Ranganathan and Hanor (1988).

about 3 km, recrystallization should yield angles below 60°. In many basins, evaporite deposits are buried to several kilometers depth (e.g., along the Gulf Coast of the United States). At such depths these salt beds could have permeabilities comparable to those of sandstones. This means that in otherwise argillaceous environments, salt beds could be acting as fluid conduits! This inference is consistent with stable isotopic evidence suggesting that deep evaporite beds have drawn in fluid from surrounding sediments (Knauth and others, 1980; Land and others, 1988).

9.3 Salt domes

Once layered beds of salt have been buried deeply enough, the contrast in density between the salt ($2,200$ kg/m^3) and the surrounding sediment ($2,500$ kg/m^3 at 3-km depth) creates an unstable condition that leads to the slight bulging of salt upward through the overlying sediment. The salt focuses itself into diapirs or *salt domes* that can rise several kilometers and eventually break through to the land surface. The upward migration of the salt is accompanied by a reciprocal downward sagging of the surrounding sediments, such that much of the relative vertical displacement is actually due to the latter (Trusheim, 1960). Salt domes often exist in great number in basins that contain deeply buried evaporites (Figure 9.7). They have been of great interest to geologists due mostly to their role in the entrapment of petroleum; salt domes arch overlying sediments as they

Evaporites

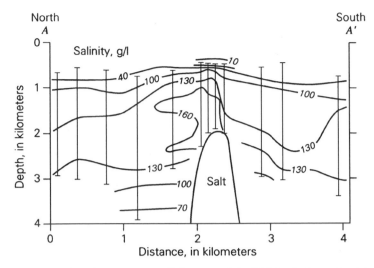

Figure 9.8 Variations in pore water salinity around the Welsh salt dome, Louisiana, as deduced from spontaneous potential measurements. From Bennett and Hanor (1987), with kind permission of Elsevier Science-NL, Amsterdam.

rise and thus create structures that can trap migrating hydrocarbons. The existence of salt domes, however, also greatly affects the quality of the surrounding groundwater (Figure 9.8) and, conversely, the surrounding groundwater greatly affects the diagenesis and dissolution of the salt. The role of groundwater in diagenesis is discussed further in Chapter 10.

9.3.1 Variable-density convection

Due to the presence and high solubility of halite in most marine salt deposits, groundwater that is in contact with the salt eventually reaches salinities greater than 300 g/liter. Brine with this salinity has a fluid density of about 1,250 kg/m³ (25% greater than dilute groundwater at standard temperature and pressure). This sets up a density contrast that has the potential to drive large-scale groundwater convection along the flanks of salt domes. An analysis of the potential for convection can be performed using the equations for variable-density solute transport in groundwater (Eqs. 2.15 and 2.26–2.28). A mathematical model can be constructed to represent a vertical cross section with one end in the center of the dome and the other end at a distance equal to half the mean separation distance between domes in the basin of interest (Ranganathan and Hanor, 1988). To reflect the radial symmetry of the dome, the numerical simulation is best done in two-dimensional radial coordinates. The boundary conditions to the problem usually involve no-flow conditions on all sides. A constant halite-equilibrium salinity is specified at the dome-sediment contact. Initial conditions elsewhere in the basin are a hydrostatic pressure gradient and freshwater salinities. Similar analyses can be done to estimate the effects of the thermal regime and geopressurized fluids on groundwater convection.

For a homogeneous basin, this problem specification yields a single convection cell with its downward flow limb wrapped around the border of the salt dome (Figure 9.9a). The salinity distribution for such a flow system develops with brine accumulating at the bottom. Because no external fluid fluxes were specified at the boundaries, the magnitude of the flow rates will be a direct and linear function of permeability. Equation (2.26) indicates that it is the vertical flux that controls density-driven flow (the density term drops out of horizontal fluxes). Thus vertical permeability is the main controlling parameter for the onset of convection cells. This implies that convection cells are not likely to develop across low-permeability argillaceous sediments, but rather within higher-permeability layers of sands or gravels. In fact, the salinity distribution for this simulation of a homogeneous system does not match the distribution seen around many salt domes (Figure 9.8).

Many basins, such as those along the Gulf of Mexico (Chapters 4.1.4 and 10.2.1) have layers of low permeability that result in elevated fluid pressures at depth, so that the overall pattern of fluid flow is up and out of the basin. A similar mathematical model of fluid flow near a salt dome can be modified to include a deep source of geopressurized fluids. In this case, if vertical faults near the salt dome act as conduits, the fluid motion may be directed upward along the edge of the dome (Figure 9.9b), and the flow pattern could create a plume of brine that extends outward from near the top of the salt dome. This salinity distribution is more like the observed patterns (Figure 9.8) than the distribution caused by free convection alone.

Because salt domes often originate at depths of several kilometers, it is possible that the thermal regime around the dome could cause variations in density that may, in turn, cause groundwater convection. Salt has a greater thermal conductivity than typical basin sediments. The implication of this is that under steady-state conditions, salt domes act as heat conduits and have lower geothermal gradients than the surrounding sediments. Analyses of constant-salinity heat transport around salt domes, analogous to the solute transport simulations described above, show that for a homogeneous basin, two cells would develop – one cell near the top of the dome with upward flow at the salt edge, and another cell below extending to the bottom of the system, with downward flow

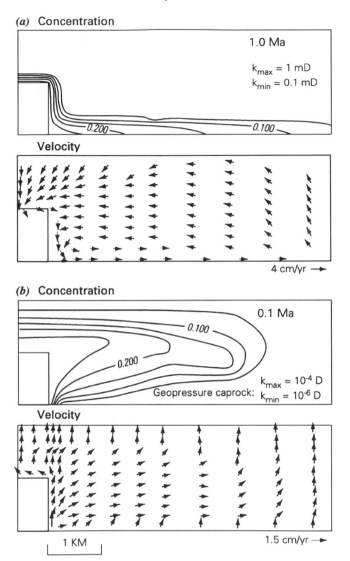

Figure 9.9 Simulated salinity and groundwater velocity distributions near a salt dome after (a) 1.0 Ma in a normally pressured system and (b) 0.1 Ma in a geopressured system. After Ranganathan and Hanor (1988).

near the salt edge (Ranganathan and Hanor, 1988). In this environment the density contrasts associated with temperature variations are smaller than those created by salinity variations, and thus the thermal effects are thought to be less important.

9.3.2 Caprock formation

At the top of many salt domes is a *caprock* of salt and rock that has formed by continuous dissolution of the mother salt and diagenesis of the adjacent rocks. Dissolution of the salt occurs as rising salt continuously comes into contact with undersaturated groundwaters. Diagenesis occurs because of contact between the salt and these undersaturated waters and brine moving up the flanks of the dome. The rising brine often includes dissolved metals that can react with anhydrite, leading to the precipitation of metal sulfide deposits. A significant part of the caprock diagenesis is controlled by bacterial action upon hydrocarbons or sulfate. A classic caprock contains zones of anhydrite (nearest the salt), gypsum, and calcite (Figure 9.10). Porous, permeable transitional areas are often present between the major zones due to the discontinuity of the ongoing

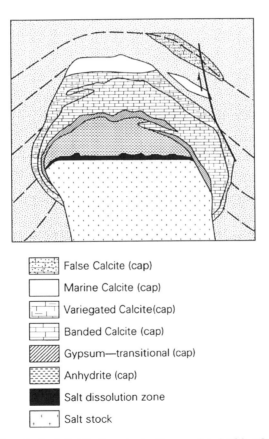

False Calcite (cap)

Marine Calcite (cap)

Variegated Calcite(cap)

Banded Calcite (cap)

Gypsum—transitional (cap)

Anhydrite (cap)

Salt dissolution zone

Salt stock

Figure 9.10 Mineralogical distributions typically associated with salt-dome caprocks in the Gulf of Mexico. After Prikryl and others (1988).

dissolution and diagenetic processes. These transitional zones are usually the most reactive, and are thus often the primary sites of the metal sulfide deposition (Kyle and others, 1987). The geochemistry, petrology, and structure of many of these caprocks have been studied in detail, for example, at Winnfield Dome, Louisiana (Kyle and others, 1987), Tatum Dome, Mississippi (Saunders and others, 1988), Richton Dome, Mississippi (Werner and others, 1988), Boling Dome, Texas (Seni, 1987), Hockley Dome, Texas (Kyle and Agee, 1988), and Damon Mound, Texas (Prikryl and others, 1988).

There is a generally accepted conceptual model for the genesis of the different zones within the caprock, at least within the geological framework of the Gulf Coast. Although the exact age of a caprock is difficult to measure, it is thought that they commonly begin to form early in the stage of salt diapirism (Posey and Kyle, 1988). This means that caprocks can move upward stratigraphically as the salt rises relative to the surrounding sediments.

Anhydrite in the caprock is mostly original salt that has been transported upward with the predominant halite. It accumulates by underplating as the halite dissolves, preferentially at the salt–caprock interface. Thus the anhydrite at the bottom of the zone is always the youngest and that at the top the oldest. Changing pressure, temperature, and chemical conditions over time often lead to recrystallization of the anhydrite. It can also contain terrigenous clastic material that has been incorporated into the salt through contact with formations at the edge of the dome (Seni, 1987). This exotic material can give evidence as to the nature of the stratigraphy below the current caprock horizon, and it can help to indicate the timing of the dome and caprock development.

Just above the anhydrite a thinner zone of gypsum often forms because of increased exposure to more dilute meteoric waters. Evidence of the role of more dilute waters can be seen in caprocks where gypsum veins are concentrated along fractures (Werner and others, 1988). Arching of the original sediments may create the initial fractures. The subsequent volume increase that occurs when anhydrite converts to gypsum then leads to additional fracturing and water inflow. Changing temperature and salinity conditions during dome and caprock growth can lead to multiple phases and textures of gypsum.

Calcite zones above the anhydrite and gypsum zones are also produced from the top down, in this case at the expense of the anhydrite. Anhydrite can convert to calcite by the replacement of sulfate with carbonate. This is believed to happen through hydrocarbon degradation by sulfate-reducing bacteria, with the sulfate being supplied by dissolution of the anhydrite (Posey and others, 1987; Prikryl and others, 1988). The hydrocarbons are likely supplied by upward-migrating methane or oil that becomes trapped along the salt dome flanks. Evidence for this transformation comes from anomalous carbon- and oxygen-isotopic signatures of the calcite and also from studies of the hydrocarbons themselves (Sassen and others, 1988). The isotopic data also suggest that the transformation occurs at temperatures between 40 and 80°C and that the waters involved could be some mixture of meteoric, formation, and sea water. Brecciation observed in the calcite zone suggests that the ongoing reactions and associated mineral-volume changes keep permeable pathways open for continual movement of the fluids that feed the reactions. "False" calcite zones can also exist some distance above the caprock. These zones are typically calcite-cemented clastics that were not formed in association with the caprock. They can be distinguished both texturally and by their isotopic signatures.

Caprocks often contain metal sulfide deposits that can include pyrite, sphalerite, galena, or pyrrhotite. Although the temperatures at which these sulfides formed have not been well constrained, they may have formed at greater than 100°C – warmer than the likely ambient temperatures. This suggests that the metals may have come from deeper brines that are known to be metal-rich. The description of fluid movement in Chapter 9.3.1 favored upward migration of deeper, overpressured brines along the flanks of domes through fracture zones. This pattern of flow agrees well with the sulfur-isotope evidence for the origin of metal sulfides in the caprock (Kyle and Agee, 1988). Sulfur from anhydrite or meteoric-water sulfate is too isotopically light to be the sole source of the sulfur in the sulfides. Hydrogen sulfide from deep basin sources has a heavier sulfur-isotope signature. It is likely that there is a combination of sulfur sources, such that when H_2S and metal-rich brines rise along the flanks of a salt dome, they encounter an environment already rich in sulfate, leading to supersaturation with respect to the metal sulfide minerals. These reactions have been modeled geochemically to reproduce both the major and minor mineralizations (Saunders, 1992).

Problems

9.1 Calculate the ratios of soluble salts that will precipitate out of seawater. One simple approximation is to start with the composition of seawater and use the chemical divide rule to eliminate each limiting element from the water as you work through the expected precipitation sequence of minerals. Fill in the chart below in this manner. The first mineral, dolomite, is already entered (bicarbonate limits the amount of dolomite that can precipitate to 1 mmole). Use the mineral formulas in Table 9.2 to determine the correct element ratios for each mineral.

Evaporation of seawater to produce $MgSO_4$-*rich deposit.*

	Ca	Mg	Na	K	Cl	SO$_4$	HCO$_3$	Precipitate in mmols
mmol/liter in seawater	10	55	457	10	535	28	2	0
dolomite	9	54	457	10	535	28	0	1
gypsum								
halite								
kieserite								
carnallite								
bischofite								

Now assume that there has been an exchange of magnesium for calcium in the input water (seawater), for example by dolomitization or mid-ocean ridge circulation. Fill in the chart again to obtain potential salt ratios in a $MgSO_4$-poor deposit.

Evaporation of seawater to produce $MgSO_4$-*poor deposit.*

	Ca	Mg	Na	K	Cl	SO$_4$	HCO$_3$	Precipitate in mmols
mmol/liter in altered seawater	41	24	457	10	535	28	2	0
dolomite								
gypsum								
halite								
sylvite								
tachyhydrite								

9.2 **(a)** Use Eq. (9.2) to calculate how many evaporated volumes it would take for salinities in a basin to reach 99% of their final value given leakage ratios of 0.1, 0.01, and 0.001. *Hint:* First rearrange the equation to give an answer in terms of evaporated volumes (dimensionless time). **(b)** Assume that the water depth in the basin is 1 meter and that the average net evaporation is 2 m/yr. What are the answers to part (a) in terms of years? **(c)** If the incoming water were seawater, and about 1% of the seawater precipitates by volume, how much salt would have been deposited in (b), given the different leakage ratios? **(d)** Evaporite basins contain thousands of meters of salt. Do you think that any given point in the sequence is more likely to

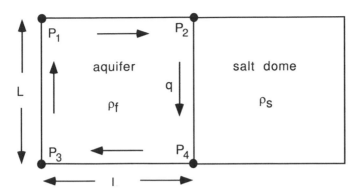

represent a transient evolving brine, or a brine under pseudo-steady-state (leaky) conditions? **(e)** For the basin in (b), calculate the average residence time of the water. Then, calculate the average residence time of the salt. (*Hint:* Use the leakage rates and concentrations relative to the lake volumes and concentrations.) Does the difference between these two residence times support or contradict the assumption of a well-mixed body of brine?

9.3 One can make a first-order approximation of steady-state convective flux next to a salt dome by approximating the system as a homogeneous equidimensional freshwater aquifer next to halite-saturated brine with higher density (see above). Fluid pressures can be calculated solely at points in the corners of the aquifer. By assuming saltwater density ρ_s along the aquifer–salt contact and freshwater density ρ_f along the opposite boundary, find algebraic expressions for the clockwise flux along the aquifer boundaries and the three unknown fluid pressures (assume $P_2 = 0$). These expressions should demonstrate that the flux is independent of the aquifer's size and that pressures are independent of the aquifer permeability. *Hint:* Write the variable-density form of Darcy's law (Eq. 2.26) for the flux between each pair of points around the boundary. Then write three equations expressing fluid continuity at three of the corner points (the fourth corner would be redundant). Given that P_2 is known, you now have seven unknowns and seven equations that can be rearranged algebraically in terms of the desired quantities.

9.4 Simple mass-flux calculations can provide insight. **(a)** For example, assume that the top of a salt dome 1 km in radius is being dissolved by groundwater flow. If flow across the top of the dome is occurring at about 1 m/yr (average linear velocity) in an aquifer 30-m thick with 30% porosity, and the increase in salinity in the aquifer water across the dome is about 300 g/liter (all due to dissolved halite), at what relative speed must the salt be rising for the caprock to maintain its vertical position? **(b)** If an anhydrite caprock at the top is 100-m thick and the mother salt contains 5% anhydrite and has 0% porosity, what total thickness of salt has risen? **(c)** At the rates calculated in (a), how long would it have taken for that much salt to rise? Is such a time span consistent with the age of the Louann salt (Jurassic) and post–salt deposition in the Gulf of Mexico?

10

Diagenesis and metamorphism

In this chapter we discuss aspects of the role of groundwater in diagenesis and metamorphism. *Diagenesis* refers to the changes undergone by sediments after initial deposition and during and after lithification. We will focus on the role of groundwater in diagenesis at the regional scale of sedimentary basins and *carbonate platforms* and then at the local scale of *pressure solutioning* and *geochemical banding*. We will not discuss karst development in carbonate terranes, an important diagenetic process that has been covered in depth by others (e.g., White, 1988). *Metamorphism* is a broad term that encompasses all adjustments of solid rocks to changes in physical and chemical conditions; here, its use is reserved for relatively deep, high-temperature processes. In this context we discuss the evidence for voluminous fluid fluxes at midcrustal depths, the nature of metamorphic permeability, and selected case studies. We do not discuss the dehydration reactions and associated metasomatism that occur in and above subduction zones, another topic that has recently been considered in detail elsewhere (e.g., Tarney and others, 1991; Manning, 1996). Fluid flow in the *accretionary prisms* of sediment that overlie subduction zones is treated briefly in Chapter 4.1.5.

10.1 Reaction-flow coupling

Before proceeding to discuss individual diagenetic processes we will first consider the means by which virtually all diagenetic and many metamorphic reactions are coupled to groundwater flow. Geochemical reactions that lead to the dissolution or precipitation of a solid mineral phase cause an associated change in the porosity of the porous media. As discussed in Chapter 1.2.1, there is usually a strong positive correlation between porosity and permeability. An increase in porosity due to dissolution will therefore lead to an increase in permeability. Permeability, in turn, affects the flow system through Darcy's

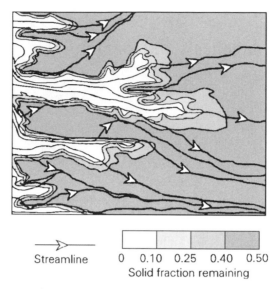

Figure 10.1 Pattern of dissolution resulting from positive feedback between fluid flow and reactive transport via the dependence of permeability upon the evolving porosity. After Lichtner (1996).

law – the result being that flow is channeled into areas of higher permeability. If the reactions are limited by the rate at which the reactants and products are being supplied and removed, then an increase in the flow rate will lead to more dissolution. A positive feedback loop will then proceed unchecked until mechanical effects begin to reduce the openings created by the dissolution (Ortoleva and others, 1987). Thus dissolution often produces a fingering phenomenon, such that discrete dissolution zones develop that channel virtually all of the groundwater flow (Figure 10.1). Limestone karstification is a prime example of this process. Mineral precipitation acts in reverse, causing negative feedback as reduction of porosity and permeability acts to channel flow away from areas of active precipitation. Under such conditions, it can be imagined that cementation of a formation would occur rather evenly across a broad area.

Because of the feedback between porosity, permeability, and fluid flow, the relation between porosity and permeability is important in the analysis of diagenesis. There are two general approaches to estimating this relation. One is to make many measurements of porosity and permeability on representative rocks and develop an empirical relation from the data. Examples of such data sets include those developed for petroleum-reservoir characterization in siliciclastic (Chilingar, 1964) or carbonate (Lucia, 1995) formations. These data sets tend to yield log–linear expressions between permeability and porosity. The second approach is to assume that a local process is occurring under predictable geo-metrical conditions. An example is uniform dissolution or precipitation along fracture walls or spherical surfaces. Such approaches usually result in expressions relating permeability to fracture or grain-size parameters. Examples are the *cubic law* developed for uniformly fractured rocks by Snow (1968),

$$k = \frac{Nb^3}{12},\qquad(10.1)$$

where N is the number of fractures per unit distance and b is the fracture aperture, and the *Kozeny–Carman equation* developed for granular media (Eq. 1.8) by Carman (1956).

Another relation that is significant to large spatial-scale diagenetic processes is that between permeability and depth (Chapter 1.2.4). Permeability generally decreases with depth over scales of 100s of meters to kilometers, due mostly to compaction and mineralization. In analyses of basin-scale behavior a log–linear relation between permeability and depth is often invoked. Such relations have been developed for intracratonic sedimentary basins such as the Denver basin (Belitz and Bredehoeft, 1988) and for carbonate platforms such as the Floridan Peninsula (Schmoker and Halley, 1982). In real systems the porosity–(or permeability–) depth curve is complicated by the fact that compaction, cementation, and dissolution can occur simultaneously and at different rates (Figure 10.2). Quantitative basin analyses that include sedimentation histories, diagenetic reactions, and transport-feedback mechanisms have been computationally prohibitive, but with recent advances in computer technology are now becoming more feasible.

10.2 Diagenesis of siliciclastic sequences

Diagenesis in layered sedimentary rocks is usually accompanied by significant fluid flow. Many of the mineral alterations (e.g., of feldspars and clays) observed with increasing depth within these systems cannot be accounted for by chemical mass balance calculations that invoke only the local mineral assemblages (Land and others, 1997). External sources and sinks of certain chemical elements are required to account for the mineralogical changes, and fluids constitute the only possible means of transport. Hundreds of pore volumes are often required to exchange adequate amounts of the elements in question. Driving mechanisms for such fluid flow include topography, compaction, and thermal convection (see Chapter 4 or Bjorlykke, 1993). These mechanisms have also been described as "meteoric," "compactional," and "thermobaric" hydrologic regimes (Galloway, 1984).

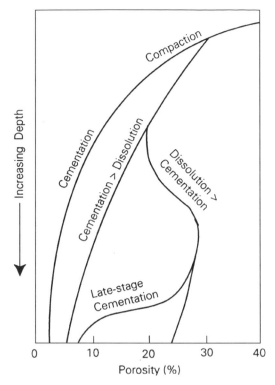

Figure 10.2 Hypothetical curves of porosity versus depth showing the effects of simultaneous compaction, cementation, and dissolution at different rates. From Loucks and others (1984), reprinted by permission of the American Association of Petroleum Geologists.

10.2.1 Diagenesis in sedimentary basins

Sedimentary basins are usually dominated by arenaceous and argillaceous (non-carbonate) sediments. As these sediments are buried to depths of many kilometers, several diagenetic reactions occur as a result of increasing temperature and pressure, including cementation by silica, calcite, iron, or clay; dissolution of these same cements; conversion of smectitic clays to illitic clays; and albitization of both potassium feldspars and plagioclase (see, e.g., Hower and others, 1976; Blatt, 1979). At greater depths, chlorite begins to form, which then undergoes changes in composition with increasing temperature (deCaritat and others, 1993). An increase in potassium content with depth is common (Land and others, 1997) and in some circumstances may be caused by downward movement of surficial brines (Leising and others, 1995). Prediction of these diagenetic changes is important for hydrocarbon reservoir exploration and production because of the associated changes in porosity and permeability. However, the complex interplay among the various relevant reactions, which occurs amidst significant gradients in temperature, pressure, and aqueous chemical concentrations, leaves a number of questions unanswered. Temperature, pressure, and fluid fluxes can often be predicted by groundwater models (e.g., Bethke and others, 1993), and chemical reactions can often be predicted by geochemical batch models (e.g., Meshri, 1990), but field data sometimes contradict predictions from these models. Fully coupled diagenetic, reactive-flow models, although potentially powerful, require accurate field-derived input parameters in order to make useful predictions.

The *Gulf Coast* basin (Chapter 4.1.4) is one of the most-studied sedimentary basins in the world. A study of diagenetic reaction paths in the Gulf sediments (Harrison and Tempel, 1993) illustrated the use of a "loose" coupling between a reaction-path model and a groundwater flow model. The first step in such analyses is to simulate groundwater flow over time, invoking estimated sedimentation rates to create a transient, basin-wide flow domain (using, e.g., BASIN2: Bethke and others, 1993). The second step is to choose a packet of sediment that will be followed throughout the history of the basin and to record the flow rates at the points in space that the moving packet occupies. These flow rates are then converted to pore volumes for each time-step interval. The final step is to input the current mineralogical assemblage for the packet into a geochemical reaction-path model and calculate the mineralogical changes that would occur due to reaction with the changing mineralogy. In this way a history of mineralogical changes can be accrued that reflects not only flow conditions but changes in temperature and pressure (Figure 10.3). Although a useful technique for qualitative assessments of diagenesis, it cannot predict true volumes of solid phases reacted because it does not consider the effect of transporting the reaction products. Water-to-rock ratios and pore volumes are batch (scalar) concepts that are not meaningful for arbitrary subvolumes within a vector (fluid flow) field. Furthermore, sparse thermodynamic data bases (for clay minerals especially) often limit the accuracy not only of this method but of more fully coupled models as well.

The *North Sea* basin has also been extensively studied. The Brent Group (Jurassic, mostly sandstone) is an important reservoir rock and has been extensively drilled and studied from a diagenetic standpoint (Bjorlykke and others, 1992). Conclusions concerning the diagenesis of the Brent Group are typical of many sedimentary basins and illustrate some major diagenetic processes. Different facies exist within the Brent Group and their distribution controls some of the diagenetic reactions (e.g., carbonate precipitation). The occurrence of facies-dependent diagenetic trends suggests that many reactions occurred

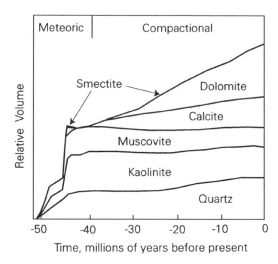

Figure 10.3 Authigenic mineral phases simulated by a diagenetic pathway model for an onshore section of the Wilcox Group sandstone of the Gulf Coast basin. After Harrison and Tempel (1993).

early in the burial history, before large changes in temperature and pressure became the dominant controlling factors. Meteoric-fluid flow has been suggested as the driving mechanism for some of the early diagenesis (McAulay and others, 1994). Potassium feldspars are depleted in zones below 2–3 km depth, but this creates only minor secondary porosity. The feldspars are converted to sodium-rich albite (*albitization*), with the potassium loss accommodated by illite formation. The presence of illite significantly reduces permeability, but illite is only pervasive below 3.5-km depth. Authigenic illite in the North Sea basin has been dated (e.g., Hogg and others, 1993), and such dating helps to elucidate the timing of diagenetic fluid migration. Quartz cement is only abundant below 4 km, where *stylolite* formation (Chapter 10.4.1) has occluded most of the primary porosity. Convection has been invoked to explain some of the deeper diagenetic patterns (Rabinowicz and others, 1985).

10.2.2 Silica cementation by thermal convection

Spatial variations in temperature and pressure are responsible for causing much of the cementation and dissolution that occurs in rocks at depth. Fluid transport by *thermal convection* (Chapters 3.5.3–3.6 and 7.2–7.3) may operate under such gradients to enhance cementation of quartz sandstones. Theoretical investigations of heat transport in sandstone layers indicate that common geometries

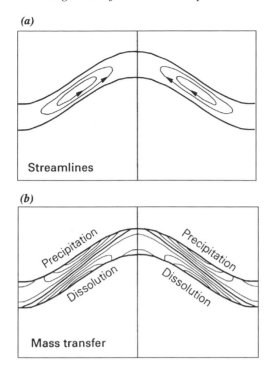

Figure 10.4 Circulation patterns in a high-permeability domed layer showing (a) streamlines during thermal convection and (b) regions of potential mineral dissolution and precipitation. After Davis and others (1985).

related to geologic structure can give rise to internal convection cells (Wood and Hewett, 1982) that take the form of isolated polyhedrons or, along the limbs of anticlines or domes (Davis and others, 1985), long two-dimensional cells (Figure 10.4a) that circulate with velocities proportional to the Rayleigh number (Chapter 3.5.3). There is no critical Rayleigh number for any nonisothermal sloping layer, and all such layers should have fluids circulating at some finite (though perhaps diagenetically insignificant) velocity. Significant temperature gradients within the circulating cells will cause differences in the equilibrium state of the fluid with respect to different minerals. Silica equilibrium, for example, is prograde with respect to temperature below about 350°C (Fournier and Potter, 1982). Under chemical equilibrium, warmer regions will be undersaturated and dissolve silica, whereas cooler regions will be supersaturated and precipitate silica (Figure 10.4b). Minerals with a retrograde equilibrium will have the reverse pattern.

Patterns of cementation such as that shown in Figure 10.4b are predicted by assuming that local equilibrium exists continually within the system. In reality, dissolution and precipitation follow kinetic rates that impose a time requirement to reach equilibrium. If fluid velocities are great enough, the equilibrium state can be out of phase with respect to the local temperature and direction of fluid migration. Investigations of the effects of kinetic rates on circulation and diagenesis reveal very complex patterns of diagenesis that are nearly chaotic and highly dependent on the feedback between porosity and permeability (Bolton and others, 1996). At any given time, any parcel of fluid within the formation could have any possible combination of relative saturation, temperature, and migration direction, making the prediction of diagenetic patterns in real situations virtually impossible for high Rayleigh number conditions. Other complications that arise under various field conditions include the dominance of *pressure solutioning* in deep sandstones (Chapter 10.4.1) and the presence of calcite cement, which will dissolve faster than silica will precipitate (Bjorlykke and Egeberg, 1993).

10.3 Diagenesis of carbonate platforms

In *carbonate platforms* the primary mineralogy is dominated by carbonates rather than silicates. Many mechanisms have been suggested to drive fluid circulation in carbonate platforms (Whitaker and Smart, 1990, 1993) at rates sufficient to cause extensive diagenesis (e.g., dolomitization). As in sedimentary basins, reactions are often controlled by gradients in temperature and pressure, so that burial depths and the temperature dependence of the mineral solubilities are still important parameters affecting diagenesis. Here we will focus mainly on dolomitization and mixing-zone dissolution. Other important diagenetic reactions in this environment include calcite cementation and aragonite-to-calcite transformation (e.g., Morse and MacKenzie, 1990).

10.3.1 Dolomitization

In spite of the ubiquitous occurrence of dolomite ($CaMg(CO_3)_2$) in the ancient rock record, sites of modern dolomite formation are scarce, and no one chemical environment can explain all of the variations in the rock record (Hardie, 1987; Budd, 1997). Although modern dolomite has been observed in some groundwater (von der Borch and others, 1975) and hypersaline (Patterson and Kinsman, 1982) environments, normal or near-normal seawater is likely the most common dolomitizing fluid because of its broad availability as a source of magnesium (Land, 1985). Although there is a considerable thermodynamic drive to precipitate dolomite from seawater, laboratory experiments have consistently failed to synthesize dolomite under reasonable near-surface conditions. This suggests either that seawater or its environment are modified somewhat

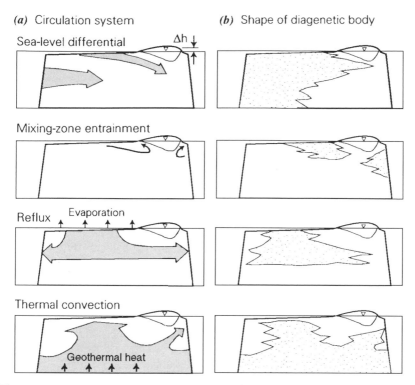

Figure 10.5 (a) Mechanisms that will force seawater to circulate through carbonate platforms and (b) hypothesized patterns of dolomitization. After Whitaker and Smart (1990).

before dolomitization occurs, or that dolomite precipitates so slowly that it fails to form measurable amounts on the time scale of laboratory experiments. In either event, a mechanism is required to deliver magnesium to the mineral sites and remove calcium, and *molecular diffusion* alone cannot transport solutes in the required time frame (see Figure 2.1). Some driving force is necessary to move seawater through large volumes of carbonate rock.

Four different mechanisms have been invoked to circulate significant quantities of seawater (Figure 10.5a). The first mechanism operates by a hydraulic forcing: a difference in the mean elevation of sea level across a carbonate platform. Given the high permeabilities of these platforms at shallow depths ($>10^{-10}$ m^2), even a small difference in sea level could result in significant lateral fluxes of seawater. The second mechanism is a result of the buoyant and hydraulic forces that operate to create the mixing zone. Meteoric recharge of freshwater sets up a shallow buoyant flow system that entrains seawater. Seawater must then circulate from the ocean to replace the entrained water. The third mechanism, *reflux*, uses a buoyant force resulting from an increase in salinity due to evaporation in shallow water on a submerged platform. The fourth also invokes a buoyant force, in this case due to the temperature (density) differences between cold water in the ocean depths and geothermally heated water within the platform. Each of these mechanisms could theoretically cause seawater to circulate through the subsurface, and the distribution of dolomite may reflect the circulation mechanism (Figure 10.5b). It is likely, however, that only under certain circumstances will the fluxes be large enough and the driving mechanism in place long enough to deliver the quantity of magnesium necessary for extensive dolomitization.

A mixing-zone model for dolomite formation was originally proposed by Hanshaw and others (1971) and Badiozamani (1973) and is based on the thermodynamic effects of mixing dissimilar waters. Certain mixtures of seawater and calcite-saturated fresh groundwater (the percentages vary depending upon the exact thermodynamic data that are used) are both undersaturated with respect to calcite and supersaturated with respect to dolomite (Plummer, 1975). The seawater end member is supersaturated with respect to both minerals, and the fresh groundwater end member is undersaturated with respect to dolomite. The transition zone between freshwater and seawater near the coastline provides an environment in which seawater will continually be supplied to the potential reaction site (though it is unclear whether the reaction could proceed at a sufficient rate under ambient temperatures). Small amounts of dolomite appear to be forming in modern mixing zones in Israel (Magaritz and others, 1980), Mexico (Ward and Halley, 1985), and Florida (Randazzo and Bloom, 1985), but the idea that extensive dolomitization occurs in the mixing zone has been contentious because many mixing zones have no dolomite (Machel and Mountjoy, 1986), and any dolomite which does precipitate could be due to the presence of organics (Whitaker and Smart, 1994).

Geothermal circulation of seawater through carbonate platforms was first suggested by Kohout (1965). Numerical analyses of groundwater flow and heat transport indicate that geothermal circulation of seawater through a carbonate platform can be substantial (Kaufman, 1994). Resulting patterns of dolomite distribution can be calculated given the relative kinetic rate of dolomite formation. By using a first-order rate equation in a solute-transport equation (as described in Chapter 2.5.1), it can be demonstrated that the pattern of dolomitization would be a function of the *Damkohler number*, with a high number (fast kinetic rates) yielding more dolomite near the inflow and a low number yielding more dolomite in the warmer, central portion of the platform (Jones and others, 1997). Increased temperatures in the center of the platform are usually one result of this flow mechanism, and at least two field areas – one modern and one ancient – possess this characteristic. The Floridan platform has elevated temperatures at its center near the Gulf coastline (Kohout and others, 1977), and the Triassic Latimar platform in Italy has geochemical signatures indicating internally elevated temperatures that mimic a similarly shaped region of dolomitization (Wilson and others, 1990).

Water-chemistry data from springs discharging from modern carbonate platforms provide strong evidence for ongoing dolomitization. Water discharging from blue holes (subsea karstic springs) at the edge of the Great Bahama bank are depleted in magnesium (Whitaker and others, 1994), and calculations suggest that dolomitization is currently occurring at a rate of 2–30% per million years. Subtidal springs around the submarine cavern of Grotto Azzurra, at Capo Palinuro, Italy, are discharging water with a magnesium depletion and a calcium excess of up to 19% with respect to chlorine (Stuben and others, 1996). These data, and the presence of dolomite in the host rocks and cave-floor sediments at the site, suggest that mixing-zone dolomitization is occurring, perhaps kinetically assisted by elevated temperatures and/or reducing conditions in the subseafloor zone. Subsea thermal springs off the west coast of Florida are discharging seawater that is relatively unaltered except for a 3 mmol depletion in magnesium and a similar excess in calcium (Fanning and others, 1981). The proximity of the thermal springs to the center of the Floridan platform is consistent with temperature patterns produced by heat-transport simulations for carbonate platforms (Jones and others, 1997) and suggests the circulation may be driven by thermally induced convection.

Laboratory experiments indicate that at temperatures in excess of $100°C$ dolomite can form relatively rapidly (Baron, 1960; Morrow, 1982; Sibley and others, 1987). Most ancient carbonate rocks have been buried to depths where temperatures exceed $100°C$ (>2–3 km). Although seawater has the high Mg/Ca ratio required for dolomitization, warm, fresh groundwater can cause dolomitization at Mg/Ca ratios as low as one (Folk and Land, 1975), and such waters are abundant in sedimentary basins. Thus the question might be posed: Why haven't all ancient carbonate rocks been dolomitized? Dolomitization of ancient massive carbonates can occur in some cases without the aid of seawater. Dolomitization by subsurface burial and hydrothermal processes has been advocated for some basin environments (Anderson and Macqueen, 1982; Gregg, 1985) where it may be driven by regional flow (Garven and Freeze, 1984a,1984b) associated with Mississippi Valley–type ore deposits (Chapter 5.1). Many other field examples exist where "warm-burial" dolomitization has been hypothesized as the driving mechanism (Mountjoy and Amthor, 1994). Geochemical calculations also indicate that anhydrite precipitation should accompany dolomitization in seawater that is warmer than $50°C$. This is consistent with the widespread

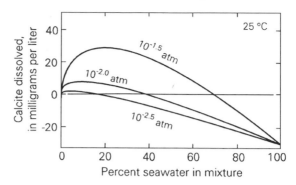

Figure 10.6 Amounts of dissolved calcite required to reequilibrate groundwaters with respect to calcite that are various mixtures of freshwater and seawater. Curves represent different freshwater end members that have equilibrated with respect to calcite and three different partial pressures of carbon dioxide. After Sanford and Konikow (1989a).

presence of anhydrite cement and its correlation with dolomite in many carbonate petroleum reservoirs (Moore, 1989).

10.3.2 Mixing-zone dissolution

Field evidence suggests that significant quantities of limestone can be dissolved along coastlines where freshwater and seawater mix. Dissolution features have been well documented in limestone along the coast of the Yucatan Peninsula, Mexico (Back and others, 1979, 1986), at coastal springs in Greece (Higgins, 1980), and at South Andros Island in the Bahamas (Smart and others, 1988). In each case, dissolution appears to be most extensive where mixtures of freshwater and seawater are discharging into the ocean. In some cases, dissolution may be partly due to increased CO_2 levels in the mixing zone that result from oxidation of organic matter in a horizon of buoyant equilibrium (Smart and others, 1988).

Undersaturation with respect to calcite in the mixing zone is due to the difference in CO_2 contents of the two end members, the nonlinear nature of chemical equilibria equations, and ionic strength effects (Wigley and Plummer, 1976). Simulations with the geochemical reaction-path model PHREEQE (Parkhurst and others, 1980) predict the amount of calcite that must dissolve into solution in order to reequilibrate mixtures of seawater and fresh groundwater with respect to calcite (Figure 10.6). In this analysis the groundwater end member is assumed to have reached equilibrium with respect to calcite along its travel path to the transition zone. Seawater remains supersaturated for kinetic reasons. Depending on the temperature, the exact $CaCO_3$ mineralogy, and the CO_2 content of the freshwater end member, undersaturation is predicted to occur for freshwater-rich mixtures.

Sanford and Konikow (1989a, 1989b) simulated the geochemical reactions and variable-density solute transport (Chapter 2.3) associated with a dynamic transition zone between seawater and fresh groundwater in a hypothetical coastal aquifer (Figure 10.7). The results illustrate the use of a fully coupled reaction-transport model for analyzing diagenetic processes. They indicate that porosity development within the mixing would not be evenly distributed (Figure 10.7a). Porosity develops on the fresher side of the transition zone, as predicted by the geochemical model (Figure 10.6). Porosity also develops faster at the base or toe of the mixing zone and at its top, in the discharge area at the coastline. This can be attributed to "virgin" mixing of end-member waters at the base, where the full potential for calcite dissolution can be realized, and to higher velocities near the coastline in the region of convergent flow, which allow more calcite to dissolve.

For typical flow conditions, porosity could be significantly enhanced over time on the order of 10^4 years. More specifically, sensitivity analyses indicated that the rate of porosity increase could be expressed as

$$R_p = 0.45 \Delta Ca_s q, \tag{10.2}$$

where R_p is the change in percent porosity per 10^4 years, ΔCa_s is the maximum amount of calcite required per liter of water in order to reach saturation (see Figure 10.6), and q is the specific discharge in the mixing zone at the point of interest (m/day). Other distributions of permeabilities and boundary conditions lead to somewhat different results. A heterogeneous permeability distribution, for example, leads to more localized porosity enhancement where flow is channeled into the higher-permeability zones (Figure 10.7b). Without a low-permeability basal unit, no zone of porosity enhancement develops at the inland toe, because inland velocities are slow enough to offset any effects from virgin mixing.

Including a porosity–permeability feedback allowed the flow system to respond to dissolution over time (Figures 10.7c, 10.7d). As porosity is enhanced on the fresher side of the mixing zone, the resulting permeability enhancement causes the transition zone to migrate landward over time (Figure 10.7d). This migration causes the dissolution to take place over a larger volume of rock, so that the largest porosity increases are less than if the mixing zone were stationary (Figure 10.7c). Over long time spans, any extensive transgressions or regressions of the coastline (e.g., during glacial periods) will cause the transition zone to migrate over significant distances, resulting in minor porosity enhancements over large areas, rather than a major porosity enhancement at any one location.

Results from the dissolution-rate study can be compared with conditions estimated at various field sites. At the Yucatan Peninsula, given a specific discharge

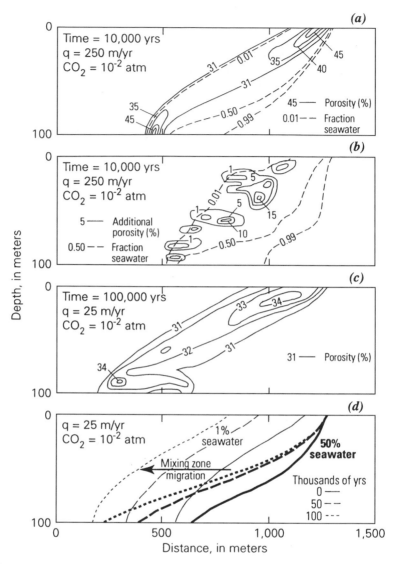

Figure 10.7 Calcite dissolution predicted in a freshwater/seawater mixing zone (a) assuming local equilibrium, homogeneous permeability, and no porosity–permeability feedback, (b) same as (a) except for heterogeneous permeability field, and (c) assuming that an initially homogeneous permeability field evolves with porosity. (d) Landward translation of the mixing zone due to uneven porosity and permeability development. After Sanford and Konikow (1989a).

of about 8 m/day and calcite dissolution of about 19 mg/liter, the porosity is estimated to increase by 68 percent in 10^4 years (Eq. 10.1). For a small island such as Bermuda, where the discharge may be only 6.5 cm/day, the porosity increase might be only 0.003% in 10^4 years. Given the magnitude of the sea level changes and coastline migrations associated with the latest ice age, it is likely that only the rates of porosity development predicted for large carbonate platforms such as the Yucatan would translate into visible localization of dissolution features. Small islands cannot produce the discharge fluxes necessary for significant porosity enhancements.

10.4 Local-scale diagenetic features

To this point we have discussed diagenetic processes as they occur at the regional scale. At this scale, diagenesis is often driven by regional variations in temperature. However, variations in pressure and concentration tend to be more important to the development of local-scale diagenetic features. Pore-scale pressure variations caused by higher pressures at grain-to-grain contacts result in differences in mineral–solution equilibria that drive reactions through *mechanochemical coupling*. Rate-limited reactions under conditions of transient solute transport can create cyclic concentration variations that lead to *mineral banding* in a wide variety of geochemical environments.

10.4.1 Mechanochemical coupling

Although thermally driven flow may control some regional-scale diagenetic processes, other flow mechanisms may dominate under certain pressure and temperature conditions in the crust (Ortoleva, 1994). Rayleigh number (Chapter 3.5.3) calculations based on density differences caused by both thermal expansion and calcite dissolution, for example, can be plotted in a pressure (depth) versus temperature-gradient field to illustrate where each process may dominate (Figure 10.8). At depths of more than 3 km in regions with a normal geothermal gradient, *mechanochemical coupling* may drive many diagenetic processes. Mechanochemical coupling is feedback between a mechanical and a chemical process, usually between pressure on mineral grains and mineral–solution equilibria. Two of these mechanochemical couplings will be discussed briefly here: pressure solution (e.g., stylolites) and concretions (e.g., geodes).

Pressure solution is an important process in reducing porosity in deep reservoirs (Angevine and Turcotte, 1983; Lemee and Gueguen, 1996). It results from the fact that the free energy of a mineral phase is a function of the pressure or stress within that mineral (Guzzetta, 1984; Tada and Siever, 1989; Gavrilenko

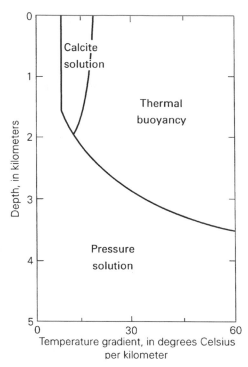

Figure 10.8 Pressure and temperature-gradient regimes in the crust where different buoyancy-driven mechanisms could operate. After Ortoleva (1994).

and Gueguen, 1993). A higher pressure, in general, leads to an increase in mineral solubility. Regions with higher porosity will have smaller grain-to-grain contact areas over which the stress of the overburden is borne, and thus the pressure within the mineral-contact zones will be higher. This will result in dissolution in higher-porosity areas and precipitation in lower-porosity areas – a positive feedback trend that will seal certain zones and leave others porous enough that physical compaction (readjustment of grains) will then act to reduce porosity. The porous zones also provide pathways for escaping fluids. In this fashion, siliciclastic or carbonate rocks can be compacted to a few percent porosity. At this point, mineral dislocations become the local stress foci and align themselves according to the stress field to become the planar, yet often jagged, features known as *stylolites*.

Concretions can also develop as a result of variations in pressure within mineral grains (Dewers and Ortoleva, 1990a), though other processes are thought to contribute to their formation under certain conditions (Coleman, 1993; Mozley, 1996). Consider the example of the growth of silica or quartz concretions (*geodes*) within a limestone formation. The groundwater should be near saturation with respect to calcite and either near saturation or under-saturated with respect to silica. If the formation is cooled (e.g., by uplift) the groundwater would become supersaturated with respect to silica due to pro-grade solubility. Silica will then begin to precipitate around nucleation seeds such as small detrital quartz grains. If there is no room for silica growth because of adjacent calcite grains, the thermodynamic drive to precipitate will create pressure from the solid-phase silica against the calcite, causing a local area of undersaturation with respect to calcite due to pressure increase. The calcite will then dissolve in a volume-for-volume exchange for silica. This process will progress at a slowing rate as the concretion increases in size and dissolved silica in the immediate vicinity is depleted.

10.4.2 Geochemical banding

Mineral banding at the local scale can often be explained by certain types of reaction-transport processes (e.g., Merino, 1990; Harder, 1993; Ortoleva, 1994; Merino and others, 1995). Examples include *banded iron oxides, zebrastone*, wood-grained chert, *banded agates*, and *Liesegang banding*. For example, when one fluid is displaced by another with a different solute composition (Figure 10.9a) and kinetic mass-transfer rates cause the continued buildup of concentrations past a *nucleation threshold* (Figure 10.9b), the in situ mineral phase can be replaced by bands of a different mineral (Figure 10.9c). The nucleation-threshold effect allows concentrations to build beyond supersaturation. Once the threshold is reached, precipitation will continue until concentrations return to saturation; the cycle then repeats. This process can be represented by two solute transport equations and two kinetic rate equations (Ortoleva, 1994, p. 148) similar to Eqs. (2.15) and (2.16):

$$n\frac{\partial X_s}{\partial t} = nD_X\nabla^2 X_s - n\nabla\cdot(X_s v) + \rho_A n_A \frac{\partial}{\partial t}\left(\frac{4}{3}\pi r_A^3\right) \quad (10.3)$$

$$- \rho_B n_B \frac{\partial}{\partial t}\left(\frac{4}{3}\pi r_B^3\right),$$

$$n\frac{\partial Y_s}{\partial t} = nD_Y\nabla^2 Y_s + n\nabla\cdot(Y_s v) - \rho_A n_A \frac{\partial}{\partial t}\left(\frac{4}{3}\pi r_A^3\right) \quad (10.4)$$

$$- \rho_B n_B \frac{\partial}{\partial t}\left(\frac{4}{3}\pi r_B^3\right),$$

$$\frac{\partial r_A}{\partial t} = -k_A[K_A X_s - Y_s], \quad (10.5)$$

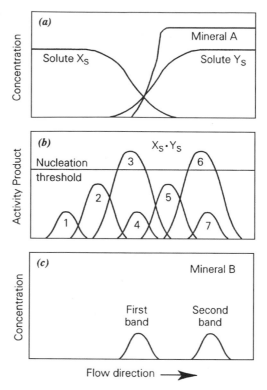

Figure 10.9 Solute transport processes that lead to *mineral banding* including (a) displacement of a fluid with solute Y_s by a fluid with solute X_s through a porous medium with mineral A and (b) temporal stages of the product of concentrations of X_s and Y_s that result in (c) the mineral B being deposited in bands. After Ortoleva (1994).

and

$$\frac{\partial r_B}{\partial t} = k_B[X_s Y_s - K_B], \tag{10.6}$$

where the subscripts A and B refer to the initial and replacement mineral phases, respectively, X_s is the concentration of an inflowing solute, Y_s is the concentration of a displaced solute, D_X and D_Y are diffusion coefficients, v is the seepage velocity, n_A and n_B are volumetric fractions, r_A and r_B are mineral-grain radii, k_A and k_B are rate constants for dissolution and precipitation, and K_A and K_B are equilibrium constants. Solutions to these equations yield different banding patterns depending on the exact parameter values. Metamorphic differentiation can also be explained by this type of cycle coupled with mechanical grain-scale effects (see, e.g., Dewers and Ortoleva, 1990b).

10.5 Metamorphism

There is no hard and fast distinction between diagenesis and metamorphism. In the preceding sections we used *diagenesis* to refer to chemical, physical, and biological changes undergone by a sediment in a typical sedimentary environment (*P* generally <100 MPa, *T* generally <100°C). We reserved the term *metamorphism* to refer to chemical and physical changes occurring in a rock mass at greater pressures and/or temperatures. There is a further distinction between *contact metamorphism*, which takes place near an igneous intrusion, and the *regional metamorphism* that occurs in orogenic belts at a wide range of pressures and temperatures. In both types of metamorphism there is generally a gain, loss, and/or exchange of chemical constituents via a fluid phase, a process sometimes termed *metasomatism*.

In this section we will review some of the evidence for voluminous fluid fluxes in metamorphic environments, speculate as to the nature of the permeability that might allow for such fluxes, and describe examples of contact and regional metamorphism. The understanding of metamorphic processes is a frontier area for hydrogeology. In the words of Rumble (1994), "We face a discrepancy between the growing body of evidence of massive fluid flow and lack of understanding of the physics of flow at middle to lower crustal depths." Our understanding of the physical dynamics of the *magmatic–hydrothermal systems* (Chapter 7.2) associated with contact metamorphism in the mid to upper crust is relatively good, but in the mid- to lower-crustal context of regional metamorphism, as at the mid-ocean ridges (Chapter 7.9), basic questions about fluid dynamics and the transient nature of permeability remain to be resolved. Furthermore, both contact and regional metamorphism often involve nonisothermal, *multicomponent reactive transport* processes (Chapter 2.5) that we as yet are unable to fully describe.

10.5.1 The evidence for voluminous fluid fluxes

In the late 1960s, oxygen- and hydrogen-isotope data from water and rocks convinced most workers that *meteoric* (atmospherically derived) fluids are or had been abundant in many metamorphic environments. That understanding depended on knowledge of the isotopic composition of precipitation (Craig, 1961a), of geothermal waters (Craig, 1963), and of igneous rocks (Taylor, 1968).

There are two stable isotopes of hydrogen: ^1H and ^2H (deuterium or D) and three of oxygen: ^{16}O, ^{17}O, and ^{18}O, of which ^{16}O and ^{18}O are the more abundant. Because the vapor pressure of water molecules is inversely proportional to their masses, water vapor is depleted in the heavier isotopes, D and ^{18}O, relative to

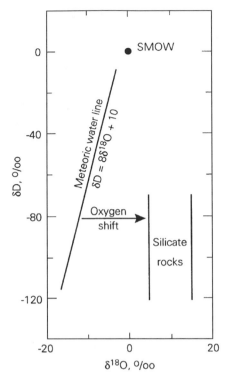

Figure 10.10 Relation between deuterium and oxygen-18 contents showing the *meteoric water line* (Craig, 1961a), a typical *oxygen shift* for high-temperature geothermal waters (Craig, 1963), and the range of isotopic compositions for silicate rocks (Faure, 1986).

coexisting liquid water (e.g., Faure, 1986). The D and ^{18}O content of water is usually reported in "δ-notation" relative to Craig's (1961b) *Standard Mean Ocean Water* (SMOW):

$$\delta = [(R_{\text{sample}}/R_{\text{SMOW}}) - 1] \times 1,000, \qquad (10.7)$$

where R is the D/H or $^{18}O/^{16}O$ ratio and the units of δ are permil (o/oo). Craig (1961a) showed that meteoric waters worldwide define the relation $\delta D = 8\delta^{18}O + 10$ (Figure 10.10), with δD and $\delta^{18}O$ values becoming increasingly more negative at higher latitudes. In general, the water vapor in an air mass becomes progressively more isotopically depleted as precipitation falls from it. The mean isotopic composition of precipitation at a particular location is approximately constant over time periods that are long enough to minimize the effect of seasonal variations and short enough to preclude significant climate change. Characteristic spatial variations in isotopic composition are often used to infer groundwater recharge areas, indicate mixing, or delineate different groundwater systems (e.g., Scholl and others, 1996).

Most groundwater samples, like precipitation, lie near the *meteoric water line* defined by $\delta D = 8\delta^{18}O + 10$, but high-temperature geothermal waters are an exception. Near-neutral-pH geothermal waters that have been heated to temperatures in excess of about 150°C frequently show an *oxygen shift* away from the meteoric water line toward an increasing abundance of the heavier isotope, ^{18}O (Figure 10.10). Craig (1963) documented this phenomenon and suggested that it was due to oxygen isotope exchange between water and rock at elevated temperatures. The hydrogen isotope values of the water would remain largely unaffected because of the low hydrogen content of most rocks. The suggested link between the oxygen isotope composition of geothermal water and altered rock was later confirmed by Taylor (1968), who found that the original $\delta^{18}O$ composition of igneous rocks is typically 5.5 to 7 o/oo (versus -2 to -17 o/oo for groundwaters) but that many hydrothermally altered igneous rocks are depleted in ^{18}O due to exchange with meteoric waters. Mainly on the basis of oxygen isotope data, it has been estimated that tertiary meteoric hydrothermal systems have altered the rocks exposed over about 5% of the land surface of the northwestern United States and southwestern Canada (Criss and others, 1991). These systems account for 25–50% of mineral production in the region. In some fossil hydrothermal systems exhumed by erosion the oxygen isotope data indicate circulation of meteoric water to depths of at least 8 to 10 km in the crust (e.g., Taylor, 1990).

The size of the oxygen shift can be used to calculate a quantitative index of the degree of water–rock interaction, a *water–rock ratio* that is an apparent mass ratio of exchanged water to exchanged rock based on a material balance (e.g., Taylor, 1971; Blattner, 1985). Open-system water–rock ratios for active and fossil hydrothermal systems are typically >1 (e.g., Blattner, 1985; Larson and Taylor, 1986; White and Peterson, 1991). Given the low porosity of many altered rocks, equal masses of exchanged water and rock may imply that the pore space was emptied and replenished 100 times or more.

Oxygen- and hydrogen-isotope data provide rather clear evidence as to the origin as well as the presence of aqueous fluids. Many other metamorphic reactions require the presence of large volumes of an H_2O-rich fluid but do not necessarily distinguish its origin. *Decarbonation* reactions, for example, require large volumes of water of unspecified origin. The progress of reactions such as

$$\text{calcite} \quad \text{quartz} \quad \text{wollastonite}$$

$$CaCO_3 + SiO_2 = CaSiO_3 + CO_2 \tag{10.8}$$

depends on continuous percolation of an H_2O-rich fluid to dilute and carry away the produced CO_2 (Ferry 1978; Rice and Ferry, 1982). On the basis of the equilibrium value for this reaction, Hoisch (1987) calculated that formation of massive wollastonite in the Big Maria Mountains of southern California required a water–rock ratio ($X_{H_2O} = 0.97$–1.0) of at least 17 at pressures of 300 ± 100 MPa (about 12 ± 4 km lithostatic depth) and temperatures of 430 to 590°C.

Other arguments for voluminous fluid fluxes in metamorphic environments are based on the solubilities and phase equilibria of minerals (see, e.g., the overview by Ferry, 1994a). Each of the three main lines of evidence (oxygen isotopes, devolatilization reactions, solubilities and phase equilibria) appears to require large volumes of water. Calculated values of time-integrated fluid flux range up to 10^6 cm^3 fluid/cm^2 rock for pervasive fluid flow and up to 10^9 cm^3/cm^2 for fluid flow in veins (Ferry, 1994a). In the context of contact metamorphism near mid- to upper-crustal plutons, such fluxes are readily explained; an intrusion can heat roughly its own mass in meteoric water (e.g., Norton and Cathles, 1979), and the resulting thermally driven circulation will be focused near the pluton. But in deeper-seated regional metamorphism the most obvious source of aqueous fluid is dehydration of underlying rock, and in many cases the calculated fluxes far exceed the amounts that could be obtained even by complete dehydration of subjacent rocks (see the discussion by Rumble, 1994).

10.5.2 The nature of permeability in metamorphic environments

The evidence from the rock record for pervasive fluid fluxes as large as 10^6 cm^3/cm^2 implies significant permeabilities deep in the crust. To calculate the required permeabilities, we must translate these time-integrated fluxes into flow rates. This requires some consideration of the time frames over which metamorphism is likely to occur. Episodes of *contact metamorphism* are likely to persist for $\ll 1$ Ma, because even the largest igneous intrusions will cool to near-ambient temperatures in about 1 Ma (e.g., Smith and Shaw, 1979). Thus in a contact-metamorphic aureole a pervasive time-integrated flux of 10^6 cm^3/cm^2 might translate to a volumetric flow rate (q) of $>3 \times 10^{-8}$ cm/s, or $>3 \times 10^{-10}$ m/s. Assuming that the maximum hydraulic gradient is on the order of 10^0 (Chapter 4.1.2), this requires a time-averaged hydraulic conductivity (K) of about 3×10^{-10} m/s or, at about 300 bars and 350°C, a permeability (k) on the order of 10^{-18} m^2. For such values of permeability, heat transfer will be conduction dominated (Chapter 7.2.3) and geologic forcing (Chapter 4.1.1) may be able to generate elevated fluid pressures. (See also Chapter 1 for discussions of *volumetric flow rate* or *Darcy velocity, hydraulic head, hydraulic conductivity,* and *intrinsic permeability*.)

In a contact-metamorphic environment, permeabilities on the order of 10^{-18} m^2 are not problematically high. Studies of plutonic environments have led researchers to infer substantially larger host-rock permeabilities on the order of 10^{-16} m^2 (Chapter 7.2.4), and heat transport in many active hydrothermal systems and fossil *epithermal systems* is (was) advection dominated, which generally requires permeabilities $>10^{-16}$ m^2.

The duration of *regional metamorphism* associated with igneous intrusions (Hanson and Barton, 1989) or crustal thickening (Thompson and England, 1984) may be on the order of 10 Ma, so that time-integrated fluid fluxes of 10^6 cm^3/cm^2 require permeabilities only on the order of 10^{-19} m^2. Such permeabilities are very low by upper-crustal standards, comparable to those measured near the surface in shales or unfractured igneous and metamorphic rocks. However, the existence of even such modest permeabilities in a midcrustal context is somewhat problematic. The nature of this permeability has been the subject of much discussion.

It has been suggested that ductile deformation should act to destroy most porosity (and therefore permeability) at temperatures above about 350 to 400°C. This inference is supported by observations in active geothermal systems; data from geothermal wells that encounter temperatures greater than about 350°C indicate elevated fluid pressures and low permeabilities (Fournier, 1991). However, layered gabbros show clear evidence of meteoric hydrothermal alteration at temperatures of 500 to 900°C (e.g., Taylor, 1990), and in hydrothermal systems associated with the mid-ocean ridges there is evidence for *supercritical phase separation* above the ~400°C critical point (Chapter 3.1.1) of seawater (Lowell and others, 1995; see also Chapter 7.9). At least in the case of these basaltic systems there is fairly clear evidence for a geochemically significant level of permeability at temperatures above 350 to 400°C.

The results of laboratory experiments suggest that this permeability is probably not due to pervasive flow along mineral-grain boundaries. Experiments that mix fine-grained mineral powders with $H_2O \pm CO_2 \pm NaCl$ fluids at elevated pressures and temperatures (e.g., Watson and Brenan, 1987) generally show rapid recrystallization, with water confined to isolated pores at grain corners (see the reviews by Brenan (1991) and Ferry (1994a)).

In contrast, there is extensive field evidence that fracturing is important to metamorphic fluid flow. The fracturing occurs on numerous scales (e.g., Titley, 1990); some of the earliest clear evidence for *microfractures* was the recognition of swarms of microscopic *fluid inclusions* (Chapter 5.1.1) in well-defined planar arrays (Roedder, 1984). In metamorphic environments, open cracks are likely to seal very rapidly (e.g., Walther, 1990; Nur and Walder, 1990). Thus permeability will be a highly transient and perhaps episodic phenomenon.

Though the physics of fracturing are poorly understood, even for simple systems (Marder and Fineberg, 1996), some level of heuristic understanding can be obtained through concepts introduced earlier in this book. In low-permeability environments, metamorphic devolatilization reactions are likely to elevate fluid pressures. For most such reactions the volume of mineral + fluid products greatly exceeds the volume of reactants (e.g., Rumble, 1994). In fact, the data in Table 4.1 suggest that the resulting fluid-production rates or *geologic forcings* (Chapter 4.1.1) in contact-metamorphic aureoles are larger than those in any other geologic environment. Furthermore, in a shallow- to midcrustal contact-metamorphic setting the thermal expansion of pore fluid may comprise a "virtual" fluid source of similar magnitude. The combined effects of devolatilization and thermal expansion should be expected to cause *hydraulic fracturing* (Chapter 4.1.3) of host rocks with intrinsic (prefracturing) permeabilities less than or equal to 10^{-17} m^2 (e.g., Eq. 4.4, Chapter 7.2.3). The geologic forcings in *regional-metamorphic* settings are about two orders of magnitude smaller, are largely due to devolatilization, and are comparable to those in subsiding sedimentary basins. Depending on the effective length scale of the process (Eq. 4.4), regional metamorphism might reasonably be expected to cause hydraulic fracturing in rocks with permeabilities $\leq 10^{-19}$ m^2. In both contact- and regional-metamorphic settings, then, the permeabilities required to accommodate the calculated flow rates (10^{-18} and 10^{-19} m^2, respectively) are quite similar to the limiting permeabilities required for hydraulic fracturing (10^{-17} and 10^{-19} m^2).

10.5.3 Contact metamorphism at Skaergaard

Field studies of contact-metamorphic aureoles typically reveal changes in mineral assemblages with increasing distance from intrusive contacts. The type and distribution of mineral zones provide clues as to the nature of the hydrothermal system associated with metamorphism.

Manning and others (1993; Manning and Bird, 1991) studied porosity evolution and fluid flow near the Skaergaard intrusion in east Greenland. At Skaergaard, a layered gabbroic intrusion was emplaced about 55 Ma ago at a depth of about 3.5 km within a 6- to 7-km-thick section of extrusive basalts. Circulation and cooling of the magma drove a meteoric hydrothermal system (Norton and Taylor, 1979) that depleted the ^{18}O content of the host rock by as much as 4.5 o/oo (Taylor and Forester, 1979). Much of the intrusion and the adjacent contact-metamorphic aureole has since been exposed by uplift and erosion (Figure 10.11).

The work by Manning and others (1993) focused on the *pyroxene* ((Mg, Fe^{2+})$_2$Si$_2$O$_6$ or Ca(Mg, Fe^{2+})Si$_2$O$_6$) and *actinolite + chlorite* (Ca$_2$(Mg, Fe^{2+})$_5$

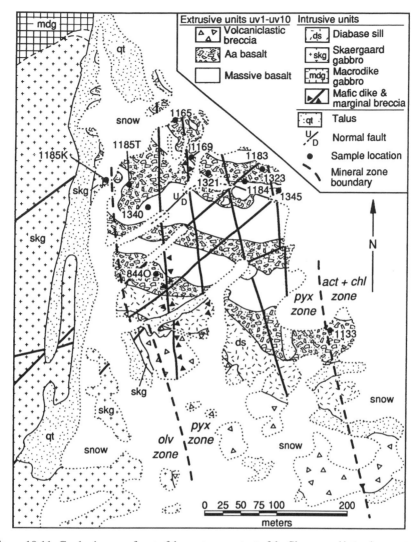

Figure 10.11 Geologic map of part of the eastern contact of the Skaergaard intrusion, east Greenland. Mineral zone boundaries are shown with heavy dashed lines. Abbreviations are olv, olivine zone; pyx, pyroxene zone; act + cl, actinolite + chlorite zone. From Manning and others (1993), reprinted by permission of American Journal of Science.

$Si_8O_{22}(OH)_2 + (Mg, Fe, Al)_6(Si, Al)_4O_{10}(OH)_8)$ mineral assemblages nearest the magma. The pyroxene zone occurs at distances of about 10 to 250 meters from the intrusive contact (Figure 10.11), and the chemical compositions of the various pyroxene phases revealed peak metamorphic temperatures ranging from about 900°C nearest the intrusion to about 800°C at 250 m distance.

The mineralogy of the actinolite + chlorite zone at >250 m from the contact required peak metamorphic temperatures in the range of 300 to 550°C. Simulations of heat and fluid flow near the intrusion (using Eqs. 3.17 and 3.18) revealed that these peak metamorphic temperatures could be matched only if (1) postemplacement temperatures in the middle layers of the intrusion were maintained above the $\sim$1,050°C solidus for some time by a process that retarded crystallization, such as magma convection, and (2) heat transport in the host rock occurred dominantly by conduction. The requirement of conduction-dominated heat transport implies host-rock permeabilities $\leq 10^{-16}$ m^2, whereas pervasive fluid flow sufficient to explain the observed ^{18}O depletion of the host rock (4×10^3 mol/cm^2) requires host-rock permeabilities $\geq 10^{-17}$ m^2. Thus the simulations, in conjunction with the field evidence, constrain host-rock permeabilities to be in the fairly narrow range of 10^{-17} to 10^{-16} m^2. This permeability is large enough that significant hydraulic fracturing is unlikely, although there is clear field evidence for some fracturing related to isolated pores (Manning and Bird, 1991; Manning and others, 1993). (Although Manning and others (1993) were interested in zones very close to the intrusive contact, they simulated a spatial domain 24-km wide by 8-km deep in order to assign reasonable boundary conditions for their flow and transport problem; see Chapter 3.2).

10.5.4 Low-pressure metamorphic belts

Regionally extensive *low-pressure metamorphism* in the upper 15 km of the crust requires temperature gradients of 35°C/km or more, significantly larger than the global average value of about 25°C/km; possible mechanisms for such elevated gradients include crustal extension, rapid uplift, anomalous heat flow at the base of the crust, and advection of heat via aqueous fluid or magma (Hanson and Barton, 1989). We suggested in Chapter 10.5.2 that the permeabilities associated with regional metamorphism in the midcrust might be on the order of 10^{-19} m^2, because values in this range would be sufficient to accommodate the estimated devolatilization fluxes during *prograde* regional metamorphism (heating) yet low enough to permit hydraulic fracturing.

For permeabilities in this range, patterns of fluid flow during prograde metamorphism will be dominated by the effects of fluid production. Flow will generally be up and away from regions of maximum fluid production (Hanson, 1992). Flow rates are such that heat transport will be dominantly by conduction, although solute transport may still be advection dominated (e.g., Chapter 1.2.6).

Regionally extensive low-pressure metamorphic belts are commonly associated with ancient magmatic arcs (Figure 10.12; Barton and Hanson, 1989). Numerical simulations indicate that felsic intrusions can elevate temperatures

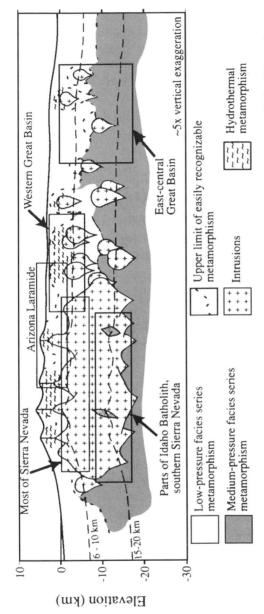

Figure 10.12 Schematic cross section showing the distribution of metamorphic facies in the western United States during the Cretaceous. From Barton and Hanson (1989), reprinted by permission of Geological Society of America.

sufficiently to induce low-pressure metamorphism where their abundance through time exceeds approximately 50% (Hanson and Barton, 1989). In a regional metamorphic setting, hydrothermal alteration may influence the distribution of contact-metamorphic aureoles in the upper crust, but it probably does not affect the total volume of rock metamorphosed.

In northern New England, mineral-chemistry data from metamorphic rocks have been used to infer the presence of large volumes of reactive aqueous fluid at paleo depths of up to 30 kilometers. Pervasive, time-integrated fluid fluxes are estimated to have been on the order of 10^6 cm^3/cm^2 in pelitic schists and micaceous sandstones, and these have been used in conjunction with Darcy's law to infer metamorphic permeabilities in the range of 10^{-21} to 10^{-16} m^2 (Ferry, 1994b). Quartz veins may have served as conduits for movement of metamorphic fluids down regional pressure and temperature gradients (Ague, 1994b). However, the evidence for mass and volume changes attending regional metamorphism in New England (Ague, 1994a) has been vigorously contested (Walther and Holdaway, 1995).

Problems

10.1 An unconfined limestone aquifer extends 10 km from the coast and has an average recharge rate of 10 cm per year. **(a)** If the freshwater lens is only 5-m thick at the coastline, what is the specific discharge, q, in the aquifer at the coastline? **(b)** If the groundwater in the aquifer has a partial pressure of carbon dioxide of 0.01 atmospheres, how much calcite dissolution could occur in the mixing zone over the next 10^4 years? *Hint:* Use Figure 10.6 and Eq. (10.2) to estimate the porosity increase.

10.2 When seawater is equilibrated with respect to both calcite and dolomite at 25°C, 14.6 mmols of calcite could theoretically be converted to dolomite per liter of water (Parkhurst and others, 1980, p. 77). **(a)** What is the *minimum* water-to-rock ratio (volumetric) required to completely convert limestone to dolomite at 25°C? Assume a reasonable density for the minerals and that equilibrium is continually and completely achieved in the throughgoing seawater. **(b)** If the limestone has 30% porosity, how many pore volumes of seawater must move through the rock? What is the final porosity of the dolomite rock, assuming that no compaction has occurred? **(c)** Geochemical calculations reveal that at 75°C one liter of seawater should simultaneously dissolve 85.1 mmols of calcite, precipitate 42.9 mmols of dolomite, and precipitate 22.6 mmols of anhydrite. The chemical formula for dolomite here is assumed to be of the form CaMg $(CO_3)_2$. Given a one-liter volume of calcite with 20% porosity at

75°C, what would the percentages of dolomite, anhydrite, and porosity be in the final rock after all of the calcite was dissolved by throughflowing seawater? **(d)** Which (if any) of the altered rocks in (a)–(c) would make a good reservoir?

10.3 Thermal convection is controlled in part by the coefficient of thermal expansion, α, of water. Assume that fluid density can be described by $\rho = \rho_o + \alpha(T - T_o) + \gamma(C - C_o)$, where $\alpha = -0.15$, $\gamma = +10.0$, ρ is in kg/m^3, T is in °C, and C is the total dissolved solids concentration in moles. **(a)** If the equilibrium constant of a mineral is equal to 0.01 at 25°C, what would that constant need to be at 100°C in order for the additional dissolved mass to counter the buoyancy effect from the heat? Assume that the mineral has one cation and one anion, that there are no other solutes present, and that the salt equilibrium can be described by the basic *law of mass action* (Eq. 2.47) with activity coefficients equal to one. **(b)** Is this type of solubility prograde or retrograde with respect to temperature? **(c)** What would the equilibrium constant need to be if the activity coefficients were equal to 0.2?

10.4 A thermal-water sample from a geothermal well in the Cascade Range of the northwestern United States has an isotopic composition (δD, $\delta^{18}O$) of $(-97, -12.0)$. Calculate the apparent mass ratio of reacted water to reacted rock (*water–rock ratio*) that can account for the composition of the sample. Assume a rock value of $\delta^{18}O = 6.0$, that there is negligible exchange of hydrogen, and that the unreacted meteoric recharge lay along the meteoric water line defined by $\delta D = 8\delta^{18}O + 10$. (See Chapter 10.5.1).

10.5 **(a)** Use Eq. (4.4) to estimate the value of hydraulic conductivity (m/s) required for thermal pressurization of a contact-metamorphic aureole. Assume the values of geologic forcing listed in Table 4.1 and a representative length of 100 meters. **(b)** Convert the solution to part (a) to a value of permeability (m^2), assuming that $P = 300$ bars and $T = 350$°C.

References

Adams, J. E., and Rhodes, M. L. (1960). Dolomitization by seepage reflux. *American Association of Petroleum Geologists' Bulletin* **44**, 1,912–1,920.

Ague, J. J. (1994a). Mass transfer during Barrovian metamorphism of pelites, south-central Connecticut. I. Evidence for changes in composition and volume. *American Journal of Science* **294**, 989–1,057.

Ague, J. J. (1994b). Mass transfer during Barrovian metamorphism of pelites, south-central Connecticut. II. Channelized fluid flow and the growth of staurolite and kyanite. *American Journal of Science* **294**, 1,061–1,134.

Ague, J. J., and Brimhall, G. H. (1989). Geochemical modeling of steady state fluid flow and chemical reaction during supergene enrichment of porphyry copper deposits. *Economic Geology* **84**, 506–528.

Agustdottir, A. M., and Brantley, S. L. (1994). Volatile fluxes integrated over four decades at Grimsvotn volcano, Iceland. *Journal of Geophysical Research* **99**, 9,505–9,522.

Allis, R. G. (1981). Changes in heat flow associated with exploitation of the Wairakei geothermal field. *New Zealand Journal of Geology and Geophysics* **24**, 1–19.

Alpers, C. N., and Brimhall, G. H (1988). Middle Miocene climatic change in the Atacama Desert, northern Chile: Evidence from supergene mineralization at La Escondida. *Geological Society of America Bulletin* **100**, 1,640–1,656.

Alpers, C. N., and Brimhall, G. H. (1989). Paleohydrologic evolution and geochemical dynamics of cumulative supergene metal enrichment at La Escondida, Atacama desert, northern Chile. *Economic Geology* **84**, 229–254.

Anderson, G. M., and MacQueen, R. W. (1982). Ore deposit models. 6. Mississippi–Valley-type lead-zinc deposits. *Geosciences Canada* **9**, 108–117.

Anderson, M. P., and Woessner, W. W. (1992). *Applied Groundwater Modeling.* San Diego: Academic Press.

Andrew, C. J., Crowe, R. W. A., Finlay, S., Pennell, W. M., and Pyne, J. F. (eds.) (1986). *The Geology and Genesis of Mineral Deposits in Ireland.* Dublin: Irish Association for Economic Geology.

Angevine, C. L., and Turcotte, D. L. (1983). Porosity reduction by pressure solution: A theoretical model for quartz arenites. *Geological Society of America Bulletin* **94**, 1,129–1,134.

Appelo, C. A. J., and Postma, D. (1993). *Geochemistry, Groundwater and Pollution.* Rotterdam: A. A. Balkema.

Armannsson, H., and Kristmannsdottir, H. (1992). Geothermal environmental impact. *Geothermics* **21**, 869–880.

301

Athy, L. F. (1930). Density, porosity, and compaction of sedimentary rocks. *American Association of Petroleum Geologists' Bulletin* **14**, 1–24.

Awad, M., and Mizoue, M. (1995). Earthquake activity in the Aswan region. *Pure and Applied Geophysics* **145**, 69–86.

Aziz, K., and Settari, A. (1979). *Petroleum Reservoir Simulation*. London: Elsevier.

Bachu, S. (1988). Analysis of heat transfer processes and geothermal pattern in the Alberta basin, Canada. *Journal of Geophysical Research* **93**, 7,767–7,781.

Back, W., Baedecker, M. J., and Wood., W. W. (1993). Scales in chemical hydrogeology: A historical perspective. In *Regional Groundwater Quality*, Alley, W. M. (ed.), pp. 111–129. New York: Van Nostrand Reinhold.

Back, W., Hanshaw, B. B., Herman, J. S., and Van Driel, J. N. (1986). Differential dissolution of a Pleistocene reef in the groundwater mixing zone of coastal Yucatan, Mexico. *Geology* **14**, 137–140.

Back, W., Hanshaw, B. B., Plummer, L. N., Rahn, P. H., Rightmire, C. T., and Rubin, M. (1983). Process and rate of dedolomitization: Mass transfer and ^{14}C dating in a regional carbonate aquifer. *Geological Society of America Bulletin* **94**, 1,415–1,429.

Back, W., Hanshaw, B. B., Pyle, T. E., Plummer, L. N., and Wedie, A. E. (1979). Geochemical significance of groundwater discharge and carbonate solution to the formation of Caleta Xel Ha, Quintana Roo, Mexico. *Water Resources Research* **15**, 1,521–1,535.

Badiozamani, K. (1973). The dorag dolomitization model – Application to the Middle Ordovician of Wisconsin. *Journal of Sedimentary Petrology* **43**, 956-984.

Baetsle, L. H. (1969). Migration of radionuclides in porous media. In *Progress in Nuclear Energy* **XII**, *Health Physics*, Duhamel, A. M. F. (ed.), pp. 707–730. Elmsford, N.Y.: Pelargonium Press.

Bamford, R. W. (1972). The Mount Fubilan (Ok Tedi) porphyry copper deposit, Territory of Papua and New Guinea. *Economic Geology* **67**, 1,019–1,033.

Bangs, N. L. B., Shipley, T. H., and Moore, G. F. (1996). Elevated fluid pressure and fault zone dilation inferred from seismic models of the northern Barbados Ridge decollement. *Journal of Geophysical Research* **101**, 627–642.

Bangs, N. L. B., and Westbrook, G. K. (1991). Seismic modeling of the decollement zone at the base of the Barbados Ridge accretionary complex. *Journal of Geophysical Research* **96**, 3,853–3,866.

Bangs, N. L. B., Westbrook, G. K., Ladd, J. W., and Buhl, P. (1990). Seismic velocities from the Barbados Ridge complex: Indicators of high pore fluid pressures in an accretionary complex. *Journal of Geophysical Research* **95**, 8,767–8,782.

Barker, C. (1972). Aquathermal pressuring – Role of temperature in development of abnormal-pressure zones. *American Association of Petroleum Geologists' Bulletin* **56**, 2,068–2,071.

Barnes, H. L. (ed.). (1979). *Geochemistry of Hydrothermal Ore Deposits*. New York: John Wiley and Sons.

Barnes, I. (1970). Metamorphic waters from the Pacific tectonic belt of the west coast of the United States. *Science* **168**, 973–975.

Barnes, I., Irwin, W. P., and Gibson, H. A. (1975). Geologic map showing springs rich in carbon-dioxide or chloride in California. *U.S. Geological Survey Water Resources Investigations Open-File Map,* scale 1:1,500,000.

Barnes, I., Irwin, W. P., and White, D. E. (1984). Map showing world-wide distribution of carbon-dioxide springs and major zones of seismicity. *U.S. Geological Survey Miscellaneous Investigations Series Map* **I-1528**, scale 1:40,000,000.

Baron, G. (1960). Sur la synthese de la dolomie. Application au phenomene de dolomitization. *Rev. Institute Francaise Petrole Annals Combustion Liquides* **15**, 3–68.

Barton, C. A., Zoback, M. D., and Moos, Daniel (1995). Fluid flow along potentially active faults in crystalline rock. *Geology* **23**, 683–686.

Barton, M. D., and Hanson, R. B. (1989). Magmatism and the development of low-pressure metamorphic belts: Implications from the western United States and modeling. *Geological Society of America Bulletin* **101**, 1,051–1,065.

Beane, R. E., and Titley, S. R. (1981). Porphyry copper deposits: Part II. Hydrothermal alteration and mineralization. *Economic Geology* 75th Anniversary Volume, 235–269.

Bear, J. (1961). On the tensor form of dispersion. *Journal of Geophysical Research* **66**, 1,185–1,197.

Bear, J. (1972). *Dynamics of Fluids in Porous Media*. New York: American Elsevier.

Bear, J. (1979). *Hydraulics of Groundwater*. New York: McGraw-Hill.

Beauheim, R. L., Roberts, R. M., Dale, T. F., Fort, M. D., and Stensrud, W. A. (1993). Hydraulic testing of Salado Formation evaporites at the waste isolation pilot plant site: Second interpretive report. *Sandia National Laboratory Report* **SAND92-0533**.

Bekins, B., McCaffrey, A. M., and Dreiss, S. J. (1994). Influence of kinetics on the smectite to illite transition in the Barbados accretionary prism. *Journal of Geophysical Research* **99**, 18,147–18,158.

Bekins, B., McCaffrey, A. M., and Dreiss, S. J. (1995). Episodic and constant flow models for the origin of low-chloride waters in a modern accretionary complex. *Water Resources Research* **31**, 3,205–3,215.

Belitz, K., and Bredehoeft, J. D. (1988). Hydrodynamics of the Denver Basin: Explanation of subnormal fluid pressure. *American Association of Petroleum Geologists' Bulletin* **72**, 1,334–1,359.

Bell, K. (1981). A review of the geochronology of the Precambrian of Saskatchewan – Some minerals from the Midwest uranium deposit, northern Saskatchewan. *Canadian Journal of Earth Sciences* **21**, 642–648.

Bemis, K. G., Von Herzen, R. P., and Mottl, M. J. (1993). Geothermal heat flux from hydrothermal plumes on the Juan de Fuca Ridge. *Journal of Geophysical Research* **98**, 6,351–6,365.

Bennett, S. C., and Hanor, J .S. (1987). Dynamics of subsurface salt dissolution at the Welsh Dome, Louisiana Gulf Coast. In *Dynamical Geology of Salt and Related Structures*, Lerche, I., and O'Brien, J. J. (eds.), pp. 653–677. Orlando: Academic Press.

Bentley, H. W., Phillips, F. M., Davis, S. N., Habermehl, M. A., Airey, P. L., and Gaeme, E. C. (1986). Chlorine 36 dating of very old groundwater. 1. The Great Artesian Basin, Australia. *Water Resources Research* **22**, 1,991–2,002.

Berry, F. A. F. (1973). High fluid potentials in California Coast Ranges and their tectonic significance. *American Association of Petroleum Geologists' Bulletin* **57**, 1,219–1,249.

Berryman, J. G. (1992). Effective stress for transport properties of inhomogeneous porous rock. *Journal of Geophysical Research* **97**, 17,409–17,424.

Bethke, C. M. (1985). A numerical model of compaction-driven groundwater flow and its application to the paleohydrology of intracratonic sedimentary basins. *Journal of Geophysical Research* **90**, 6,817–6,828.

Bethke, C. M. (1986). Inverse hydrologic analysis of the distribution and origin of Gulf Coast-type geopressured zones. *Journal of Geophysical Research* **91**, 6,535–6,545.

Bethke, C. M. (1996). *Geochemical Reaction Modeling*. New York: Oxford University Press.

Bethke, C. M., Lee, M. K., Quinodoz, H. A. M., and Kreiling, W. N. (1993). *Basin Modeling with BASIN2*. Urbana: University of Illinois Hydrogeology Program.

Bethke, C. M., and Marshak, S. (1990). Brine migrations across North America – The plate tectonics of groundwater. *Annual Review of Earth and Planetary Sciences* **18**, 287–315.

Biddle, K. T. (1991). The Los Angeles basin, an overview. In *Active Margin Basins*, Biddle, K. T. (ed.). *American Association of Petroleum Geologists' Memoir* **52**, 5–24.

Biot, M. A. (1941). General theory of three-dimensional consolidation. *Journal of Applied Physics* **12**, 155–164.

Birch, F., and Clark, H. (1940). The thermal conductivity of rocks and its dependence on temperature and composition. *American Journal of Science* **238**, 613–635.

Bischoff, J. L., and Dickson, F. W. (1975). Seawater-basalt interaction at 200°C and 500 bars: Implications for origin of sea-floor heavy-metal deposits and regulation of seawater chemistry. *Earth and Planetary Science Letters* **25**, 385–397.

Bischoff, J. L., and Pitzer, K. S. (1989). Liquid-vapor relations for the system $NaCl-H_2O$: Summary of the P-T-x surface from 300° to 500°C. *American Journal of Science* **289**, 217–248.

Bischoff, J. L., and Rosenbauer, R. J. (1984). The critical point and two-phase boundary of seawater, 200–500°C. *Earth and Planetary Science Letters* **68**, 172–180.

Bjorlykke, K. (1993). Fluid flow in sedimentary basins. In *Basin Analysis and Dynamics of Sedimentary Basin Evolution*. Cloetingh, S., Sassi, W., Horvath, F., and Puigdefabregas, C. (eds.). *Sedimentary Geology* **86**, 137–158.

Bjorlykke, K., and Egeberg, P. K. (1993). Quartz cementation in sedimentary basins. *American Association of Petroleum Geologists' Bulletin* **77**, 1,538–1,548.

Bjorlykke, K., Nedkvitne, T., Ramm, M., and Saigal, G. (1992). Diagenetic processes in the Brent Group (Middle Jurassic) reservoirs of the North Sea – An overview. In *Geology of the Brent Group*, Morton, A. C., Haszeldine, R. S., Giles, M. R., and Brown, S. (eds.). *Geological Society Special Publication* **61**, 263–287.

Blackwell, D. D., and Baker, S. L. (1988). Thermal analysis of the Austin and Breitenbush geothermal systems, Western Cascades, Oregon. In *Geology and Geothermal Resources of the Breitenbush-Austin Hot Springs Area, Clackamas and Marion Counties, Oregon,* Sherrod, D. R. (ed.). *Oregon Department of Geology and Mineral Industries Open-File Report* **O-88-5**, 47–62.

Blackwell, D. D., and Steele, J. L. (1989). Thermal conductivity of rocks: Measurement and significance. In *Thermal History of Sedimentary Basins: Methods and Case Histories*, Naeser, N. D., and McCulloh, T. H. (eds.), pp. 13–36. New York: Springer-Verlag.

Blackwell, D. D., Steele, J. L., and Brott, C. A. (1980). The terrain effect on terrestrial heat flow. *Journal of Geophysical Research* **85**, 4,757–4,772.

Blackwell, D. D., Steele, J. L., Frohme, M. K., Murphy, C. F., Priest, G. R., and Black, G. L. (1990). Heat flow in the Oregon Cascade Range and its correlation with regional gravity, Curie point depths, and geology. *Journal of Geophysical Research* **95**, 19,475–19,494.

Blanc, P., and Connan, J. (1994). Preservation, degradation, and destruction of trapped oil. In *The Petroleum System – From Source to Trap*, Magoon, L. B., and Dow, D. G. (eds.). *American Association of Petroleum Geologists' Memoir* **60**, 237–247.

Blatt, H. (1979). Diagenetic processes in sandstones. In *Aspects of Diagenesis*, Scholle, P. A., and Schluger, P. R. (eds.). *Society of Economic Paleontologists and Mineralogists Special Publication* **26**, 141–158.

Blattner, P. (1985). Isotope shift data and the natural evolution of geothermal systems. *Chemical Geology* **49**, 187–203.

Bodine, M. W., and Jones, B. F. (1986). The salt norm: A quantitative chemical-mineralogical characterization of natural waters. *U.S. Geological Survey Water Resources Investigations Report* **86-4086**.

Bodvarsson, G. S., Benson, S. M., and Witherspoon, P. A. (1982). Theory of development of geothermal systems charged by vertical faults. *Journal of Geophysical Research* **87**, 9,317–9,328.

Bodvarsson, G. S., Pruess, K., Stefansson, V., and Eliasson, E. T. (1984). The Krafla geothermal field, Iceland 2. The natural state of the system. *Water Resources Research* **20**, 1,531–1,544.

Bolt, B. A. (1993). *Earthquakes*. New York: W.H. Freeman.

Bolton, E. W., Lasaga, A. C., and Rye, D. M. (1996). A model for the kinetic control of quartz dissolution and precipitation in porous media flow with spatially variable permeability: Formulation and examples of thermal convection. *Journal of Geophysical Research* **101**, 22,157–22,187.

Borchert, H., and Muir, R. O. (1964). *Salt Deposits: The Origin, Metamorphism and Deformation of Evaporites*. Princeton: Van Nostrand-Reinhold.

Brace, W. F. (1980). Permeability of crystalline and argillaceous rocks. *International Journal of Rock Mechanics and Mining Sciences and Geomechanics Abstracts* **17**, 241–251.

Brace, W. F. (1984). Permeability of crystalline rocks: New in situ measurements. *Journal of Geophysical Research* **89**, 4,327–4,330.

Brace, W. F., and Kohlstedt, D. L. (1980). Limits on lithospheric stress imposed by laboratory experiments. *Journal of Geophysical Research* **85**, 6,248–6,252.

Braitsch, O. (1971). *Salt Deposits. Their Origin and Composition*. New York: Springer-Verlag.

Bredehoeft, J. D. (1967). Response of well-aquifer systems to Earth tides. *Journal of Geophysical Research* **72**, 3,075–3,087.

Bredehoeft, J. D. (1988). Will salt repositories be dry? *Eos, Transactions American Geophysical Union* **69**, 121–131.

Bredehoeft, J. D., and Hanshaw, B. B. (1968). On the maintenance of anomalous fluid pressures: I. Thick sedimentary sequences. *Geological Society of America Bulletin* **79**, 1,097–1,106.

Bredehoeft, J. D., and Ingebritsen, S. E. (1990). Degassing of carbon dioxide as a possible source of high pore pressures in the crust. In *The Role of Fluids in Crustal Processes*, Bredehoeft, J. D., and Norton, D. L. (eds.), pp. 158–164. Washington, DC: National Academy Press.

Bredehoeft, J. D., Neuzil, C. E., and Milly, P. C. D. (1983). Regional flow in the Dakota aquifer: A study of the role of confining layers. *U.S. Geological Survey Water Supply Paper* **2237**.

Bredehoeft, J. D., and Norton, D. (eds.) (1990). *The Role of Fluids in Crustal Processes*. Washington, DC: National Academy Press.

Bredehoeft, J. D., and Papadopolous, I. S. (1965). Rates of vertical groundwater movement estimated from the Earth's thermal profile. *Water Resources Research* **1**, 325–328.

Bredehoeft, J. D., Wesley, J. B., and Fouch, T. D. (1994). Simulation of the origin of fluid pressure, fracture generation, and the movement of fluids in the Uinta basin,

Utah. *American Association of Petroleum Geologists' Bulletin* **78**, 1,729–1,747.

Bredehoeft, J. D., Wolff, R. G., Keys, W. S., and Shuter, E. (1976). Hydraulic fracturing to determine the regional in situ stress, Piceance Basin, Colorado. *Geological Society of America Bulletin* **87**, 250–258.

Brenan, J. M. (1991). Development and maintenance of metamorphic permeability: Implications for fluid transport. In *Contact Metamorphism*, Kerrick, D. M. (ed.). *Reviews in Mineralogy* **26**, 291–319.

Brimhall, G. H. (1979). Lithologic determination of mass transfer mechanisms of multiple stage porphyry copper mineralization at Butte, Montana: Vein formation by hypogene leaching and enrichment of potassium silicate protore. *Economic Geology* **74**, 556–589.

Brimhall, G. H. (1980). Deep hypogene oxidation of porphyry copper potassium-silicate protore at Butte, Montana: A theoretical evaluation of the copper remobilization hypothesis. *Economic Geology* **75**, 384–409.

Brune, J. N., Henyey, T. L., and Roy, R. F. (1969). Heat flow, stress, and rate of slip along the San Andreas fault, California. *Journal of Geophysical Research* **74**, 3,821–3,827.

Bryan, T. S. (1995). *The Geysers of Yellowstone.* 3rd ed. Niwot: University of Colorado Press.

Budd, D. A. (1997). Cenozoic dolomites of carbonate islands: Their attributes and origin. *Earth-Science Reviews* **42**, 1–47.

Burnham, A. K., and Sweeney, J. J. (1989). A chemical kinetic model of vitrinite maturation and reflectance. *Geochimica et Cosmochimica Acta* **53**, 2,649–2,657.

Burrus, J., Kuhfuss, A., Doligez, B., and Ungerer, P. (1992). Are numerical models useful in reconstructing the migration of hydrocarbons? A discussion based on the northern Viking graben. In *Petroleum Migration*, England, W. A., and Fleet, A. J. (eds.). *Geological Society Special Publication* **59**, 89–109.

Byerlee, J. D. (1968). Brittle-ductile transition in rocks. *Journal of Geophysical Research* **73**, 4,741–4,750.

Byerlee, J. D. (1990). Friction, overpressure, and fault normal compression. *Geophysical Research Letters* **17**, 2,109–2,203.

Byerlee, J. D. (1993). Model for episodic flow of high-pressure water in fault zones before earthquakes. *Geology* **21**, 303–306.

Campbell, J. H., Koskinas, G. J., and Stout, N. D. (1978). Kinetics of oil generation from Colorado oil shale. *Fuel* **57**, 372–376.

Cappetti, G., Celati, R., Cigni, U., Squarci, P., Stefani, G., and Taffi, L. (1985). Development of deep exploration in the geothermal areas of Tuscany, Italy. *Geothermal Resources Council Transactions* **9** (intl. vol.), 303–309.

Carman, P. C. (1956). *The Flow of Gases through Porous Media.* New York: Academic Press.

Carslaw, H. S., and Jaeger, J. C. (1959). *Conduction of Heat in Solids.* Oxford: Clarendon Press.

Carson, B., Seke, E., Paskevich, V., and Holmes, M. L. (1994). Fluid expulsion sites on the Cascadia accretionary prism: Mapping diagenetic deposits with processed GLORIA imagery. *Journal of Geophysical Research* **99**, 11,959–11,969.

Casas, E., Lowenstein, T. K., Spencer, R. J., and Zhang, P. (1992). Carnallite mineralization in the nonmarine Qaidam basin, China: Evidence for the early diagenetic origin of potash evaporites. *Journal of Sedimentary Petrology* **62**, 881–898.

Cathles, L. M. (1977). An analysis of the cooling of intrusives by ground-water convection which includes boiling. *Economic Geology* **72**, 804–826.

Cathles, L. M., and Smith, A. T. (1983). Thermal constraints on the formation of Mississippi Valley–type lead-zinc deposits and their implications for episodic basin dewatering and deposit genesis. *Economic Geology* **78**, 983–1,002.

Chapman, D. S., Keho, T. H., Bauer, M. S., and Picard, M. D. (1984). Heat flow in the Uinta basin determined from bottom hole temperature (BHT) data. *Geophysics* **49**, 453–466.

Chapman, D. S., and Pollack, H. N. (1975). Global heat flow: A new look. *Earth and Planetary Science Letters* **28**, 23–32.

Chapman, R. E. (1981). *Geology and Water*. The Hague: Nijhoff/Junk Publishers.

Chenoweth, W. L., and McLemore, V. T. (1989). Uranium resources on the Colorado Plateau. In *Energy Frontier in the Rockies*, Lorenz, J. C., and Lucas, S. G. (eds.). Albuquerque: Albuquerque Geological Society.

Chilingar, G. V. (1964). Relationship between porosity, permeability, and grain-size distribution of sands and sandstones. In *Deltaic and Shallow Marine Deposits: Proceedings of the Sixth International Sedimentological Congress*, van Straatan, L. M. J. U. (ed.). *Developments in Sedimentology* **1**, 71–75. New York: Elsevier Science.

Clauser, C. (1988). Opacity – The concept of radiative thermal conductivity. In *Handbook of Terrestrial Heat-Flow Density Determinations*, Haenel, R., Rybach, L., and Stegena, L. (eds.), pp. 143–165. Dordrecht: Kluwer Academic Publishers.

Clauser, C. (1992). Permeability of crystalline rocks. *Eos, Transactions American Geophysical Union* **73**, 233, 237.

Cochrane, G. R., Moore, J. C., MacKay, M. E., and Moore, G. F. (1994). Velocity and inferred porosity model of the Oregon accretionary prism from multichannel seismic reflection data: Implications on sediment dewatering and overpressure. *Journal of Geophysical Research* **99**, 7,033–7,043.

Coleman, M. L. (1993). Microbiological processes: controls on the shape and composition of carbonate concretions. In *Marine Sediments, Burial, Porewater Chemistry, Microbiology and Diagenesis*, Parkes, R. J., Westbrook, P., and de Leeuw, J. W. (eds.). *Marine Geology* **113**, 127–140.

Combarnous, M. A., and Bories, S. A. (1975). Hydrothermal convection in saturated porous media. *Advances in Hydroscience* **10**, 231–307.

Cooley, R. L., Konikow, L. F., and Naff, R. L. (1986). Non-linear regression groundwater flow modeling of a deep regional aquifer system. *Water Resources Research* **22**, 1,759–1,778.

Cooper, H. H., Jr. (1966). The equation of groundwater flow in fixed and deforming coordinates. *Journal of Geophysical Research* **71**, 4,785–4,790.

Cooper, H. H., Jr., Bredehoeft, J. D., Papadopulos, I. S., and Bennett, R. R. (1965). The response of well-aquifer systems to seismic waves. *Journal of Geophysical Research* **70**, 3,915–3,926.

Corbet, T. F., and Bethke, C. M. (1992). Disequilibrium fluid pressures and groundwater flow in the Western Canada sedimentary basin. *Journal of Geophysical Research* **97**, 7,203–7,217.

Corey, A. T. (1957). Measurement of water and air permeabilities in unsaturated soil. *Soil Science Society Proceedings* **21**, 7–10.

Cox, B. L., and Pruess, K. (1990). Numerical experiments on convective heat transfer in water-saturated porous media at near-critical conditions. *Transport in Porous Media* **5**, 299–323.

Craig, H. (1961a). Isotopic variations in meteoric waters. *Science* **133**, 1,702–1,703.

Craig, H. (1961b). Standard for reporting concentrations of deuterium and oxygen-18 in natural waters. *Science* **133**, 1,833–1,834.

Craig, H. (1963). The isotopic geochemistry of water and carbon in geothermal areas. In *Nuclear Geology on Geothermal Areas*, Tongiorgi, E. (ed.), pp. 17–53. Pisa: Consiglio Nazionale delle Richerche.

Criss, R. E., Fleck, R. J., and Taylor, H. P., Jr. (1991). Tertiary meteoric hydrothermal systems and their relation to ore deposition, northwestern United States and southern British Columbia. *Journal of Geophysical Research* **96**, 13,335–13,356.

Criss, R. E., and Hofmeister, A. M. (1991). Application of fluid dynamics principles in tilted permeable media to terrestrial hydrothermal systems. *Geophysical Research Letters* **18**, 199–202.

Dahlberg, E. C. (1994). *Applied Hydrodynamics in Petroleum Exploration*. 2nd ed. New York: Springer-Verlag.

Dahlkamp, F. J. (1978). Geologic appraisal of the Key Lake U-Ni deposits, northern Saskatchewan. *Economic Geology* **73**, 1,430–1,449.

Darcy, H. (1856). *Les Fontaines Publiques de la Ville de Dijon*. Paris: Victor Dalmont.

Darnley, A. G. (1981). The relationship between uranium distribution and some major crustal features in Canada. *Mineralogical Magazine* **44**, 425–436.

Davies, P. B. (1989). Assessing deep-seated dissolution-subsidence hazards at radioactive-waste repository sites in bedded salt. *Engineering Geology* **27**, 467–487.

Davis, S. H., Rosenblat, S., Wood, J. R., and Hewett, T. A. (1985). Convective fluid-flow and diagenetic patterns in domed sheets. *American Journal of Science* **285**, 207–223.

DeGroot, S. R., and Mazur, P. (1962). *Non-Equilibrium Thermodynamics*. Amsterdam: North-Holland.

de Josselin de Jong, G. (1969). Generating functions in the theory of flow through porous media. In *Flow Through Porous Media*, De Wiest, R. J. M. (ed.), pp. 377–400. San Diego: Academic Press.

deCaritat, P., Hutcheon, I., and Walshe, J. L. (1993). Chlorite geothermometry: A review. *Clays and Clay Minerals* **41**, 219–239.

Deming, D. (1992). Catastrophic release of heat and fluid flow in the continental crust. *Geology* **20**, 83–86.

Deming, D. (1993). Regional permeability estimates from investigations of coupled heat and groundwater flow, North Slope of Alaska. *Journal of Geophysical Research* **98**, 16,271–16,286.

Deming, D., and Nunn, J. A. (1991). Numerical simulations of brine migration by topographically driven recharge. *Journal of Geophysical Research* **96**, 2,485–2,499.

Deming, D., Sass, J. H., Lachenbruch, A. H., and DeRito, R. F. (1992). Heat flow and subsurface temperature as evidence for basin-scale groundwater flow. *Geological Society of America Bulletin* **104**, 528–542.

Dewers, T., and Ortoleva, P. (1990a). Force of crystallization during the growth of siliceous concretions. *Geology* **18**, 204–207.

Dewers, T., and Ortoleva, P. (1990b). Geochemical organization. III. A mechano-chemical model of metamorphic differentiation. *American Journal of Science* **290**, 473–521.

De Wiest, R. J. M. (1966). On the storage coefficient and the equations of groundwater flow. *Journal of Geophysical Research* **71**, 1,117–1,122.

Dickey, P. A. (1979). *Petroleum Development Geology*. Tulsa: PPC Books.

Dickinson, G. (1953). Geological aspects of abnormal reservoir pressures in Gulf

Coast Louisiana. *American Association of Petroleum Geologists' Bulletin* **37**, 410–432.

Diment, W. H., and Pratt, H. R. (1988). Thermal conductivity of some rock-forming minerals: A tabulation. *U.S. Geological Survey Open-File Report* **88-690**.

Doe, B. R., and Zartman, R. E. (1979). Plumbotectonics, the Phanerozoic. In *Geochemistry of Hydrothermal Ore Deposits*, Barnes, H. L. (ed.), pp. 22–70. New York: John Wiley and Sons.

Doebl, F., Heling, D., Homann, W., Karweil, J., Teichmuller, M., and Welte, D. (1974). Diagenesis of Tertiary clayey sediments and included dispersed organic matter in relationship to geothermics in the Upper Rhine graben. In *Approaches to Taphrogenesis*, Illies, J. and Fuchs, K. (eds.), pp. 192–207. Stuttgart: Schweizbard.

Domenico, P. A., and Palciauskas, V. V. (1973). Theoretical analysis of forced convective heat transfer in regional groundwater flow. *Geological Society of America Bulletin* **84**, 3,803–3,814.

Domenico, P. A., and Schwartz, F. W. (1990). *Physical and Chemical Hydrogeology*. New York: John Wiley and Sons.

Drever, J. I. (1982). *The Geochemistry of Natural Waters*. Englewood Cliffs: Prentice-Hall.

Driscoll, F. G. (1986). *Groundwater and Wells*. 2nd ed. Saint Paul: Johnson Division.

Du Commun, J. (1828). On the cause of freshwater springs, fountains, and c.. *American Journal of Science* **14**, 174–176.

Duffy, C. J., and Al-Hassan, S. (1988). Groundwater circulation in a closed desert basin: Topographic scaling and climatic forcing. *Water Resources Research* **24**, 1,675–1,688.

Dunn, J. C., and Hardee, H. C. (1981). Superconvecting geothermal zones. *Journal of Volcanology and Geothermal Research* **11**, 189–201.

Earlougher, R. C., Jr. (1977). *Advances in Well Test Analysis*. New York: Society of Petroleum Engineers of AIME.

Eastoe, C. J. (1982). Physics and chemistry of the hydrothermal system at the Panguna porphyry copper deposit, Bougainville, Papau New Guinea. *Economic Geology* **77**, 127–153.

Economides, M. J., and Miller, F. G. (1985). The effects of adsorption phenomena in the evaluation of vapour-dominated geothermal reservoirs. *Geothermics* **14**, 3–27.

Edmond, J. M., Measures, C., McDuff, R. E., Chan, L. H., Collier, R., Grant, B., Gordon, L. I., and Corliss, J. B. (1979). Ridge crest hydrothermal activity and the balances of the major and minor elements in the ocean: The Galapagos data. *Earth and Planetary Science Letters* **46**, 1–18.

Elsworth, D., and Voight, B. (1992). Theory of dike intrusion in a saturated porous solid. *Journal of Geophysical Research* **97**, 9,105–9,117.

Eugster, H. P., and Jones, B. F. (1979). Behavior of major solutes during closed-basin brine evolution. *American Journal of Science* **279**, 609–631.

Evans, D. G., and Raffensperger, J. P. (1992). On the stream function for variable-density groundwater flow. *Water Resources Research* **28**, 2,141–2,145.

Fanning, K. A., Byrne, R. H., Breland, J. A., II, and Betzer, P. R. (1981). Geothermal springs of the West Florida continental shelf: Evidence for dolomitization and radionuclide enrichment. *Earth and Planetary Science Letters* **52**, 345–354.

Faure, G. (1986). *Principles of Isotope Geology.* 2nd ed. New York: John Wiley and Sons.

Faust, C. R., and Mercer, J. W. (1979a). Geothermal reservoir simulation. 1. Mathematical models for liquid- and vapor-dominated hydrothermal systems. *Water Resources Research* **15**, 23–30.

Faust, C. R., and Mercer, J. W. (1979b). Geothermal reservoir simulation. 2. Numerical solution techniques for liquid- and vapor-dominated hydrothermal systems. *Water Resources Research* **15**, 31–46.

Felmy, A. R., and Weare, J. H. (1991). Calculation of multicomponent ionic diffusion from zero to high concentration. I. The system $Na-K-Ca-Mg-Cl-SO_4-H_2O$ at $25°C$. *Geochimica et Cosmochimica Acta* **55**, 113–131.

Ferguson, I. J., Westbrook, G. K., Langseth, M. G., and Thomas, G. P. (1993). Heat flow and thermal models of the Barbados Ridge accretionary complex. *Journal of Geophysical Research* **98**, 4,121–4,142.

Ferry, J. M. (1978). Fluid interaction between granite and sediment during metamorphism. *American Journal of Science* **278**, 1,025–1,056.

Ferry, J. M. (1994a). A historical review of metamorphic fluid flow. *Journal of Geophysical Research* **99**, 15,487–15,498.

Ferry, J. M. (1994b). Overview of the petrologic record of fluid flow during regional metamorphism in northern New England. *American Journal of Science* **294**, 905–988.

Fetter, C. W. (1994). *Applied Hydrogeology*. 3rd ed. New York: Macmillan College.

Finlayson, B. A. (1992). *Numerical Methods for Problems with Moving Fronts*. Seattle: Ravenna Park Publishing.

Folk, R. L., and Land, L. S. (1975). Mg/Ca ratio and salinity: Two controls over crystallization of dolomite. *American Association of Petroleum Geologists' Bulletin* **59**, 60–68.

Force, E. R., Back, W., and Spiker, E. C. (1986). A groundwater mixing model for the origin of the Imini manganese deposit (Cretaceous) of Morocco. *Economic Geology* **81**, 65–79.

Forster, C., and Smith, L. (1989). The influence of groundwater flow on thermal regimes in mountainous terrain: A model study. *Journal of Geophysical Research* **97**, 9,439–9,451.

Fournier, R. O. (1969). Old Faithful: A physical model. *Science* **163**, 304–305.

Fournier, R. O. (1981). Application of water geochemistry to geothermal exploration and reservoir engineering. In *Geothermal Systems: Principles and Case Histories*, Rybach, L., and Muffler, L. J. P. (eds.), pp. 109–144. New York: John Wiley and Sons.

Fournier, R. O. (1983). Self-sealing and brecciation resulting from quartz deposition within hydrothermal systems. *Proceedings of the International Symposium on Water-Rock Interaction* **4**, 137–140.

Fournier, R. O. (1987). Conceptual models of brine evolution in magmatic-hydrothermal systems. In *Volcanism in Hawaii*, Decker, R. W., Wright, T. L., and Stauffer, P. H. (eds.). *U.S. Geological Survey Professional Paper* **1350**, 1,487–1,506.

Fournier, R. O. (1989). Geochemistry and dynamics of the Yellowstone National Park hydrothermal system. *Annual Review of Earth and Planetary Sciences* **17**, 13–53.

Fournier, R. O. (1991). The transition from hydrostatic to greater than hydrostatic fluid pressure in presently active continental hydrothermal systems in crystalline rock. *Geophysical Research Letters* **18**, 955–958.

Fournier, R. O., and Potter, R. W. (1982). A revised and expanded silica (quartz) geothermometer. *Geothermal Resources Council Bulletin* **11(10)**, 3–9.

Freeze, R. A., and Back, W. (eds.) (1983). *Physical Hydrogeology*. Stroudsburg: Hutchinson Ross.

Freeze, R. A., and Cherry, J. A. (1979). *Groundwater*. Englewood Cliffs: Prentice-Hall.

Freeze, R. A., and Witherspoon, P. A. (1966). Theoretical analysis of regional groundwater flow. 1. Analytical and numerical solutions to the mathematical model. *Water Resources Research* **2**, 641–656.

Fridleifsson, I. B., and Freeston, D. H. (1994). Geothermal energy research and development. *Geothermics* **23**, 175–214.

Furlong, K. P., Hanson, R. B., and Bowers, J. R. (1991). Modeling thermal regimes. In *Contact Metamorphism*, Kerrick, D. M. (ed.). *Reviews in Mineralogy* **26**, 437–505.

Fyfe, W. S., Price, N. J., and Thompson, A. B. (1978). *Fluids in the Earth's Crust*. New York: Elsevier Scientific.

Gaines, G. L., and Thomas, H. C. (1953). Adsorption studies on clay minerals. II. A formulation of the thermodynamics of exchange adsorption. *Journal of Chemical Physics* **21**, 714–718.

Galloway, W. E. (1984). Hydrogeologic regimes of sandstone diagenesis. In *Clastic Diagenesis*, McDonald, D. A., and Surdam, R. C. (eds.). *American Association of Petroleum Geologists' Memoir* **37**, 3–13.

Gangi, A. F. (1978). Variation of whole and fractured porous rock permeability with confining pressure. *International Journal of Rock Mechanics and Mining Sciences and Geomechanics Abstracts* **15**, 249–257.

Gapon, E. N. (1933). Theory of exchange adsorption. *Journal of General Chemistry* (USSR) **3**, 667–669.

Garabedian, S. P., LeBlanc, D. R., Gelhar, L. W., and Celia, M. A. (1991). Large-scale natural gradient tracer test in sand and gravel, Cape Cod, Massachusetts. 2. Analysis of spatial moments for a nonreactive tracer. *Water Resources Research* **27**, 911–924.

Garven, G. (1989). A hydrogeologic model for the formation of the giant oil sands deposits of the Western Canada sedimentary basin. *American Journal of Science* **289**, 105–166.

Garven, G., and Freeze, R. A. (1984a). Theoretical analysis of the role of groundwater flow in the genesis of stratabound ore deposits 1. Mathematical and numerical model. *American Journal of Science* **284**, 1,085–1,124.

Garven, G., and Freeze, R. A. (1984b). Theoretical analysis of the role of groundwater flow in the genesis of stratabound ore deposits 2. Quantitative results. *American Journal of Science* **284**, 1,125–1,174.

Garven, G., Ge, S., Person, M. A., and Sverjensky, D. A. (1993). Genesis of stratabound ore deposits in the midcontinent basins of North America. 1. The role of regional groundwater flow. *American Journal of Science* **293**, 497–568.

Gavrilenko, P., and Gueguen, Y. (1993). Fluid overpressures and pressure solution in the crust. *Tectonophysics* **217**, 91–110.

Gelhar, L. W. (1986). Stochastic subsurface hydrology from theory to applications. *Water Resources Research* **22**, 135S–145S.

Gelhar, L. W., Welty, C., and Rehfeldt, K. R. (1992). A critical review of data on field-scale dispersion in aquifers. *Water Resources Research* **28**, 1,955–1,974.

Germanovich, L. N., and Lowell, R. P. (1992). Percolation theory, thermoelasticity, and discrete hydrothermal venting in the Earth's crust. *Science* **255**, 1,564–1,567.

Goldhaber, M. B., Rowan, E. L., Hatch, J., Zartman, R., Pitman, J., and Reynolds, R. (1994). The Illinois Basin as a flow path for low temperature hydrothermal fluids. In *Proceedings of the Illinois Basin Energy and Mineral Workshop, Evansville, IN*, Ridgely, J. L, Drahovzal, J. A., Keith, B. D., and Kolata, D. R. (eds.). *U.S. Geological Survey Open-File Report* **94-298**, 10–12.

Goode, D. J., and Konikow, L. F. (1989). Modification of a method-of-characteristics solute-transport model to incorporate decay and equilibrium-controlled sorption or ion exchange. *U.S. Geological Survey Water Resources Investigations Report* **89-4030**.

Grant, M. A. (1977). Permeability reduction factors at Wairakei. Paper presented at the AIChE-ASME Heat Transfer Conference, Salt Lake City, August 15–17, 1977.

Grant, M. A. (1979). Interpretation of downhole pressure measurements at Baca. *Proceedings of the Workshop on Geothermal Reservoir Engineering* **5**, 261–268.

Grant, M. A., and Sorey, M. L. (1979). The compressibility and hydraulic diffusivity of a water-steam flow. *Water Resources Research* **15**, 684–686.

Gray, W. G., and Pinder, G. F. (1976). An analysis of the numerical solution to the transport equation. *Water Resources Research* **12**, 547–555.

Green, D. H., and Wang, H. F. (1990). Specific storage as an elastic coefficient. *Water Resources Research* **26**, 1,631–1,637.

Greenberg, J. A., Mitchell, J. K., and Witherspoon, P. A. (1973). Coupled salt and water flows in a groundwater basin. *Journal of Geophysical Research* **78**, 6,341–6,353.

Greenkorn, R. A., and Kesslar, D. P. (1972). *Transfer Operations*. New York; McGraw-Hill.

Gregg, J. M. (1985). Regional epigenetic dolomitization in the Bonneterre dolomite (Cambrian), southeastern Missouri. *Geology* **13**, 503–506.

Gustafson, L. B., and Curtis, L. W. (1983). Post-Kombolgie metasomatism at Jabiluka, Northern Territory, Australia, and its significance in the formation of high-grade uranium mineralization in Lower Proterozoic rocks. *Economic Geology* **78**, 26–56.

Guzzetta, G. (1984). Kinematics of stylolite formation and physics of the pressure-solution process. *Tectonophysics* **101**, 383–394.

Haar, L., Gallagher, J. S., and Kell, G. S. (1984). *NBS/NRC Steam Tables: Thermodynamic and Transport Properties and Computer Programs for Vapor and Liquid States of Water in SI Units*. New York: Hemisphere Publishing.

Haase, R. (1969). *Thermodynamics of Irreversible Processes*. Reading: Addison-Wesley.

Hanor, J. S. (1994a). Physical and chemical controls on the composition of waters in sedimentary basins. *Marine and Petroleum Geology* **11**, 31–45.

Hanor, J. S. (1994b). Origin of saline fluids in sedimentary basins. In *Geofluids: Origin of Fluids in Sedimentary Basins*, Parnell, J. (ed.). *Geological Society of London Special Publication* **78**, 151–174.

Hanshaw, B. B., Back, W., and Deike, R. G. (1971). A geochemical hypothesis for dolomitization by groundwater. *Economic Geology* **66**, 710–724.

Hanshaw, B. B., and Zen, E. (1965). Osmotic equilibrium and overthrust faulting. *Geological Society of America Bulletin* **76**, 1,379–1,386.

Hansley, P. L., and Spirakis, C. S. (1992). Organic matter diagenesis as the key to a unifying theory for the genesis of tabular uranium-vanadium deposits in the Morrison Formation, Colorado Plateau. *Economic Geology* **87**, 352–365.

Hanson, R. B. (1992). Effects of fluid production on fluid flow during regional and contact metamorphism. *Journal of Metamorphic Geology* **10**, 87–97.

Hanson, R. B., and Barton, M. D. (1989). Thermal development of low-pressure metamorphic belts: Results from two-dimensional numerical models. *Journal of Geophysical Research* **94**, 10,363–10,377.

Harder, H. (1993). Agates – Formation as a multi component colloid chemical precipitation at low temperatures. *Neues Jahrbook fuer Mineralogie* **1**, 31–48.

Hardie, L. A. (1983). Origin of $CaCl_2$ brines by basalt-seawater interaction: Insights provided by some simple mass balance calculations. *Contributions to Mineralogy and Petrology* **82**, 205–213.

Hardie, L. A. (1984). Evaporites: Marine or nonmarine? *American Journal of Science* **284**, 193–240.

Hardie, L. A. (1987). Dolomitization: A critical view of some current views. *Journal of Sedimentary Petrology* **57**, 166–183.

Hardie, L. A. (1990). The roles of rifting and hydrothermal CaCl$_2$ brines in the origin of potash evaporites: An hypothesis. *American Journal of Science* **290**, 43–106.

Hardie, L. A. (1996). Secular variation in seawater chemistry: An explanation for the coupled secular variation in the mineralogies of marine limestones and potash evaporites over the past 600 m.y. *Geology* **24**, 279–283.

Hardie, L. A., and Eugster, H. P. (1970). The evolution of closed basin brines. *Mineralogical Society of America Special Paper* **3**, 273–290.

Harris, K. R., Hertz, H. G., and Mills, R. (1978). The effect of structure on self-diffusion in concentrated electrolytes: Relationship between water and ionic self-diffusion coefficients for structure-forming salts. *Journal of Chemical Physics* **75**, 391–396.

Harris, P. G., Kennedy, W. Q., and Scarfe, C. M. (1970). Volcanism versus plutonism – The effect of chemical composition. In *Mechanism of Igneous Intrusion*, Newall, G. and Rast, N. (eds.). *Geological Journal Special Issue* **2**, 187–200.

Harrison, W. J., and Summa, L. L. (1991). Paleohydrology of the Gulf of Mexico basin. *American Journal of Science* **291**, 109–176.

Harrison, W. J., and Tempel, R. N. (1993). Diagenetic pathways in sedimentary basins. In *Diagenesis and Basin Development*, Horbury, A. D., and Robinson, A. G. (eds.). *American Association of Petroleum Geologists' Studies in Geology* **36**, 69–86.

Harvie, C. E., Eugster, H. P., and Weare, J. H. (1982). Mineral equilibria in the six-component seawater system Na-K-Mg-Ca-SO$_4$-Cl-H$_2$O at 25°C. II. Composition of the saturated solutions. *Geochimica et Cosmochimica Acta* **46**, 1,603–1,618.

Harvie, C. E., Moller, N., and Weare, J. H. (1984). The prediction of mineral solubilities in natural waters: The Na-K-Mg-Ca-Cl-SO$_4$-OH-HCO$_3$-CO$_3$-CO$_2$-H$_2$O system to high ionic strengths at 25°C. *Geochimica et Cosmochimica Acta* **48**, 723–751.

Harvie, C. E., and Weare, J. H. (1980). The prediction of mineral solubilities in natural waters: The Na-K-Mg-Ca-SO$_4$-Cl-H$_2$O system from zero to high concentrations at 25°C. *Geochimica et Cosmochimica Acta* **44**, 981–997.

Harvie, C. E., Weare, J. H., Hardie, L. A., and Eugster, H. P. (1980). Evaporation of seawater: Calculated mineral sequences. *Science* **208**, 498–500.

Hayba, D. O. (1993). Numerical hydrologic modeling of the Creede epithermal ore-forming system. Ph.D. thesis, University of Illinois.

Hayba, D. O., and Bethke, C. M. (1995). Timing and velocity of oil migration in the Los Angeles basin. *Journal of Geology* **103**, 33–49.

Hayba, D. O., and Ingebritsen, S. E. (1994). The computer model HYDROTHERM, a three-dimensional finite-difference model to simulate ground-water flow and heat transport in the temperature range of 0 to 1, 200°C. *U.S. Geological Survey Water-Resources Investigations Report* **94-4045**.

Hayba, D. O., and Ingebritsen, S. E. (1997). Multiphase groundwater flow near cooling plutons. *Journal of Geophysical Research* **102**, 12,235–12,252.

Healy, J. H., Rubey, W. W., Griggs, D. T., and Raleigh, C. B. (1968). The Denver earthquakes. *Science* **161**, 1,301–1,310.

Hebein, J. J. (1985). Historical evolutionary facets revealed within The Geysers steam field. *Geothermal Resources Council Bulletin* **14(6)**, 13–16.

Hedenquist, J. W. (ed.) (1992). Magmatic contributions to hydrothermal systems and the behavior of volatiles in magma. *Geological Survey of Japan Report* **279**.

Henry, P., and Le Pichon, X. (1991). Fluid flow along a decollement layer: A model applied to the 16°N section of the Barbados accretionary wedge. *Journal of Geophysical Research* **96**, 6,507–6,528.

Henry, P., and Wang, C. Y. (1991). Modeling of fluid flow and pore pressure at the toe of Oregon and Barbados accretionary wedges. *Journal of Geophysical Research* **96**, 20,109–20,130.

Hickman, S., Sibson, R., and Bruhn, R. (1995). Introduction to special section: Mechanical involvement of fluids in faulting. *Journal of Geophysical Research* **100**, 12,831–12,840.

Higgins, C. G. (1980). Nips, notches, and the solution of coastal limestone: An overview of the problem with examples from Greece. *Estuarine and Coastal Marine Science* **10**, 15–30.

Hildreth, W. (1981). Gradients in silicic magma chambers: Implications for lithospheric magmatism. *Journal of Geophysical Research* **86**, 10,153–10,192.

Hill, D. P., and others (1993). Seismicity remotely triggered by the magnitude 7.3 Landers, California earthquake. *Science* **260**, 1,617–1,626.

Hitchon, B. (1984). Geothermal gradients, hydrodynamics, and hydrocarbon occurences, Alberta, Canada. *American Association of Petroleum Geologists' Bulletin* **68**, 713–743.

Hite, R. J. (1961). Potash-bearing evaporite cycles in the salt anticlines of the Paradox Basin, Colorado and Utah. *U.S. Geological Survey Professional Paper* **424-D**, 135–138.

Hite, R. J., and Japakasetr, T. (1979). Potash deposits of the Khorat Plateau, Thailand and Laos. *Economic Geology* **74**, 448–458.

Hoeve, J., and Quirt, D. (1984). Mineralization and host-rock alteration in relation to clay mineral diagenesis and evolution of the Middle-Proterozoic, Athabasca Basin, northern Saskatchewan, Canada. *Saskatchewan Research Council Technical Report* **187**.

Hoeve, J., and Quirt, D. (1987). A stationary redox front as a critical factor in the formation of high-grade, unconformity-type uranium ores in the Athabasca Basin, Saskatchewan, Canada. *Bulletin de Mineralogie* **110**, 157–171.

Hoeve, J., and Sibbald, T. I. I. (1978). On the genesis of Rabbit Lake and other unconformity-type uranium deposits in northern Saskatchewan, Canada. *Economic Geology* **73**, 1,450–1,473.

Hogg, A. J. C., Hamilton, P. J., and Macintyre, R. M. (1993). Mapping diagenetic fluid flow within a reservoir: K-Ar dating in the Alwyn area (UK North Sea). *Marine and Petroleum Geology* **10**, 279–294.

Hoisch, T. D. (1987). Heat transport by fluids during Late Cretaceous regional metamorphism in the Big Maria Mountains, southeastern California. *Geological Society of America Bulletin* **98**, 549–553.

Holland, H. H. (1984). *The Chemical Evolution of the Atmosphere and Oceans*. Princeton: Princeton University Press.

Horne, R. N., and Ramey, H. J., Jr. (1978). Steam/water relative permeabilities from production data. *Geothermal Resources Council Transactions* **2**, 291–293.

Hower, J., Eslinger, E. V., Hower, M. E., and Perry, E. A. (1976). Mechanism of burial metamorphism of argillaceous sediment. I. Mineralogical and chemical evidence. *Geological Society of America Bulletin* **87**, 725–737.

Hsieh, P. A., and Bredehoeft, J. D. (1981). A reservoir analysis of the Denver earthquakes: A case of induced seismicity. *Journal of Geophysical Research* **86**, 903–920.

Hubbert, M. K. (1940). The theory of ground-water motion. *Journal of Geology* **48**, 785–944.

Hubbert, M. K. (1953). Entrapment of petroleum under hydrodynamic conditions. *American Association of Petroleum Geologists' Bulletin* **37**, 1,954–2,026.

Hubbert, M. K. (1956). Darcy's law and the field equations of the flow of underground fluids. *Transactions of the American Institute of Mining, Metallurgical, and Petroleum Engineers* **207**, 222–239.

Hubbert, M. K., and Rubey, W. W. (1959). Role of fluid pressure in mechanics of overthrust faulting: I. Mechanics of fluid-filled porous solids and its application to overthrust faulting. *Geological Society of America Bulletin* **70**, 115–166.

Hubbert, M. K., and Willis, D. G. (1957). Mechanics of hydraulic fracturing. *Transactions of the American Institute of Mining, Metallurgical, and Petroleum Engineers* **210**, 153–168.

Huenges, E., Erzinger, J., Kuck, J., Engeser, B., and Kessels, W. (1997). The permeable crust: Geohydraulic properties down to 9101 m depth. *Journal of Geophysical Research* **102**, 18,255–18,265.

Hutchinson, R. A. (1985). Hydrothermal changes in the Upper Geyser Basin, Yellowstone National Park, after the 1983 Borah Peak earthquake. In *Proceedings of Workshop XXVIII on the Borah Peak, Idaho, Earthquake*, Stein, R. S., and Bucknam, R. C. (eds.). *U.S. Geological Survey Open-File Report* **85-290**, 612–624.

Huyakorn, P. S., and Pinder, G. F. (1983). *Computational Methods in Subsurface Flow*. New York: Academic Press.

Hyndman, R. D., Wang, K., and Yamano, M. (1995). Thermal constraints on the seismogenic portion of the southwestern Japan subduction thrust. *Journal of Geophysical Research* **100**, 15,373–15,392.

Hyndman, R. D., Wang, K., Yaun, T., and Spence, G. D. (1993). Tectonic sediment thickening, fluid expulsion, and the thermal regime of subduction zone accretionary prisms: The Cascadia margin off Vancouver Island. *Journal of Geophysical Research* **98**, 21,865–21,876.

Igarashi, G., Saeki, S., Takahata, N., Sumikawa, K., Tasaka, S., Sasaki, Y., Takahashi, M., and Sano, Y. (1995). Groundwater radon anomaly before the Kobe earthquake in Japan. *Science* **269**, 60–61.

Ingebritsen, S. E., and Hayba, D. O. (1994). Fluid flow and heat transport near the critical point of H_2O. *Geophysical Research Letters* **21**, 2,199–2,203.

Ingebritsen, S. E., Mariner, R. H., and Sherrod, D. R. (1994). Hydrothermal systems of the Cascade Range, north-central Oregon. *U.S. Geological Survey Professional Paper* **1044-L**.

Ingebritsen, S. E., and Rojstaczer, S. A. (1993). Controls on geyser periodicity. *Science* **262**, 889–892.

Ingebritsen, S. E., and Rojstaczer, S. A. (1996). Geyser periodicity and the response of geysers to deformation. *Journal of Geophysical Research* **101**, 21,891–21,905.

Ingebritsen, S. E., and Scholl, M. A. (1993). The hydrogeology of Kilauea volcano. *Geothermics* **22**, 255–270.

Ingebritsen, S. E., Scholl, M. A., and Sherrod, D. R. (1993). Heat flow from four new research drill holes in the Western Cascades, Oregon. *Geothermics* **22**, 151–163.

Ingebritsen, S. E., Sherrod, D. R., and Mariner, R. H. (1989). Heat flow and hydrothermal circulation in the Cascade Range, north-central Oregon. *Science* **243**, 1,458–1,462.

Ingebritsen, S. E., Sherrod, D. R., and Mariner, R. H. (1992). Rates and patterns of groundwater flow in the Cascade Range volcanic arc, and the effect on subsurface temperatures. *Journal of Geophysical Research* **97**, 4,599–4,627.

Ingebritsen, S. E., and Sorey, M. L. (1985). A quantitative analysis of the Lassen hydrothermal system. *Water Resources Research* **21**, 853–868.

Ingebritsen, S. E., and Sorey, M. L. (1988). Vapor-dominated zones within hydrothermal systems: Evolution and natural state. *Journal of Geophysical Research* **93**, 13,635–13,655.

Irwin, W. P., and Barnes, I. (1975). Effect of geologic structure and metamorphic fluids on seismic behavior of the San Andreas fault system in central and northern California. *Geology* **3**, 713–716.

Jacob, C. E. (1940). On the flow of water in an elastic artesian aquifer. *Transactions American Geophysical Union* **21**, 574–586.

Jaeger, J. C., and Cook, N. G. W. (1979). *Fundamentals of Rock Mechanics*. 3rd ed. New York: Chapman and Hall.

Jankowski, J., and Jacobson, G. (1990). Hydrochemical processes in groundwater-discharge playas, central Australia. *Hydrological Processes* **4**, 59–70.

Jessop, A. M., Hobart, M. A., and Schlater, J. G. (1976). The world heat flow data collection – 1975. *Geothermal Service of Canada Geothermal Series* **5**.

Johnson, J. W., and Norton, D. (1991). Critical phenomena in hydrothermal systems: State, thermodynamic, electrostatic, and transport properties of H_2O in the critical region. *American Journal of Science* **291**, 541–648.

Jones, B. F., and Bodine, M. W. (1987). Normative salt characterization of natural waters. In *Saline Water and Gases in Crystalline Rocks*, Fritz, P., and Frape, S. K., (eds.). *Geological Society of Canada Special Paper* **33**, 5–18.

Jones, B. F., Eugster, H. P., and Rettig, S. L. (1977). Hydrochemistry of the Lake Magadi basin, Kenya. *Geochimica et Cosmochimica Acta* **41**, 53–72.

Jones, C. L. (1972). Permian basin potash deposits, southwestern United States. In *Geology of Saline Deposits*, Richter-Bernburg, G. (ed.), pp.191–201. Paris: UNESCO.

Jones, F. W., Majorowicz, J. A., and Lam, H. L. (1985). The variation of heat flow density with depth in the prairies basin of western Canada. *Tectonophysics* **121**, 35–44.

Jones, G., Whitaker, F. F., Smart, P. L., and Sanford, W. E. (1997). Dolomitization of carbonate platforms by saline ground water: Coupled numerical modelling of thermal and reflux circulation mechanisms. In *Geofluids II: 2nd International Conference on Fluid Evolution, Migration, and Interaction in Sedimentary Basins and Orogenic Belts*, Henry, J., Carey, P., Parnell, J., Ruffel, A., and Worden, R., (eds.), 378–381.

Kanamori, H., and Anderson, D. L. (1975). Theoretical basis of some empirical relations in seismology. *Bulletin of the Seisomological Society of America* **65**, 1,073–1,095.

Kauahikaua, J. (1993). Geophysical characteristics of the hydrothermal systems of Kilauea volcano, Hawai'i. *Geothermics* **22**, 271–299.

Kaufman, J. (1994). Numerical models of fluid flow in carbonate platforms: Implications for dolomitization. *Journal of Sedimentary Research* **64**, 128–139.

Kimura, S., Schubert, G., and Straus, J. M. (1986). Route to chaos in porous-medium thermal convection. *Journal of Fluid Mechanics* **166**, 305–324.

King, F. H. (1899). Principles and conditions of the movement of groundwater. *U.S. Geological Survey Annual Report* **19** (Part II), 59–294.

King, R. H. (1947). Sedimentation in Permian Castile Sea. *American Association of Petroleum Geologists' Bulletin* **26**, 470–477.

Kipp, K. L. (1987). HST3D: A computer code for simulation of heat and solute transport in three-dimensional groundwater flow systems. *U.S. Geological Survey Water-Resources Investigations Report* **86-4095**.

Knauth, L. P. (1979). A model for the origin of chert in limestone. *Geology* **7**, 274–277.

Knauth, L. P., Kumar, M. B., and Martinez, J. D. (1980). Isotope geochemistry of water in Gulf Coast salt domes. *Journal of Geophysical Research* **85**, 4,863–4,871.

Kohout, F. A. (1965). A hypothesis concerning the cyclic flow of salt water related to geothermal heating in the Floridan aquifer. *Transactions of the New York Academy of Sciences Series II* **28**, 249–271.

Kohout, F. A., Henry, H. R., and Banks, J. E. (1977). Hydrogeology related to geothermal conditions of the Floridan Plateau. In *The Geothermal Nature of the Floridan Plateau*, Smith, D. L., and Griffin, G. M. (eds.). Bureau of Geology, Division of Resource Management, *Florida Department of Natural Resources Special Publication* **21**, ix–41.

Konikow, L. F. (1985). Process and rate of dedolomitization: Mass transfer and ^{14}C dating in a regional aquifer: extended interpretation. *Geological Society of America Bulletin* **96**, 1,096–1,098.

Konikow, L. F., and Bredehoeft, J. D. (1978). Computer model of two-dimensional solute transport and dispersion in groundwater. *U.S. Geological Survey Techniques of Water-Resources Investigations* **Book 7, Chapter C2**.

Konikow, L. F., and Grove, D. B. (1977). Derivation of equations describing solute transport in groundwater. *U.S. Geological Survey Water Resources Investigations Report* **77-19**.

Krauskopf, K. B. (1979). *Introduction to Geochemistry*. New York: McGraw-Hill.

Kuehn, R., and Hsu, K. J. (1978). Chemistry of halite and potash salt cores, D.S.D.P. sites 374 and 376, leg 42A, Mediterranean Sea. *Initial Reports of the Deep Sea Drilling Projects* **V42A**, 613–619.

Kyle, J. R., and Agee, W. N., Jr. (1988). Evolution of metal ratios and δ^{34}S composition of sulfide mineralization during anhydrite cap rock formation, Hockley Dome, Texas, U.S.A. *Chemical Geology* **74**, 37–55.

Kyle, J. R., Ulrich, M. R., and Gose, W. A. (1987). Textural and paleomagnetic evidence for the mechanism and timing of anhydrite cap rock formation, Winnfield salt dome, Louisiana. In *Dynamical Geology of Salt and Related Structures*, Lerche, I., and O'Brien, J. J. (eds.), pp. 497–542. Orlando: Academic Press.

Lachenbruch, A. H. (1970). Crustal temperature and heat production: Implications of the linear heat-flow relation. *Journal of Geophysical Research* **75**, 3,291–3,300.

Lachenbruch, A. H., and Marshall, B. V. (1986). Changing climate: Geothermal evidence from permafrost in the Alaskan Arctic. *Science* **234**, 689–696.

Lachenbruch, A. H., and Sass, J. H. (1980). Heat flow and energetics of the San Andreas fault zone. *Journal of Geophysical Research* **85**, 6,185–6,222.

Lachenbruch, A. H., and Sass, J. H. (1992). Heat flow from Cajon Pass, fault strength, and tectonic implications. *Journal of Geophysical Research* **97**, 4,995–5,015.

Lachenbruch, A. H., Sass, J. H., Munroe, R. J., and Moses, T. H., Jr. (1976). Geothermal setting and simple heat-conduction models for the Long Valley caldera. *Journal of Geophysical Research* **81**, 769–784.

Land, L. S. (1985). The origin of massive dolomite. *Journal of Geological Education* **33**, 112–125.

Land, L. S., Kupecz, J. A., and Mack, L. E. (1988). Louann salt geochemistry (Gulf of Mexico sedimentary basin, USA): A preliminary synthesis. *Chemical Geology* **47**, 25–35.

Land, L. S., Mack, L. E., Milliken, K. L., and Lynch, F. L. (1997). Burial diagenesis of argillaceous sediment, south Texas Gulf of Mexico sedimentary basin: A reexamination. *Geological Society of America Bulletin* **109**, 2–15.

Langford, F. F. (1974). A supergene origin for vein-type uranium ores in the light of the western Australian calcrete-carnotite deposits. *Economic Geology* **69**, 516–526.

Langmuir, D. (1997). *Aqueous Environmental Chemistry*. Upper Saddle River, NJ: Prentice-Hall.

Lapwood, E. R. (1948). Convection of a fluid in a porous media. *Proceedings Cambridge Philosophical Society* **44**, 508–521.

Larson, P. B., and Taylor, H. P., Jr. (1986). An oxygen isotope study of hydrothermal alteration in the Lake City caldera, San Juan Mountains, Colorado. *Journal of Volcanology and Geothermal Research* **30**, 47–82.

Lasaga, A. C. (1979). The treatment of multi-component diffusion and ion pairs in diagenetic fluxes. *American Journal of Science* **279**, 324–346.

Last, W. M. (1989). Continental brines and evaporites of the Northern Great Plains of Canada. *Sedimentary Geology* **64**, 207–221.

Leach, D. L. (1979). Temperature and salinity of the fluids responsible for minor occurrences of sphalerite in the Ozark region of Missouri. *Economic Geology* **74**, 931–937.

Leach, D. L., and Rowan, E. L. (1986). Genetic link between Ouachita foldbelt tectonism and the Mississippi Valley–type lead-zinc deposits of the Ozarks. *Geology* **14**, 931–935.

LeBlanc, D. R., Garabedian, S. P., Hess, K. M., Gelhar, L. W., Quadri, R. D., Stollenwerk, K. G., and Wood, W. W. (1991). Large-scale natural gradient tracer test in sand and gravel, Cape Cod, Massachusetts. 1. Experimental design and observed tracer movement. *Water Resources Research* **27**, 895–910.

LeHuray, A. P., Caulfield, J. B. D., Rye, D. M., and Dixon, P. R. (1987). Basement controls on sediment-hosted Zn-Pb deposits: A Pb isotope study of Carboniferous mineralization in central Ireland. *Economic Geology* **82**, 1,695–1,709.

Leising, J. F., Tyler, S. W., and Miller, W. W. (1995). Convection of saline brines in enclosed lacustrine basins: A mechanism for potassium metasomatism. *Geological Society of America Bulletin* **107**, 1,157–1,163.

Lemee, C., and Gueguen, Y. (1996). Modeling of porosity loss during compaction and cementation of sandstones. *Geology* **24**, 875–878.

Levinson, A. A. (1980). *Introduction to Exploration Geochemistry*. Wilmette, IL: Applied Publishing.

Levorsen, A. I. (1967). *Geology of Petroleum*. San Francisco: W.H. Freeman.

Lewan, M. D. (1987). Petrographic study of primary petroleum migration in the Woodford Shale and related rock units. In *Migration of Hydrocarbons in Sedimentary Basins*, Doligez, B. (ed.), pp. 113–130. Paris: Editions Technip.

Lewan, M. D. (1994). Assessing natural oil expulsion from source rocks by laboratory pyrolysis. In *The Petroleum System – From Source to Trap*, Magoon, L. B., and Dow, W. G. (eds.). *American Association of Petroleum Geologists' Memoir* **60**, 201–210.

Lewis, S., and Holness, M. (1996). Equilibrium halite-H_2O dihedral angles: High rock-salt permeability in the shallow crust? *Geology* **24**, 431–434.

Li, Y. H., and Gregory, S. (1974). Diffusion of ions in sea water and in deep-sea sediments. *Geochimica et Cosmochimica Acta* **38**, 703–714.

Lichtner, P. C. (1992). Time-space continuum description of fluid/rock interaction in permeable media. *Water Resources Research* **28**, 3,135–3,156.

Lichtner, P. C. (1996). Continuum formulation of multiphase multicomponent reactive transport. In *Reactive Transport in Porous Media*, Lichtner, P. C., Steefel, C. I., and Oelkers, E. H. (eds.). *Reviews in Mineralogy* **34**, 1–81.

Lichtner, P. C., and Biino, G. G. (1992). A first principles approach to supergene enrichment of a porphyry copper protore: I. Cu-Fe-S subsystem. *Geochimica et Cosmochimica Acta* **56**, 3,987–4,013.

Lister, C. R. B. (1974). On the penetration of water into hot rock. *Geophysical Journal of the Royal Astronomical Society* **39**, 465–509.

Loucks, R. G., Dodge, M. M., and Galloway, W. E. (1984). Regional controls of diagenesis and reservoir quality in Lower Tertiary sandstones along the Texas Gulf Coast. In *Clastic Diagenesis*, McDonald, D. A., and Surdam, R. C. (eds.). *American Association of Petroleum Geologists' Memoir* **37**, 15–46.

Lowell, R. P., and Germanovich, L. N. (1994). On the temporal evolution of high-temperature hydrothermal systems at ocean ridge crests. *Journal of Geophysical Research* **99**, 565–575.

Lowell, R. P., Rona, P. A., and Von Herzen, R. P. (1995). Seafloor hydrothermal systems. *Journal of Geophysical Research* **100**, 327–352.

Lowell, R. P., Van Cappellen, P., and Germanovich, L. N. (1993). Silica precipitation in fractures and the evolution of permeability in hydrothermal upflow zones. *Science* **260**, 192–194.

Lucia, F. J. (1968). Recent sediments and diagenesis of South Bonaire, Netherlands Antilles. *Journal of Sedimentary Petrology* **38**, 845–858.

Lucia, F. J. (1995). Rock-fabric/petrophysical classification of carbonate pore space for reservoir characterization. *American Association of Petroleum Geologists' Bulletin* **79**, 1,275–1,300.

Ludwig, K. R., Grauch, R. I., Nutt, C. J., Nash, J. T., Frishman, D., and Simmons, K. R. (1987). Age of uranium mineralization at the Jabiluka and Ranger deposits, Northern Territory, Australia: New U-Pb isotope evidence. *Economic Geology* **82**, 857–874.

Lutz, R. A., and Kennish, M. J. (1993). Ecology of deep-sea hydrothermal vent communities: A review. *Reviews of Geophysics* **31**, 211–242.

Maasland, M. (1957). Soil anisotropy and soil drainage. In *Drainage of Agricultural Lands*, Maddock, J. N. (ed.), pp. 216–285. Madison: American Society of Agromony.

Machel, H. G., and Mountjoy, E. W. (1986). Chemistry and environments of dolomitization – A reappraisal. *Earth-Science Reviews* **23**, 175–222.

MacKenzie, F. T., and Garrels, R. M. (1966). Chemical mass balance between rivers and oceans. *American Journal of Science* **264**, 507–525.

Magaritz, M., Goldenberg, L., Kafri, U., and Arad, A. (1980). Dolomite formation in the seawater-freshwater interface. *Nature* **287**, 622–624.

Majorowicz, J. A., and Jessop, A. M. (1981). Regional heat flow patterns in the Western Canada sedimentary basin. *Tectonophysics* **74**, 209–238.

Majorowicz, J. A., Jones, F. W., Lam, H. L., and Jessop, A. M. (1984). The variability of heat flow both regional and with depth in southern Alberta, Canada: Effect of groundwater flow. *Tectonophysics* **106**, 1–29.

Majorowicz, J. A., Jones, F. W., MacQueen, R. W., and Ertman, M. E. (1989). The bearing of heat-flow data on the nature of fluid flow in the Keg River-Pine Point oil field region, Canada. *Economic Geology* **84**, 708–714.

Manning, C. E. (1996). Coupled reaction and flow in subduction zones: Silica metasomatism in the mantle wedge. In *Fluid Flow and Transport in Rocks*, Jamtveit, B., and Yardley, B. W. D. (eds.), pp. 139–148. London: Chapman and Hall.

Manning, C. E., and Bird, D. K. (1991). Porosity evolution and fluid flow in the basalts of the Skaergaard magmatic-hydrothermal system, East Greenland. *American Journal of Science* **291**, 201–257.

Manning, C. E., Ingebritsen, S. E., and Bird, D. K. (1993). Missing mineral zones in contact metamorphosed basalts. *American Journal of Science* **293**, 894–938.

Marder, M., and Fineberg, J. (1996). How things break. *Physics Today* **49(9)**, 24–29.

Marine, I. W., and Fritz, S. J. (1981). Osmotic model to explain anomalous hydraulic heads. *Water Resources Research* **17**, 73–82.

Marler, G. D., and White, D. E. (1977). Evolution of Seismic Geyser, Yellowstone National Park. *Earthquake Information Bulletin* **9(2)**, 21–25.

Marmont, S. (1987). Ore deposit models # 13. Unconformity-type uranium deposits. *Geoscience Canada* **14**, 219–229.

Marsh, B. D. (1989). Magma chambers. *Annual Review of Earth and Planetary Sciences* **17**, 439–474.

Mase, C. W., Sass, J. H., Lachenbruch, A. H., and Munroe, R. J. (1982). Preliminary heat-flow investigations of the California Cascades. *U.S. Geological Survey Open-File Report* **82-150**.

McAulay, G. E., Burley, S. D., Fallick, A. E., and Kusznir, N. J. (1994). Paleohydrodynamic fluid flow regimes during diagenesis of the Brent Group in the Hutton-NW Hutton reservoirs: Constraints from oxygen isotope studies of authigenic kaolin and reverse flexural modelling. *Clay Minerals* **29**, 609–626.

McDonald, M. G., and Harbaugh, A. W. (1988). A modular three-dimensional finite-difference groundwater flow model. *U.S. Geological Survey Techniques of Water Resources Investigations* **Book 6, Chapter A1**.

McGee, H. W., Meyer, H. J., and Pringle, T. R. (1989). Shallow geothermal anomalies overlying deeper oil and gas deposits in Rocky Mountain region. *American Association of Petroleum Geologists' Bulletin* **73**, 576–597.

McPherson, B. J. O. L. (1996). A three-dimensional model of the geologic and hydrodynamic history of the Uinta basin, Utah: Analysis of overpressures and oil migration. Ph.D. thesis, University of Utah.

Melchior, P. J. (1978). *The Tides of the Planet Earth*. Oxford: Pergamon Press.

Mellinger, M., Quirt, D., and Hoeve, J. (1987). Geochemical signatures of uranium deposition in the Athabasca Basin of Saskatchewan, Canada. *Uranium* **3**, 187–209.

Meriaux, M., and Gannat, E. (1980). Connaissances actuelles sur la potasse en France. *Annales des Mines* **186**, 167–176.

Merino, E. (1990). The geochemistry of habits and textures of authigenic quartz. In *Geochemistry of the Earth's Surface and Mineral Formations*, Noack, Y. and Nahon, D. (eds.). *Chemical Geology* **84**, 233–234.

Merino, E., Wang, Y., and Deloule, E. (1995). Genesis of agates in flood basalts: Twisting of chalcedony fibers and trace-element geochemistry. *American Journal of Science* **295**, 1,156–1,176.

Meshri, I. D. (1990). An overview of chemical models and their relationship to porosity prediction in the subsurface. In *Prediction of Reservoir Quality Through*

Chemical Modeling, Meshri, I. D., and Ortoleva, P. (eds.). *American Association of Petroleum Geologists' Memoir* **49**, 45–53.

Miller, D. G. (1960). Thermodynamics of irreversible processes: The experimental verification of the Onsager reciprocal relations. *Chemical Reviews* **60**, 15–37.

Miller, D. G., Rard, J. A., Eppstein, L. B., and Albright, J. G. (1984). Mutual diffusion coefficients and ionic transport coefficients l_{ij} of $MgCl_2$-H_2O at 25°C. *Journal of Physical Chemistry* **88**, 5,739–5,748.

Minshull, T., and White, R. (1989). Sediment compaction and fluid migration in the Makran accretionary prism. *Journal of Geophysical Research* **94**, 7,387–7,402.

Mitchell, J. K. (1993). *Fundamentals of Soil Behavior*. 2nd ed. New York: John Wiley and Sons.

Moore, C. H. (1989). Carbonate diagenesis and porosity. *Developments in Sedimentology* **46**, Amsterdam, Elsevier.

Moore, D. E., Lockner, D. A., Summers, R. S., Shengli, M., and Byerlee, J. D. (1996). Strength of chrysotile-serpentinite gouge under hydrothermal conditions: Can it explain a weak San Andreas fault?: *Geology* **24**, 1,041–1,044.

Moore, J. C., Mascle, A., and many others (1988). Tectonics and hydrogeology of the northern Barbados Ridge: Results from the Ocean Drilling Program Leg 100. *Geological Society of America Bulletin* **100**, 1,578–1,593.

Morgan, J. K., and Karig, D. E. (1995). Decollement processes at the Nankai accretionary margin, southeast Japan: Propagation, deformation, and dewatering. *Journal of Geophysical Research* **100**, 15,221–15,231.

Morgan, J. K., Karig, D. E., and Maniatty, A. (1994). The estimation of diffuse strains in the toe of the western Nankai accretionary prism. *Journal of Geophysical Research* **99**, 7,019–7,032.

Morrow, C. A., and Byerlee, J. D. (1992). Permeability of core samples from Cajon Pass scientific drill hole: Results from 2100 to 3500 m depth. *Journal of Geophysical Research* **97**, 5,145–5,151.

Morrow, C. A., and Lockner, D. A. (1994). Permeability differences between surface-derived and deep drillhole core samples. *Geophysical Research Letters* **21**, 2,151–2,154.

Morrow, C. A., Lockner, D., Hickman, S., Rusanov, M., and Rockel, T. (1994). Effects of lithology and depth on the permeability of core samples from the Kola and KTB drill holes. *Journal of Geophysical Research* **99**, 7,263–7,274.

Morrow, C. A., Radney, B., and Byerlee, J. D. (1992). Frictional strength and the effective pressure law of montmorillonite and illite clays. In *Fault Mechanics and Transport Properties of Rocks*, Evans, B., and Wong, T.-F. (eds.), pp. 69–88. London: Academic.

Morrow, D. W. (1982). Diagenesis I. Dolomite–Part I. The chemistry of dolomitization and dolomite precipitation. *Geoscience Canada* **9**, 5–13.

Morse, J. W., and MacKenzie, F. T. (1990). *Geochemistry of Sedimentary Carbonates*. Amsterdam: Elsevier.

Mountjoy, E. W., and Amthor, J. E. (1994). Has burial dolomitization come of age? Some answers from the Western Canada Sedimentary Basin. In *Dolomites – A Volume in Honour of Dolomieu*, Purser, B., Tucker, M., and Zenger, D. (eds.). *International Association of Sedimentologists Special Publication* **21**, 203–229. Oxford: Blackwell Scientific Publications.

Mozley, P. S. (1996). The internal structure of carbonate concretions in mudrocks: A critical evaluation of the conventional concentric model of concretion growth. *Sedimentary Geology* **103**, 85–91.

Muir-Wood, R., and King, G. C. P. (1993). Hydrological signatures of earthquake strain. *Journal of Geophysical Research* **98**, 22,035–22,068.

Naeser, N. D., Naeser, C. W., and McCulloh, T. H. (1989). The application of fission-track dating to the depositional and thermal history of rocks in sedimentary basins. In *Thermal History of Sedimentary Basins: Methods and Case Histories*, Naeser, N. D., and McCulloh, T. H. (eds.), pp. 157–180. New York: Springer Verlag.

Narasimhan, T. N., and Witherspoon, P. A. (1977). Numerical model for saturated-unsaturated flow in deformable porous media. 1. Theory. *Water Resources Research* **14**, 1,017–1,034.

Needham, R. S., and Stuart-Smith, P. G. (1976). The Cahill formation-host to uranium deposits in the Alligator Rivers Uranium Field, Australia. *BMR Journal of Australian Geology and Geophysics* **1**, 321–333.

Neglia, S. (1979). Migration of fluids in sedimentary basins. *American Association of Petroleum Geologists' Bulletin* **63**, 573–597.

Neuman, S. P. (1979). Perspective on 'delayed yield.' *Water Resources Research* **15**, 899–908.

Neuzil, C. E. (1986). Groundwater flow in low-permeability environments. *Water Resources Research* **22**, 1,163–1,195.

Neuzil, C. E. (1994). How permeable are clays and shales? *Water Resources Research* **30**, 145–150.

Neuzil, C. E. (1995). Abnormal pressures as hydrodynamic phenomena. *American Journal of Science* **295**, 742–786.

Norton, D., and Cathles, L. M. (1979). Thermal aspects of ore deposition. In *Geochemistry of Hydrothermal Ore Deposits*, Barnes, H. L. (ed.), pp. 611–631. New York: John Wiley and Sons.

Norton, D., and Knight, J. (1977). Transport phenomena in hydrothermal systems: Cooling plutons. *American Journal of Science* **277**, 937–981.

Norton, D., and Taylor, H. P., Jr. (1979). Quantitative simulation of the hydrothermal systems of crystallizing magmas on the basis of transport theory and oxygen isotope data: An analysis of the Skaergaard intrusion. *Journal of Petrology* **20**, 421–486.

Nur, A., and Byerlee, J. D. (1971). An exact effective stress law for elastic deformation of rock with fluids. *Journal of Geophysical Research* **76**, 6,414–6,419.

Nur, A., and Walder, J. (1990). Time-dependent permeability of the Earth's crust. In *The Role of Fluids in Crustal Processes*, Bredehoeft, J. D., and Norton, D. (eds.), pp. 113–127. Washington, DC: National Academy Press.

Nurmi, R. D., and Freedman, G. M. (1977). Sedimentology and depositional environments of basin-center evaporites, Lower Salina Group (Upper Silurian), Michigan Basin. In *Reefs and Evaporites – Concepts and Depositional Models*, Fisher, J. H. (ed.). *American Association of Petroleum Geologists' Studies in Geology* **5**, 23–52.

Nutting, P. G. (1930). Physical analysis of oil sands. *American Association of Petroleum Geologists' Bulletin* **14**, 1,337–1,349.

Okada, Y. (1992). Internal deformation due to shear and tensile faults in a half-space. *Bulletin of the Seismological Society of America* **82**, 1,018–1,040.

Oliver, J. (1986). Fluids expelled tectonically from orogenic belts: Their role in hydrocarbon migration and other geologic phenomena. *Geology* **14**, 99–102.

Olsen, H. W. (1965). Deviations from Darcy's law in saturated clays. *Soil Science Society of America Proceedings* **29**, 135–140.

Olsen, H. W. (1969). Simultaneous fluxes of liquid and charge in saturated kaolinite. *Soil Science Society of America Proceedings* **33**, 338–344.

Ortiz, N. V., Ferentchak, J. A., Ethridge, F. G., Granger, H. C., and Sunada, D. K. (1980). Ground-water flow and uranium in Colorado Plateau. *Groundwater* **18**, 596–606.

Ortoleva, P. (1994). *Geochemical Self-Organization*. New York: Oxford University Press.

Ortoleva, P., Merino, E., Moore, C., and Chadam, J. (1987). Geochemical self-organization. I. Reaction-transport feedbacks and modeling approach. *American Journal of Science* **287**, 979–1,007.

Ottaway, T. L., Wicks, F. J., Bryndzia, L. T., Kyser, T. K., and Spooner, E. T. C. (1994). Formation of the Muzo hydrothermal emerald deposit in Colombia. *Nature* **369**, 552–554.

Parker, J. C., Lenhard, R. J., and Kuppusamy, T. (1987). A parametric model for constitutive properties governing multiphase flow in porous media. *Water Resources Research* **23**, 618–624.

Parkhurst, D. L. (1995). User's guide to PHREEQC – A computer program for speciation, reaction-path, advective transport, and inverse geochemical calculations. *U.S. Geological Survey Water-Resources Investigations Report* **95-4227**.

Parkhurst, D. L., Thorstenson, D. C., and Plummer, L. N. (1980). PHREEQE – A computer program for geochemical calculations. *U.S. Geological Survey Water Resources Investigations Report* **80-96**.

Patterson, R. J., and Kinsman, D. J. J. (1982). Formation of diagenetic dolomite in coastal sabkha along Arabian (Persian) Gulf. *American Association of Petroleum Geologists' Bulletin* **66**, 28–43.

Perkins, T. K., and Johnston, O. C. (1963). A review of diffusion and dispersion in porous media. *Society of Petroleum Engineering Journal* **3**, 70–83.

Person, M. A., and Garven, G. (1992). Hydrologic constraints on petroleum generation within continental rift basins: Theory and application to the Rhine graben. *American Association of Petroleum Geologists' Bulletin* **76**, 468–488.

Person, M. A., and Garven, G. (1994). A sensitivity study of the driving forces on fluid flow during continental-rift basin evolution. *Geological Society of America Bulletin* **106**, 461–475.

Person, M. A., Raffensperger, J. P., Ge, S., and Garven, G. (1996). Basin-scale hydrogeologic modeling. *Reviews of Geophysics* **34**, 61–87.

Person, M. A., Toupin, D., and Eadington, P. (1995). One-dimensional models of groundwater flow, sediment thermal history and petroleum generation within continental rift basins. *Basin Research* **7**, 81–96.

Phillips, F. M., Bentley, H. W., Davis, S. N., Elmore, D., and Swanick, G. B. (1986). Chlorine 36 dating of very old groundwater. 2. Milk River Aquifer, Alberta, Canada. *Water Resources Research* **22**, 2,003–2,016.

Phillips, O. M. (1991). *Flow and Reactions in Permeable Rocks*. Cambridge: Cambridge University Press.

Piquemal, J. (1994). Saturated steam relative permeabilities of unconsolidated porous media. *Transport in Porous Media* **17**, 105–120.

Pitzer, K. S. (1973). Thermodynamics of electrolytes. 1. Theoretical basis and general equations. *Journal of Physical Chemistry* **77**, 268–277.

Pitzer, K. S. (1975). Thermodynamics of electrolytes. 5. Effects of higher-order electrostatic terms. *Journal of Solution Chemistry* **4**, 249–265.

Plumlee, G. S. (1989). Processes controlling epithermal mineral distribution in the Creede Mining District, Colorado. Ph.D. thesis, Harvard University.

Plumlee, G. S., Goldhaber, M. B., and Rowan, E. L. (1995). The potential role of magmatic gases in the genesis of Illinois-Kentucky fluorspar deposits: Implications from chemical reaction path modeling. *Economic Geology* **90**, 999–1,011.

Plummer, L. N. (1975). Mixing of seawater with calcium carbonate groundwater. In *Quantitative Studies in the Geological Sciences*, Whitten, E. H. T. (ed.). *Geological Society of America Memoir* **142**, 219–236.

Plummer, L. N., Busby, J. F., Lee, R. W., and Hanshaw, B. B. (1990). Geochemical modeling of the Madison aquifer in parts of Montana, Wyoming, and South Dakota. *Water Resources Research* **26**, 1,981–2,014.

Plummer, L. N., Michel, R. L., Thurman, E. M., and Glynn, P. D. (1993). Environmental tracers for age dating young ground water. In *Regional Ground-Water Quality*, Alley, W. M. (ed.), pp. 255–294. New York: van Norstrand Reinhold.

Plummer, L. N., Parkhurst, D. L., Fleming, G. W., and Dunkle, S. A. (1988). A computer program incorporating Pitzer's equations for calculation of geochemical reactions in brines. *U.S. Geological Survey Water-Resources Investigations Report* **88-4153**.

Plummer, L. N., Prestemon, E. C., and Parkhurst, D. L. (1994). An interactive code (NETPATH) for modeling net chemical reactions along a flow path version 2.0. *U.S. Geological Survey Water-Resources Investigations Report* **94-4169**.

Pollack, H. N. Hurter, S. J., and Johnson, J. R. (1993). Heat Flow from the Earth's interior: Analysis of the global data set. *Reviews of Geophysics* **31**, 267–280.

Pollock, D. W. (1989). Documentation of computer programs to compute and display pathlines using results from the U.S. Geological Survey modular three–dimensional finite-difference ground-water flow model. *U.S. Geological Survey Open-File Report* **89-381**.

Posey, H. H., and Kyle, J. R. (1988). Fluid-rock interactions in the salt dome environment: An introduction and review. *Chemical Geology* **74**, 1–24.

Posey, H. H., Price, P. E., and Kyle, J. R. (1987). Mixed carbon sources for calcite cap rocks of gulf coast salt domes. In *Dynamical Geology of Salt and Related Structures*, Lerche, I., and O'Brien, J. J. (eds.), pp. 593–630. Orlando: Academic Press.

Powers, M. C. (1967). Fluid-release mechanisms in compacting marine mudrocks and their importance in oil exploration. *American Association of Petroleum Geologists' Bulletin* **51**, 1,240–1,254.

Press, W. H., Flannery, B. P., Teukolsky, S. A., and Vetterling, W. T. (1986). *Numerical Recipes: The Art of Scientific Computing*. New York: Cambridge University Press.

Prikryl, J. D., Posey, H. H., and Kyle, J. R. (1988). A petrographic and geochemical model for the origin of calcite cap rock at Damon Mound salt dome, Texas, U.S.A. *Chemical Geology* **74**, 67–97.

Pruess, Karsten (1991). TOUGH2 – A general-purpose numerical simulator for multiphase fluid and heat flow. *Lawrence Berkeley Laboratory Report* **LBL-29400**.

Quilty, E. G., Farrar, C. D., Galloway, D. L., Hamlin, S. N., Laczniak, R. J., Roeloffs, E. A., Sorey, M. L., and Woodcock, D. E. (1995). Hydrologic effects associated with the January 17, 1994 Northridge, California, earthquake. *U.S. Geological Survey Open-File Report* **95-813**.

Rabinowicz, M., Dandurand, J. L., Jakubowski, J., Schott, J., and Cassam, J. P. (1985). Convection in a North Sea oil reservoir: Inferences on diagenesis and hydrocarbon migration. *Earth and Planetary Science Letters* **74**, 387–404.

Raffensperger, J. P., and Garven, G. (1995a). The formation of unconformity-type uranium ore deposits 1. Coupled groundwater flow and heat transport modeling. *American Journal of Science* **295**, 581–636.

Raffensperger, J. P., and Garven, G. (1995b). The formation of unconformity-type uranium deposits 2. Coupled hydrochemical modeling. *American Journal of Science* **295**, 639–696.

Raleigh, C. B., Healy, J. H., and Bredehoeft, J. D. (1976). An experiment in earthquake control at Rangely, Colorado. *Science* **191**, 1,230–1,236.

Randazzo, A. F., and Bloom, J. I. (1985). Mineralogical changes along the freshwater/saltwater interface of a modern aquifer. *Sedimentary Geology* **43**, 219–239.

Ranganathan, V., and Hanor, J. S. (1988). Density-driven groundwater flow near salt domes. *Chemical Geology* **74**, 173–188.

Ranganathan, V., and Hanor, J. S. (1989). Perched brine plumes above salt domes and dewatering of geopressured sediments. *Journal of Hydrology* **110**, 63–86.

Renshaw, C. E. (1996). Influence of subcritical fracture growth on the connectivity of fracture networks. *Water Resources Research* **32**, 1,519–1,530.

Rice, J. M., and Ferry, J. M. (1982). Buffering, infiltration, and the control of intensive variables during infiltration. In *Characterization of Metamorphism through Mineral Equilibria*, Ferry, J. M. (ed.). *Reviews in Mineralogy* **10**, 263–326.

Rice, J. R. (1992). Fault stress states, pore pressure distributions, and the weakness of the San Andreas fault. In *Fault Mechanics and Transport Properties of Rocks*, Evans, B., and Wong, T.-F. (eds.), pp. 475–503. London: Academic.

Rice, J. R., and Cleary, M. P. (1976). Some basic stress diffusion solutions for fluid-saturated elastic porous media with compressible constituents. *Reviews of Geophysics and Space Physics* **14**, 227–241.

Rinehart, J. S. (1972). Fluctuations in geyser activity caused by variations in Earth tidal forces, barometric pressure, and tectonic stresses. *Journal of Geophysical Research* **77**, 342–350.

Rinehart, J. S. (1980). *Geysers and Geothermal Energy*. New York: Springer-Verlag.

Robertson, J. B. (1974). Digital modeling of radioactive and chemical waste transport in the Snake River Plain aquifer at the National Reactor Testing Station, Idaho. *U.S. Geological Survey Water Resources Division Report* **IDO-22054**.

Roedder, E. (1984). Fluid inclusions. *Reviews in Mineralogy* **12**.

Roeloffs, E. (1988a). Fault stability changes induced beneath a reservoir with cyclic variations in water level. *Journal of Geophysical Research* **93**, 2,107–2,124.

Roeloffs, E. (1988b). Hydrologic precursors to earthquakes: A review. *Pure and Applied Geophysics* **126**, 171–209.

Roeloffs, E. (1996). Poroelastic techniques in the study of earthquake-related hydrologic phenomena. *Advances in Geophysics* **37**, 135–195.

Roeloffs, E., Danskin, W. R., Farrar, C. D., Galloway, D. L., Hamlin, S. N., Quilty, E. G., Quinn, H. M., Schaefer, D. H., Sorey, M. L., and Woodcock, D. E. (1995). Hydrologic effects associated with the June 28, 1992 Landers, California, earthquake sequence. *U.S. Geological Survey Open-File Report* **95-42**.

Rojstaczer, S. A. (ed.) (1994). The Loma Prieta, California earthquake of October 17, 1989 – Hydrologic disturbances. *U.S. Geological Survey Professional Paper* **1551-E**.

Rojstaczer, S. A., and Bredehoeft, J. D. (1988). Groundwater and fault strength. In *The*

Geology of North America v. O-2: Hydrogeology, Back, W., Rosenshein, J. S., and Seaber, P. R. (eds.), pp. 447–460. Boulder: The Geological Society of America.

Rojstaczer, S. A., and Wolf, S. (1992). Permeability changes associated with large earthquakes: An example from Loma Prieta, California, 10/17/89. *Geology* **20**, 211–214.

Rojstaczer, S. A., and Wolf, S. (1994). Hydrologic changes associated with the earthquake in the San Lorenzo and Pescadero drainage basins. In *The Loma Prieta, California Earthquake of October 17, 1989 – Hydrologic Disturbances*, Rojstaczer, S. A. (ed.). *U.S. Geological Survey Professional Paper* **1551-E**, E51-E64.

Rojstaczer, S. A., Wolf, S. and Michel, R. (1995). Permeability enhancement in the shallow crust as a cause of earthquake-induced hydrological changes. *Nature* **373**, 237–239.

Rowan, E. L., and Goldhaber, M. B. (1996). Fluid inclusions and biomarkers in the Upper Mississippi Valley zinc-lead district – Implications for the fluid-flow and thermal history of the Illinois Basin. *U.S. Geological Survey Bulletin* **2094-F**.

Rubey, W. W., and Hubbert, M. K. (1959). Role of fluid pressure in mechanics of overthrust faulting. II. Overthrust belt in geosynclinal area of western Wyoming in light of fluid-pressure hypothesis. *Geological Society of America Bulletin* **70**, 167–206.

Rubin, J. (1983). Transport of reacting solutes in porous media: Relation between mathematical nature of problem formulation and chemical nature of reactions. *Water Resources Research* **19**, 1,231–1,252.

Rubin, J. (1990). Solute transport with multisegment, equilibrium-controlled reactions: A feed-forward simulation method. *Water Resources Research* **26**, 2,029–2,055.

Rumble, D., III (1994). Water circulation in metamorphism. *Journal of Geophysical Research* **99**, 15,499–15,502.

Runnels, D. D. (1969). Diagenesis, chemical sediments, and the mixing of natural waters. *Journal of Sedimentary Petrology* **39**, 1,188–1,201.

Russell, M. J. (1986). Extension and convection: A genetic model for the Irish Carboniferous base metal and barite deposits. In *The Geology and Genesis of Mineral Deposits in Ireland*, Andrew, C. J., Crowe, R. W. A., Finlay, S., Pennell, W. M., and Pyne, J. F. (eds.), pp. 545–554. Dublin: Irish Association for Economic Geology.

Ryan, G. R. (1979). The genesis of Proterozoic uranium deposits in Australia. *Royal Society (London) Philosophical Transactions* **A291**, 339–353.

Sammel, E. A., Ingebritsen, S. E., and Mariner, R. H. (1988). The hydrothermal system at Newberry volcano, Oregon. *Journal of Geophysical Research* **93**, 10,149–10,162.

Sanford, R. F. (1990). Hydrogeology of an ancient arid closed basin: Implications for tabular sandstone-hosted uranium deposits. *Geology* **18**, 1,099–1,102.

Sanford, R. F. (1992). A new model for tabular-type uranium deposits. *Economic Geology* **87**, 2,041–2,055.

Sanford, W. E. (1997). Correcting for diffusion in carbon-14 dating of groundwater. *Ground Water* **35**, 357–361.

Sanford, W. E., and Buapeng, S. (1996). Assessment of a groundwater flow model of the Bangkok Basin, Thailand, using carbon-14-based ages and paleohydrology. *Hydrogeology Journal* **4**, 26–40.

Sanford, W. E., and Konikow, L. F. (1989a). Simulation of calcite dissolution and porosity changes in salt water mixing zones in coastal aquifers. *Water Resources Research* **25**, 655–667.

Sanford, W. E., and Konikow, L. F. (1989b). Porosity development in coastal carbonate aquifers. *Geology* **17**, 249–252.

Sanford, W. E., and Wood, W. W. (1991). Brine evolution and mineral deposition in hydrologically open evaporite basins. *American Journal of Science* **291**, 687–710.

Sanford, W. E., and Wood, W. W. (1995). Paleohydrologic record from lake brine on the southern High Plains, Texas. *Geology* **23**, 229–232.

Sanford, W. E., Wood, W. W., and Councill, T. B. (1992). Calcium chloride-dominated brines: An ion-exchange model. In *Proceedings of the 7th International Symposium on Water-Rock Interaction*, Kharaka, Y. K., and Maest, A. S. (eds.), pp. 669–672. Rotterdam: A.A. Balkema.

Sass, J. H., and Galanis, S. P., Jr. (1983). Temperatures, thermal conductivity, and heat flow from a well in Pierre Shale near Hayes, South Dakota. *U.S. Geological Survey Open-File Report* **83-25**.

Sass, J. H., Lachenbruch, A. H., Moses, T. H., Jr., and Morgan, P. (1992). Heat flow from a scientific research well at Cajon Pass, California. *Journal of Geophysical Research* **97**, 5,017–5,030.

Sass, J. H., Lachenbruch, A. H., and Munroe, R. J. (1971). Thermal conductivity of rocks from measurements on fragments and its application to heat-flow determinations. *Journal of Geophysical Research* **76**, 3,391–3,401.

Sassen, R., Chinn, E. W., and McCabe, C. (1988). Recent hydrocarbon alteration, sulfate reduction and formation of elemental sulfur and metal sulfides in salt dome cap rock. *Chemical Geology* **74**, 57–66.

Saunders, J. A. (1992). Geochemical modeling of salt dome cap rock-brine reactions. In *Proceedings of the 7th International Symposium on Water-Rock Interaction*, Kharaka, Y. K., and Maest, A. S. (eds.), pp. 1,121–1,124. Rotterdam: A.A. Balkema.

Saunders, J. A., Prikryl, J. D., and Posey, H. H. (1988). Mineralogic and isotopic constraints on the origin of strontium-rich cap rock, Tatum Dome, Mississippi, U.S.A. *Chemical Geology* **74**, 1–24.

Sawkins, F. J. (1990). *Metal Deposits in Relation to Plate Tectonics*. New York: Springer-Verlag.

Scheidegger, A. E. (1961). General theory of dispersion in porous media. *Journal of Geophysical Research* **66**, 3,273–3,278.

Scheidegger, A. E. (1974). *The Physics of Flow Through Porous Media*. 3rd ed. Toronto: University of Toronto Press.

Schmoker, J. W., and Halley, R. B. (1982). Carbonate porosity versus depth: A predictable relation for South Florida. *American Association of Petroleum Geologists' Bulletin* **66**, 2,561–2,570.

Scholl, M. A., Ingebritsen, S. E., Janik, C. J., and Kauahikaua, J. P. (1996). Use of precipitation and groundwater isotopes to interpret regional hydrology on a tropical volcanic island: Kilauea volcano area, Hawaii. *Water Resources Research* **32**, 3,525–3,537.

Scholz, C. H. (1990). *The Mechanics of Earthquakes and Faulting*. Cambridge: Cambridge University Press.

Sclater, J. G., Jaupart, C., and Galson, D. (1980). The heat flow through oceanic and continental crust and the heat loss of the Earth. *Reviews of Geophysics and Space Physics* **18**, 269–311.

Screaton, E. J., Carson, B., and Lennon, G. P. (1995). Hydrogeologic properties of a thrust fault within the Oregon accretionary prism. *Journal of Geophysical Research* **100**, 20,025–20,035.

Screaton, E. J., Wuthrich, D. R., and Dreiss, S. J. (1990). Permeabilities, fluid pressures, and flow rates in the Barbados Ridge complex. *Journal of Geophysical Research* **95**, 8,997–9,007.

Segall, P., and Rice, J. R. (1995). Dilatancy, compaction, and slip instability of a fluid-infiltrated fault. *Journal of Geophysical Research* **100**, 22,155–22,171.

Seni, S. J. (1987). Evolution of Boling Dome cap rock with emphasis on included terrigenous clastics, Fort Bend and Wharton counties, Texas. In *Dynamical Geology of Salt and Related Structures*, Lerche, I. and O'Brien, J. J. (eds.), pp. 543–591. Orlando: Academic Press.

Sharp, J. M., Jr., and Domenico, P. A. (1976). Energy transport in thick sequences of compacting sediment. *Geological Society of America Bulletin* **87**, 390–400.

Sibley, D. F., Dedoes, R. E., and Bartlett, T. R. (1987). Kinetics of dolomitization. *Geology* **15**, 1,112–1,114.

Silver, P. G., and Vallette-Silver, N. J. (1992). Detection of hydrothermal precursors to large northern California earthquakes. *Science* **257**, 1,363.

Silver, P. G., and Wakita, H. (1996). A search for earthquake precursors. *Science* **273**, 77–78.

Skinner, B. J. (1979). The many origins of hydrothermal ore deposits. In *Geochemistry of Hydrothermal Ore Deposits*, Barnes, H. L. (ed.), pp. 1–21. New York: John Wiley and Sons.

Slichter, C. S. (1899). Theoretical investigations of motions of groundwater flow. *U.S. Geological Survey Annual Report* **19** (Part II), 295–380.

Slujik, D., and Nederlof, M. H. (1984). Worldwide geologic experience as a systematic basis for prospect appraisal. In *Petroleum Geochemistry and Basin Evaluation*, Demaison, G., and Murris, R. J. (eds.). *American Association of Petroleum Geologists' Memoir* **35**, 15–26.

Smart, P. L., Dawans, J. M., and Whitaker, F. F. (1988). Carbonate dissolution in a modern mixing zone. *Nature* **335**, 811–813.

Smith, G. D. (1978). *Numerical Solutions to Partial Differential Equations: Finite Difference Methods*. Oxford: Oxford University Press.

Smith, L., and Chapman, D. S. (1983). On the thermal effects of groundwater flow. 1. Regional scale systems. *Journal of Geophysical Research* **88**, 593–608.

Smith, R. L., and Shaw, H. R. (1979). Igneous-related geothermal systems. In *Assessment of Geothermal Resources of the United States* – 1978, Muffler, L. J. P. (ed.). *U.S. Geological Survey Circular* **790**, 12–17.

Smoluchowski, M. S. (1909). Some remarks on the mechanics of overthrusts. *Geological Magazine* **Decade V, Volume VI**, 204–205.

Snow, D. T. (1968). Rock fracture spacings, openings, and porosities. *Proceedings American Society of Civil Engineers* **94**, 73–91.

Sonnenfeld, Peter (1984). *Brines and Evaporites*. Orlando: Academic Press.

Sorey, M. L. (1978). Numerical modeling of liquid geothermal systems. *U.S. Geological Survey Professional Paper* **1044-D**.

Sorey, M. L., and Clark, M. D. (1981). Changes in the discharge characteristics of thermal springs and fumaroles in the Long Valley caldera, California, resulting from earthquakes on May 25–27, 1980. *U.S. Geological Survey Open-File Report* **81-203**.

Sorey, M. L., and Colvard, E. M. (1994). Measurements of heat and mass flow from thermal areas in Lassen Volcanic National Park, California, 1984–93. *U.S. Geological Survey Water Resources Investigations Report* **94-4180-A**.

Sorey, M. L., Grant, M. A., and Bradford, E. (1980). Nonlinear effects in two-phase flow to wells in geothermal reservoirs. *Water Resources Research* **16**, 767–777.

Spalding, D. B. (1972). A novel finite difference formulation for differential expressions involving both first and second derivatives. *International Journal for Numerical Methods in Engineering* **4**, 551–559.

Spencer, R. J., Eugster, H. P., Jones, B. F., and Rettig, S. L. (1985). Geochemistry of Great Salt Lake, Utah. I. Hydrochemistry since 1850. *Geochimica et Cosmochimica Acta* **49**, 727–737.

Spencer, R. J., and Hardie, L. A. (1990). Control of seawater composition by mixing of river waters and mid-ocean ridge hydrothermal brines. In *Fluid-Mineral Interactions: A Tribute to H. P. Eugster*, Spencer, R. J., and Chou, I. M. (eds.). *Geochemical Society Special Publication* **19**, 409–419.

Stein, C. A., and Stein, S. (1994). Constraints on hydrothermal heat flux through the oceanic lithosphere from global heat flow. *Journal of Geophysical Research* **99**, 3,081–3,095.

Steinberg, G. S., Merzhavov, A. G., and Steinberg, A. S. (1982). Geyser process: Its theory, modeling, and field experiment. 3. On metastability of water in geysers. *Modern Geology* **8**, 75–78.

Stober, I. (1996). Researchers study conductivity of crystalline rock in proposed radioactive waste site. *Eos, Transactions American Geophysical Union* **77**, 93–94.

Straus, J. M., and Schubert, G. (1981). One-dimensional model of vapor-dominated geothermal systems. *Journal of Geophysical Research* **86**, 9,433–9,438.

Stuben, D., Sedwick, P., and Colantoni, P. (1996). Geochemistry of submarine warm springs in the limestone cavern of Grotta Azzurra, Capo Palinuro, Italy: Evidence for mixing-zone dolomitisation. *Chemical Geology* **131**, 113–125.

Stumm, W., and Morgan, J. J. (1981). *Aquatic Chemistry*. New York: Wiley Interscience.

Sudicky, E. A. (1986). A natural gradient experiment on solute transport in a sand aquifer: Spatial variability of hydraulic conductivity and its role in the dispersion process. *Water Resources Research* **22**, 2,069–2,082.

Sugisaki, R., Ito, T., Nagamine, K., and Kawabe, I. (1996). Gas geochemical changes at mineral springs associated with the 1995 southern Hyogo earthquake ($M = 7.2$), Japan. *Earth and Planetary Science Letters* **139**, 239–249.

Summer, N. S., and Verosub, K. L. (1987). Maturation anomalies in sediments underlying thick volcanic strata, Oregon: Evidence for a thermal event. *Geology* **15**, 30–33.

Summerfield, M. A., and Hulton, N. J. (1994). Natural controls of fluvial denudation rates in major world drainage basins. *Journal of Geophysical Research* **99**, 13,871–13,883.

Swanberg, C. A., Walkey, W. C., and Combs, J. (1988). Core hole drilling and the "rain curtain" phenomenon at Newberry volcano, Oregon. *Journal of Geophysical Research* **93**, 10,163–10,173.

Swanson, V. E., and Palacas, J. G. (1965). Humate in coastal sands of northwest Florida. *U.S. Geological Survey Bulletin* **1214-B**.

Sweeney, J. J., Burnham, A. K., and Braun, R. L. (1987). A model of hydrocarbon generation from Type I kerogen: Application to the Uinta basin, Utah. *American Association of Petroleum Geologists' Bulletin* **71**, 967–985.

Tada, R., and Siever, R. (1989). Pressure solution during diagenesis. *Annual Review of Earth and Planetary Sciences* **17**, 89–118.

Tarney, J., Pickering, K. T., Knipe, R. J., and Dewey, J. F. (1991). *The Behaviour and Influence of Fluids in Subduction Zones*. London: The Royal Society.

Taylor, H. P., Jr. (1968). The oxygen isotope geochemistry of igneous rocks. *Contributions to Mineralogy and Petrology* **19**, 1–71.

Taylor, H. P., Jr. (1971). Oxygen isotope evidence for large-scale interaction between meteoric groundwaters and Tertiary granodiorite intrusions, Western Cascade Range, Oregon. *Journal of Geophysical Research* **76**, 7,855–7,874.

Taylor, H. P., Jr. (1979). Oxygen and hydrogen isotope relationships in hydrothermal mineral deposits. In *Geochemistry of Hydrothermal Ore Deposits*, Barnes, H. L. (ed.), pp. 173–235. New York: John Wiley and Sons.

Taylor, H. P., Jr. (1990). Oxygen and hydrogen isotope constraints on the deep circulation of surface waters into zones of hydrothermal metamorphism and melting. In *The Role of Fluids in Crustal Processes*, Bredehoeft, J. D., and Norton, D. L. (eds.), pp. 72–95, Washington, DC: National Academy Press.

Taylor, H. P., Jr., and Forester, R. W. (1979). An oxygen isotope study of the Skaergaard intrusion and its country rocks: A description of a 55-m.y. old fossil hydrothermal system. *Journal of Petrology* **20**, 355–419.

Terzaghi, K. (1925). *Erdbaummechanic*. Vienna: Franz Deuticke.

Thompson, A. B., and England, P. C. (1984). Pressure-temperature-time paths of regional metamorphism. II. Their inference and interpretation using mineral assemblages in igneous rocks. *Journal of Petrology* **25**, 929–955.

Tissot, B. P., Durand, B., Espitalie, J., and Combaz, A. (1974). Influence of nature and diagenesis of organic matter in formation of petroleum. *American Association of Petroleum Geologists' Bulletin* **58**, 499–506.

Tissot, B. P., Pelet, R., and Ungerer, P. (1987). Thermal kinetics of oil and gas generation. *American Association of Petroleum Geologists' Bulletin* **71**, 1445–1466.

Titley, S. R. (1978). Geologic history, hypogene features, and processes of secondary sulfide enrichment at the Pleysumi copper prospect, New Britain, Papua New Guinea. *Economic Geology* **73**, 768–784.

Titley, S. R. (1990). Evolution and style of fracture permeability in intrusion-centered hydrothermal systems. In *The Role of Fluids in Crustal Processes*, Bredehoeft, J. D., and Norton, D. L. (eds.), pp. 50–63. Washington, DC: National Academy Press.

Titley, S. R., and Beane, R. E. (1981). Porphyry copper deposits. Part I. Geologic settings, petrology and tectogenesis. *Economic Geology* 75th Anniversary Volume, 214–235.

Toth, J. (1963). A theoretical analysis of groundwater flow in small drainage basins. *Journal of Geophysical Research* **68**, 4,795–4,812.

Toupin, D., Eadington, P. J., Person, M. A., Morin, P., Wieck, J., and Warner, D. (1997). Petroleum hydrogeology of the Cooper and Eromanga basins, Australia. *American Association of Petroleum Geologists' Bulletin* **81**, 577–603.

Tremblay, L. P. (1982). Geology of the uranium deposits related to the sub-Athabasca unconformity, Saskatchewan. *Geological Survey of Canada Paper* **81-20**.

Trusheim, F. (1960). Mechanisms of salt migration in northern Germany. *American Association of Petroleum Geologists' Bulletin* **44**, 1,519–1,540.

Tsunogai, U., and Wakita, H. (1995). Precursory chemical changes in ground water: Kobe earthquake, Japan. *Science* **269**, 61–63.

Turcotte, D. L., and Schubert, G. (1982). *Geodynamics*. New York: John Wiley and Sons.

Turner-Peterson, C. E., and Hodges, C. A. (1986). Descriptive model of sandstone U. In *Mineral Deposit Models*, Cox, D. P., and Singer, D. A. (eds.). *U.S. Geological Survey Bulletin* **1693**, 209–210.

Usiglio, J. (1849). Analyse de l'eau de la Mediterannee sur les cotes de France. *Annals de Chimica Physica* **27**, 92–107.

Vanselow, A. P. (1932). Equilibria of the base-exchange reactions of bentonites, permutites, soil colloids and zeolites. *Journal of Colloidal Interface Science* **109**, 219–228.

Vennard, J. K., and Street, R. L. (1975). *Elementary Fluid Mechanics*. New York: John Wiley and Sons.

Verhoogen, J. (1980). *Energetics of the Earth*. Washington, DC: National Academy Press.

Verma, A. K. (1986). Effects of phase transformation on steam-water relative permeabilities. *Lawrence Berkeley Laboratory Report* **LBL-20594**.

Vineyard, J. D. (1977). Preface to the Viburnum Trend issue. *Economic Geology* **72**, 337–338.

von der Borch, C. C., Lock, D. E., and Schwebel, D. (1975). Ground-water formation of dolomite in the Coorong region of South Australia. *Geology* **3**, 283–285.

Voss, C. I. (1984). A finite-element simulation model for saturated-unsaturated, fluid-density-dependent groundwater flow with energy transport or chemically reactive single-species solute transport. *U.S. Geological Survey Water Resources Investigations Report* **84-4369**.

Wallace, R. H., Jr., Wesselman, J. B., and Kraemer, T. F. (1981). Occurrence of geopressure in the northern Gulf of Mexico basin. *Proceedings of the Geopressured-Geothermal Energy Conference* **5**, 200–220.

Waller, R. M. (1966). Effects of the March 1964 Alaska earthquake on the hydrology of south-central Alaska. *U.S. Geological Survey Professional Paper* **544-A**.

Walters, M. A., and Combs, J. (1992). Heat flow in The Geysers–Clear Lake area of northern California, U.S.A. In *The Geysers Geothermal Field*, Stone, C. (ed.). *Geothermal Resources Council Special Report* **17**, 43–58.

Walther, J. V. (1990). Fluid dynamics during progressive regional metamorphism. In *The Role of Fluids in Crustal Processes*, Bredehoeft, J. D., and Norton, D. (eds.), pp. 64–71. Washington, DC: National Academy Press.

Walther, J. V., and Holdaway, M. J. (1995). Comment: Mass transfer during Barrovian metamorphism. *American Journal of Science* **295**, 1,020–1,025.

Wang, C. Y., Hwang, W.-T., and Cochrane, G. B. (1994). Tectonic dewatering and mechanics of protothrust zones: Example from the Cascadia accretionary margin. *Journal of Geophysical Research* **99**, 20,043–20,050.

Wang, K. (1994). Kinematic models of dewatering accretionary prisms. *Journal of Geophysical Research* **99**, 4,429–4,438.

Wang, H. F., and Anderson, M. P. (1982). *Introduction to Groundwater Modeling: Finite Difference and Finite Element Methods*. San Francisco: W. H. Freeman and Company.

Waples, D. W. (1994). Modeling of sedimentary basins and petroleum systems. In *The petroleum system – From Source to Trap*, Magoon, L. B., and Dow, W. G. (eds.). *American Association of Petroleum Geologists' Memoir* **60**, 307–322.

Ward, J. D. (1964). Turbulent flow in porous media. *Proceedings American Society of Civil Engineers* **90**, 1–12.

Ward, W. C., and Halley, R. B. (1985). Dolomitization in a mixing zone of near-seawater composition, late Pleistocene, northern Yucatan Peninsula. *Journal of Sedimentary Petrology* **55**, 407–420.

Wardlaw, N. C. (1968). Carnallite-sylvite relationships in the Middle Devonian Prairie Evaporite Formation, Saskatchewan, Canada. *Geological Society of America Bulletin* **79**, 1,273–1,294.

Wardlaw, N. C. (1972). Unusual marine evaporites with salts of calcium and magnesium chloride in Cretaceous basins of Sergipe, Brazil. *Economic Geology* **67**, 156–168.

Watson, E. B., and Brenan, J. M. (1987). Fluids in the lithosphere. 1. Experimentally determined wetting characteristics of CO_2–H_2O fluids and their implications for fluid transport, host-rock physical properties, and fluid inclusion formation. *Earth and Planetary Science Letters* **85**, 497–515.

Wendebourg, J. (1994). Simulating hydrocarbon migration and stratigraphic traps. Ph.D. thesis, Stanford University.

Werner, M. L., Feldman, M. D., and Knauth, L. P. (1988). Petrography and geochemistry of water-rock interactions in Richton Dome cap rock. *Chemical Geology* **74**, 113–135.

Whitaker, F. F., and Smart, P. L. (1990). Active circulation of saline groundwaters in carbonate platforms: Evidence from the Great Bahama Bank. *Geology* **18**, 200–203.

Whitaker, F. F., and Smart, P. L. (1993). Circulation of saline groundwater in carbonate platforms. A review and case study from the Bahamas. In *Diagenesis and Basin Development*, Horbury, A. D., and Robinson, A. G. (eds.). *American Association of Petroleum Geologists' Studies in Geology* **36**, 113–132.

Whitaker, F. F., and Smart, P. L. (1994). Bacterially mediated oxidation of organic matter: A major control on groundwater geochemistry and porosity generation in oceanic carbonate terrains. In *Breakthroughs in Karst Geomicrobiology and Redox Geochemistry*, Sasowsky, I. D. and Palmer, M. V. (eds.). *Karst Waters Institute Special Publication* **1**, 72–73.

Whitaker, F. F., Smart, P. L., Vahrenkamp, V. C., Nicholson, H., and Wogelius, R. A. (1994). Dolomitization by near-normal seawater? Field evidence from the Bahamas. In *Dolomites – A Volume in Honour of Dolomieu*, Purser, B., Tucker, M., and Zenger, D. (eds.), pp. 111–132. *International Association of Sedimentologists Special Publication* **21**. Oxford: Blackwell Scientific Publications.

White, A. F., and Peterson, M. L. (1991). Chemical equilibrium and mass balance relationships associated with the Long Valley hydrothermal system, California. *Journal of Volcanology and Geothermal Research* **27**, 371–397.

White, D. E. (1967). Some principles of geyser activity, mainly from Steamboat Springs, Nevada. *American Journal of Science* **265**, 644–684.

White, D. E. (1992). The Beowawe Geysers, Nevada, before geothermal development. *U.S. Geological Survey Bulletin* **1998**.

White, D. E., Fournier, R. O., Muffler, L. J. P., and Truesdell, A. H. (1975). Physical results of research drilling in thermal areas of Yellowstone National Park, Wyoming. *U.S. Geological Survey Professional Paper* **892**.

White, D. E., Hem, J. D., and Waring, G. A. (1963). Chemical composition of subsurface waters. *U.S. Geological Survey Professional Paper* **440-F**.

White, D. E., and Marler, G. D. (1972). Comments on paper by John S. Rinehart, 'Fluctuations in geyser activity caused by Earth tidal forces, barometric pressure, and tectonic stresses.' *Journal of Geophysical Research* **77**, 5,825–5,829.

White, D. E., Muffler, L. J. P., and Truesdell, A. H. (1971). Vapor-dominated hydrothermal systems compared with hot-water systems. *Economic Geology* **66**, 75–97.

White, W. B. (1988). *Geomorphology and Hydrology of Karst Terrains*. New York: Oxford University Press.

Whitehead, H. C., and Feth, J. H. (1961). Recent chemical analyses of waters from several closed-basin lakes and their tributaries in the western United States. *Geological Society of America Bulletin* **72**, 1,421–1,426.

Wigley, T. M. L., and Plummer, L. N. (1976). Mixing of carbonate water. *Geochimica et Cosmochimica Acta* **40**, 989–995.

Wilde, A. R., and Wall, V. J. (1987). Geology of the Nabarlek uranium deposit, Northern Territory, Australia. *Economic Geology* **82**, 1,152–1,168.

Willet, S. D., and Chapman, D. S. (1987). Temperatures, fluid flow, and the thermal history of the Uinta basin. In *Migration of Hydrocarbons in Sedimentary Basins*, Doligez, B. (ed.), pp. 533–551. Paris: Editions Technip.

Williams, B., and Brown, C. (1986). A model for the genesis of Zn-Pb deposits in Ireland. In *The Geology and Genesis of Mineral Deposits in Ireland*, Andrew, C. J., Crowe, R. W. A., Finlay, S., Pennell, W. M., and Pyne, J. F. (eds.), pp. 579–590. Dublin: Irish Association for Economic Geology.

Williams, C. F., and Narasimhan, T. N. (1989). Hydrogeologic constraints on heat flow along the San Andreas fault: A testing of hypotheses. *Earth and Planetary Science Letters* **92**, 131–143.

Williams, D. L., and Von Herzen, R. P. (1974). Heat loss from the Earth: New estimate. *Geology* **2**, 327.

Williams-Stroud, S. C. (1994). Solution to the paradox? Results of some chemical equilibrium and mass balance calculations applied to the Paradox Basin evaporite deposit. *American Journal of Science* **294**, 1,189–1,228.

Wilson, E. N., Hardie, L. A., and Phillips, O. M. (1990). Dolomitization front geometry, fluid flow patterns, and the origin of massive dolomite: The Triassic Latemar buildup, northern Italy. *American Journal of Science* **290**, 741–796.

Winograd, I. J. (1971). Hydrogeology of ash-flow tuff: A preliminary statement. *Water Resources Research* **7**, 994–1,006.

Wolery, T. J. (1979). Calculation of chemical equilibrium between aqueous solution and minerals: The EQ3/6 software package. *University of California Lawrence Livermore Laboratory Bulletin* **UCRL-52658**.

Wolery, T. J. (1992). EQ3/6, A software package for geochemical modeling of aqueous systems. *Lawrence Livermore National Laboratory Report* **UCRL-MA-110662-PT-I**.

Wolery, T. J., and Sleep, N. H. (1976). Hydrothermal circulation and geochemical flux at mid-ocean ridges. *Journal of Geology* **84**, 249–275.

Wood, J. R., and Hewett, T. A. (1982). Fluid convection and mass transfer in porous sandstones – A theoretical model. *Geochimica et Cosmochimica Acta* **46**, 1,707–1,713.

Wood, W. W., and Jones, B. F. (1990). Origin of solutes in saline lakes and springs on the southern High Plains of Texas and New Mexico. In *Geologic Framework and Regional Hydrology: Upper Cenozoic Blackwater Draw and Ogallala Formations, Great Plains*, Gustavson, T. C. (ed.), pp. 193–208. Austin: Texas Bureau of Economic Geology.

Wood, W. W., and Sanford, W. E. (1990). Groundwater control of evaporite deposition. *Economic Geology* **85**, 1,226–1,235.

Woodbury, A. P., and Smith, L. (1988). Simultaneous inversion of hydrogeologic and thermal data. 2. Incorporation of thermal data. *Water Resources Research* **24**, 356–372.

Ypma, P. J. M., and Fusikawa, K. (1980). Fluid inclusion and oxygen isotope studies of the Nabarlek and Jabiluka uranium deposits, Northern Territory, Australia. In *Uranium in the Pine Creek Geosyncline*, Ferguson, J., and Goleby, A. B. (eds.), 375–395. Sydney: International Atomic Energy Agency.

Yuan, T., Spence, G. D., and Hyndman, R. D. (1994). Seismic velocities and inferred porosities in the accretionary wedge sediments at the Cascadia Margin. *Journal of Geophysical Research* **99**, 4,413–4,427.

Zak, I. (1974). Sedimentology and bromine geochemistry of marine and continental evaporites in the Dead Sea Basin. In *Fourth Symposium on Salt*, Coogam, A. H. (ed.), pp. 349–361. Cleveland: Northern Ohio Geological Society.

Zharkov, M. A. (1984). *Paleozoic Salt-Bearing Formations of the World.* New York: Springer-Verlag.

Zhao, G., Chen, H., Ma, S., and Zhang, D. (1995). Research on earthquakes induced by water injection in China. *Pure and Applied Geophysics* **145**, 60–68.

Ziagos, J. P., and Blackwell, D. D. (1986). A model for the transient temperature effects of horizontal fluid flow in geothermal systems. *Journal of Volcanology and Geothermal Research* **27**, 371–397.

Zoback, M. D., and Healy, J. H. (1984). Friction, faulting, and in situ stress. *Annales Geophysicae* **2**, 689–698.

Zoback, M. D., and Healy, J. H. (1992). In situ stress measurements to 3.5 km depth in the Cajon Pass scientific borehole: Implications for the mechanics of crustal faulting. *Journal of Geophysical Research* **97**, 5,039–5,057.

Zoback, M. D., and Lachenbruch, A. H. (1992). Introduction to special section on the Cajon Pass scientific drilling project. *Journal of Geophysical Research* **97**, 4,991–4,994.

Zoback, M. D., Zoback, M. L., Mount, V. S., Suppe, J., Eaton, J. P., Healy, J. H., Oppenheimer, D. H., Reasenberg, P. A., Jones, L., Raleigh, C. B., Wong, I. G., Scotti, O., and Wentworth, C. M. (1987). New evidence on the state of stress of the San Andreas fault system. *Science* **238**, 1,105–1,111.

Zoback, M. L., and Zoback, M. D. (1997). Crustal stress and intraplate deformation. *Geowissenschaften* **15**,116–123.

Index